Essays in the History
and Philosophy of Artificial Life

EDITED BY JESSICA RISKIN

GENESIS REDUX

THE UNIVERSITY OF
CHICAGO PRESS *Chicago and London*

Jessica Riskin is associate professor of history at Stanford
University. She is the author of *Science in the Age of Sensibility:
The Sentimental Empiricists of the French Enlightenment*, also published
by the University of Chicago Press.

The University of Chicago Press, Chicago 60637
The University of Chicago Press, Ltd., London
© 2007 by The University of Chicago
All rights reserved. Published 2007
Printed in the United States of America

16 15 14 13 12 11 10 09 08 07   1 2 3 4 5

ISBN-13: 978-0-226-72080-7 (cloth)
ISBN-10: 0-226-72080-2 (cloth)
ISBN-13: 978-0-226-72081-4 (paper)
ISBN-10: 0-226-72081-0 (paper)

Library of Congress Cataloging-in-Publication Data

Genesis redux: essays in the history and philosophy of
artificial life / edited by Jessica Riskin.
     p.   cm.
   Includes index.
   ISBN-13: 978-0-226-72080-7 (cloth: alk. paper)
   ISBN-10: 0-226-72080-2 (cloth: alk. paper)
   ISBN-13: 978-0-226-72081-4 (pbk.: alk. paper)
   ISBN-10: 0-226-72081-0 (pbk.: alk. paper)
   1. Artificial life. I. Riskin, Jessica.
   BD418.8.G46   2007
   113'.8—dc22

                                    2007003437

∞ The paper used in this publication meets the minimum
requirements of the American National Standard for
Information Sciences—Permanence of Paper for Printed
Library Materials, ANSI Z39.48-1992.

For Madeleine, author of rare and fascinating creatures,
and for Oliver, researcher into life in all its forms

CONTENTS

CONTRIBUTORS

DAVID BATES teaches intellectual history in the Department of Rhetoric at the University of California, Berkeley. He has published *Enlightenment Aberrations: Error and Revolution in France* and is now completing a book on the history of human insight from Descartes to Artificial Intelligence.

BERNADETTE BENSAUDE-VINCENT is professor of history and philosophy of science at the Université de Paris X. Her research has focused on the history and philosophy of chemistry and on the public roles of natural science. Her current work examines ethical and philosophical issues surrounding research in nanotechnology and its applications. She is a member of the Académie des technologies and the Comité d'éthique du CNRS.

SYLVIA BERRYMAN is associate professor in the Department of Philosophy at the University of British Columbia. She studies ancient Greek natural philosophy and the reception of ancient science in philosophy. She is currently working on a book on the concept of mechanical explanation in ancient Greek natural philosophy.

JUSTINE CASSELL is a full professor in the Department of Communication Studies and the Department of Computer Science at Northwestern University, the director of the ArticuLab research group, and the graduate director of the interdisciplinary Technology and Social Behavior Ph.D. program. Her research focuses on better understanding everyday kinds of conversation and narrative as practiced by children and adults, and on building computational systems that simulate, mediate, and facilitate those everyday kinds of talk.

DENNIS DES CHENE is professor of philosophy at Washington University in St. Louis. He is the author of *Physiologia*, *Life's Form*, and *Spirits and Clocks*, and is now preparing works on the passions in the seventeenth century and mathematics in the twentieth.

ANTHONY GRAFTON teaches European intellectual and cultural history at Princeton. His interests include the history of scholarship, the history of science, and the history of books and readers from antiquity to the present.

STEFAN HELMREICH is associate professor of anthropology at MIT. His research examines the work of biologists concerned with the conceptual and material limits of the category of "life." His book *Silicon Second Nature: Culturing Artificial Life in a Digital World* is an ethnography of Artificial Life scientists at the Santa Fe Institute for the Sciences of Complexity. Helmreich is presently at work on a book about marine microbiologists who work in extreme environments such as deep-sea vents.

EVELYN FOX KELLER received her Ph.D. in theoretical physics at Harvard University, worked for a number of years at the interface of physics and biology, and is now professor of history and philosophy of science in the Program in Science, Technology, and Society at MIT. She is the author of many articles and books, including *A Feeling for the Organism: The Life and Work of Barbara McClintock*; *Reflections on Gender and Science*; *Secrets of Life, Secrets of Death: Essays on Language, Gender and Science*; *Refiguring Life: Metaphors of Twentieth-Century Biology*; *The Century of the Gene*; and *Making Sense of Life: Explaining Biological Development with Models, Metaphors, and Machines*. She is currently Chaire Blaise Pascal at Recherches Epistémologiques et Historiques sur les Sciences Exactes et les Institutions Scientifiques in Paris.

ELIZABETH KING is a sculptor interested in early clockwork automata, the history of the puppet, and literature's host of legends in which the artificial figure comes to life. Her sculpture is represented by Kent Gallery in New York, and she holds an appointment as School of the Arts research professor in the Department of Sculpture and Extended Media at Virginia Commonwealth University. She is the author of *Attention's Loop (A Sculptor's Reverie on the Coexistence of Substance and Spirit)*.

JOAN B. LANDES is the Walter L. and Helen Ferree Professor of Early Modern History and Women's Studies at the Pennsylvania State University. Her research engages questions of gender, political culture, and visual cognition in Old Regime and Revolutionary France. She is currently investigating eighteenth-century designs for artificial life. Her books include *Monstrous Bodies/Political Monstrosities in Early Modern Europe*; *Visualizing the Nation: Gender, Representation, and Revolution in Eighteenth-Century France*; and *Women and the Public Sphere in the Age of the French Revolution*.

TIMOTHY LENOIR is the Kimberly Jenkins Chair for New Technologies and Society at Duke University. He has published several books and articles on the history of biomedical science from the nineteenth century to the present. His more recent work has focused on the introduction of computers into biomedical research from the early 1960s to the present.

SCOTT MAISANO is assistant professor of English literature at the University of Massachusetts–Boston, where he teaches courses on Shakespeare, the English Renaissance, early modern theater, and artificial life. His current project, *Shakespeare's Fourth Dimension: The Scientific Romances*, positions the dramatist's last plays alongside the works of his contemporaries—Bacon, Brahe, Galileo, and Descartes—in order to reserve a place for Shakespeare not only in the history of science but also in the history of cinematic science fiction.

WILLIAM R. NEWMAN is the Ruth Halls Professor in the History and Philosophy of Science Department at Indiana University. Much of Newman's work has focused on the multiple intersections between alchemy, art, technology, and natural philosophy. His most recent books are *Atoms and Alchemy: Chymistry and the Experimental Origins of the Scientific Revolution*; *Promethean Ambitions: Alchemy and the Quest to Perfect Nature*; *George Starkey: Alchemical Laboratory Notebooks and Correspondence* (with L. M. Principe); and *Alchemy Tried in the Fire* (with L. M. Principe).

JESSICA RISKIN is associate professor of history at Stanford University. Her research interests include Enlightenment science, politics and culture, and the history of scientific explanation. She is the author of *Science in the Age of Sensibility: The Sentimental Empiricists of the French Enlightenment*. She is writing a book about the history of the animal-machine from Descartes to Darwin.

ELLIOTT SOBER is the Hans Reichenbach Professor of Philosophy and William Vilas Research Professor at the University of Wisconsin–Madison. His interests in the philosophy of science include the use of probability and statistics in scientific reasoning and conceptual issues in evolutionary biology.

ADELHEID VOSKUHL is an assistant professor in the Department of the History of Science at Harvard University. She is working on a book about the intersection of automaton-making and sentimental bourgeois culture in late eighteenth-century Switzerland and Germany as well as on the emergence of "philosophies of technology" among members of the Conservative Revolution in the Weimar Republic.

M. NORTON WISE is professor of history at the University of California, Los Angeles, where he also codirects the Center for Society and Genetics. His interests include the history of physics, technologies of knowledge, science and political economy, and landscape gardens. He is completing a book entitled *Bourgeois Berlin and Laboratory Science.*

ILLUSTRATIONS

ACKNOWLEDGMENTS

It is a great pleasure to thank the following people and institutions for their help, both in orchestrating the symposium that began our conversation and also during the collaborative process of making this book. At Stanford University, the dean of Humanities and Sciences; the Program in History and Philosophy of Science; the Program in Science, Technology, and Society; the Department of History; the Hind Fund; and the Humanities Center granted the funding for the symposium. The Stanford Humanities Center also gave us our excellent venue.

Oussama Khatib generously took time away from his research in robotics to give the workshop participants a tour of his laboratory and an explanation of the work he does there. Casey Alt designed the Web site for the symposium and served as audiovisual consultant during the proceedings. Rosemary Rogers brought her precious expertise to the administration of both the symposium and the compilation of the volume. Robert Scafe provided keen and efficient research assistance. Lucy Suchman gave an incisive and invaluable reading of the entire manuscript, as did two anonymous readers for the University of Chicago Press. Finally, I have relied with continual gratitude upon Paula Findlen's advice and guidance.

1   Introduction: The Sistine Gap JESSICA RISKIN

In Michelangelo's portrait of the creation of life, Adam reclines on a rocky ledge while God floats, supported by his angels in a mussel-shaped billow of fabric. God's arm stretches firmly from the right, while Adam's is draped, extended, waiting, from the left. The two arms converge in index fingers, Adam's swooning, God's pointing. Craning your neck to look up at the painting, at the off-center focus of the Sistine Chapel's ceiling, your eye is drawn from either side along the arm, down the finger, to the painting's own off-center focus. There, between the two fingers, one aiming, the other waiting, is Michelangelo's representation of life itself, reaching from God to Adam. It is, quite simply, a gap. The gap is powerful in its understated drama. Surrounded by flamboyance at the hub of the teeming ceiling, it is empty and open, an airy mystery.

This book is about a long tradition of attempts to sound the Sistine gap. Aiming to reenact Creation, at least in part, to produce life or some of its aspects artificially, philosophers and engineers have hoped to understand the connection between spirit and body, mind and matter, the subjective and objective experiences of life. Together, the essays collected here offer an unexpected and far-reaching result: they find that attempts at artificial life have rarely been driven by an impulse to reduce life and mind to machinery. On the contrary, designers of synthetic creatures have generally assumed a role for something nonmechanical, something other than a self-contained system of interacting parts. These designers and the respondents to their work have produced many and varied accounts of what the nonmechanical something might be, and of how it might engage with bodily mechanism. Thus, the history of attempts at artificial life is also a history of theories of soul and intellect, as these have been tested against an array of material arrangements. The essays that follow describe efforts to simulate life in machinery, to synthesize life out of material parts, and to measure living beings against inanimate mechanisms. These endeavors have not reduced living creatures to brute matter. Rather they have continually translated the meaning of vitality and sentience into new terms, and the process of translation continues. The essays ahead chart this process of transformation.

In October 2003, a group of historians, philosophers, computer scientists, and engineers came together at the Humanities Center on the campus of Stanford University to discuss the centuries-long project of understanding life by reproducing it. This volume has grown from their conversation. Our subject is artificial life in the general sense; the current fields of Artificial Life (with capital letters) and Artificial Intelligence are only instances in our examination. Some of the essays treat recent and current work in Artificial Life and Artificial Intelligence. But they all study experiments in artificial life (not capitalized, so as to encompass both past and current work): attempts to understand life and mind by reproducing them or some of their aspects artificially. They return to a common set of themes and problems; for example, how to reconcile the subjective experience of consciousness with the objective viewpoint of a mechanist explanation. This thematic coherence is no surprise, since it corresponds with a kind of historical and intellectual coherence: many of our more recent protagonists have studied and responded to the work of earlier ones.

My central purpose in putting together this collection has been to draw historical and philosophical connections among different moments and areas in the history of artificial life. I mean to treat current projects on a par with earlier developments: that is, as having a history. Such a volume cannot be comprehensive. Moreover, since there is an extensive literature—technical, philosophical, anthropological, sociological—on recent and current work in Artificial Life and Artificial Intelligence, I have been even less concerned to be comprehensive in these areas than in the others treated in our volume. Our original contribution with regard to Artificial Life and Artificial Intelligence is to treat these fields as occupying a moment in a much longer history.

Several tendencies characterize most of the recent writing in this area. First, authors have been concerned primarily to identify likenesses between historical and current attitudes toward the possibility of manufacturing living beings. Tom Standage's *The Turk*,[1] for example, emphasizes the similarities between discussions surrounding an eighteenth-century ostensible chess-playing automaton and modern conversations about machine intelligence. Second, the history of artificial life often comes across as the history of a determinedly, not to say dogmatically, reductive approach to life and consciousness. Gaby Wood's *Edison's Eve*,[2] which details a series of robot-related vignettes from the seventeenth to the twenty-first century, finds the same impulse at work in each, a sort of mechanist rationalism gone mad. Finally, recent writing has generally focused on the devices themselves, offering technological lineages from the early ancestry of electronic computing.[3]

This book, in contrast, is not primarily about continuities. We are as interested in the discontinuities and transformations in this history as in the elements that remain stable. Nor is our book mainly about devices, though it does feature some very beautiful and interesting ones. It is especially not about a centuries-long, unchanging drive to reduce life to a device. The theories of life and intelligence described in these essays include mechanical systems, but rarely, if ever, assign them full responsibility for all aspects of living beings. And, although the essays presented here establish that philosophers and engineers have been manufacturing lifelike entities and using them to take life's measure since antiquity, these essays highlight change: radical transformations, not limited to the devices themselves, but extending down into the underlying theories of life and mind, and out into the surrounding cultural, social, economic, and political terrain.

In a word, this book presents a historical approach to artificial life. Even where it considers recent and current developments—and even where its authors adopt philosophical, anthropological, and other modes of analysis—the overall framework is historical in a particular and crucial sense. That is, we are interested, not in timeless compulsions, transcendent forces, or inevitable trajectories, but in the changefulness and contingency of all views of life and mind, including current ones.

A historical assessment of artificial life will be of interest to historians, philosophers, and social scientists. We believe it can also help to inform the work of current practitioners of research in artificial life and anyone engaged in evaluating their projects. One cannot judge current intuitions about the nature of life and cognition without taking into account where these intuitions have come from; how they have depended upon myriad elements in natural science, engineering, culture, and politics; and how they have, in consequence, repeatedly and fundamentally changed.

To be sure, there are elements of continuity, but these generally take the form of questions: can one design a machine able to talk, write, reason, play chess, make music, have feelings, express emotions? These queries go back at least to the seventeenth century. What about a machine that can interact, learn, grow, evolve? These questions have generated vigorous debate since the mid-nineteenth century. As for the idea of synthesizing living creatures as we do chemical compounds, that, as William Newman's essay on the homunculus reveals, was a fond ambition of medieval and early modern alchemists.

While the questions have persisted, their answers have undergone profound renovation at each stage. Indeed, the terms in which the designers of artificial creatures, and their respondents, have defined the central problem—the very meanings they have ascribed to the key words—have transformed over time, so that what we currently mean by such words as *alive* and *conscious* is markedly different from what people have meant at various moments in the past. To Aristotle, a living being was one possessing an internal principle of change, able to grow and, in the case of animals, to move at will.[4] Early modern philosophers, doctors, and physiologists made life a matter of bodily machinery and often distinguished it, following Descartes, from the functions of mind and consciousness, which they took to be the province of an immaterial soul.[5] Enlightenment and Romantic natural philosophers deemed life and consciousness alike to be neither mechanisms nor souls but substances: animating spirits, vital fluids, animal electricity and magnetism, forms of energy.[6] Following the publication of Charles Darwin's theory of evolution by natural selection, new features became definitive of life: the ability to interact with an environment, to respond, to self-regulate, to adapt.[7]

In each period, the dominant theories of life have included mechanical and nonmechanical elements working in concert. Indeed, the nonmechanical aspects of these theories—and the ways in which they engage with the mechanical aspects—have often undergone the most striking changes. For example, during the early seventeenth century, the idea of the soul gave ground to a new notion that philosophers began to develop: an immaterial "consciousness" not reducible to bodily mechanism. John Locke, a principal author of this idea, wrote that an animal or human body was very much like a watch, a system of moving parts. On the other hand, "*personal identity*"—that by which "everyone is to himself that which he calls *self*"—was something different. Selfhood resided in "that consciousness which is inseparable from thinking." And consciousness, Locke reasoned, could not be a "substance" or arrangement of material parts, for it could flicker in and out of existence, could be interrupted by forgetfulness or deep sleep, without ever being lost.[8] Although people's selves attached to their watchwork bodies, Locke argued that the self was not made of matter and could never be accounted for by any kind of machinery. Consciousness was different from its ancestor, soul, carrying with it a certain secularism, a whiff of materialism, a philosophical novelty, by virtue of the fact that it came in a package with a mechanist view of life.

Around the middle of the twentieth century, philosophers, scientists, and engineers developed another new concept with which they invested some of the key functions of consciousness: the concept of information. The central originators of this idea were working at the Massachusetts Institute of Technology to establish a new field of inquiry called *cybernetics*. They took the name from the Greek word for "steersman," and steering represented the founding principle of cybernetics, which was that information—giving, receiving, and acting upon it, as a helmsman does—was the crux of animal life and of automatic mechanism alike. Both animals and machines maintained their internal functions and performed tasks by processing information, according to the mathematician Norbert Wiener, the leading founder of cybernetics and the term's coiner.[9] Cyberneticists and their fellow travelers established the now-ubiquitous idea of another realm of existence, apart from the realm of machines and bodies, the realm of information. Wiener wrote, "The mechanical brain does not secrete thought 'as the liver does bile,' as the earlier materialists claimed, nor does it put it out in the form of energy, as the muscle puts out activity. Information is information, not matter or energy. No materialism which does not admit this can survive at the present day."[10] Information was to supply the gap not only between organism and environment, but between animate and inert, intelligent and rote. Instead, it reproduced the old chasm within itself. For, as Wiener defined it, information was a hybrid, half physical phenomenon—light, heat, electricity—and the other half meaning: the "content" of the signal.[11]

Over the next few decades, the idea that information comprised a realm apart took firm hold. In 1979, the cognitive and computer scientist Douglas Hofstadter predicted that intelligence would turn out to be "a property that can be 'lifted' right out of the hardware in which it resides—or in other words, intelligence will be a *software* property."[12] Rodney Brooks, director of the Computer Science and Artificial Intelligence Laboratory at MIT, likes to invoke an as yet unknown "new stuff," which he also calls "the juice," and which he predicts will be a necessary element in the explanation and production of life. The new stuff in question is actually not very stuff-like; it is a "new mathematics" that will explain perception, evolution, cognition, and consciousness.[13] And the virtual-life programmer Steve Grand has written, "Although the materialist viewpoint is undoubtedly the truth, it is not the whole truth. . . . I believe that the computer, if interpreted correctly, can be the saviour of the soul rather than its executioner. It can show us how to transcend mere mechanism, and reinstate a respect for the spiritual domain that materialism has so cruelly . . . destroyed." The spiritual domain that the computer will restore to us, according to Grand, turns out to be "the world of software."[14] Thus, for Hofstadter (a computer scientist), Brooks (a roboticist), and Grand (a virtual-life programmer) alike, software assumes the role of spirit.

The idea that mathematics is the soul of the world goes back at least to the fourth century BCE and Plato's view that the material world of daily life was to the ideal world of geometrical forms as shadows cast on the wall of a fire-lit cave. On the other hand, the notion that there is some other source of life and mind, something beyond interacting bits of matter, has, under the pressure of attempts to undermine it, taken radically new forms at each juncture. Efforts to create life and mind in machinery have spawned ever-new renditions of that other something, from consciousness in the mid-seventeenth century, to information in the mid-twentieth. From soul to consciousness to software, successive theories of life have retained central, nonmechanical elements, and these frequently steal center stage in this collection.

By chronicling the history of attempts to understand life by reproducing it, or some of its aspects, artificially, the authors of this book lend depth and complexity to current conventions regarding the nature of animal life, sentience, and human cognition. Two such conventions became conspicuous during the symposium from which this volume arose and are prominent in the papers collected here. Let us consider them under the headings of "Emergence" and "Interaction."

EMERGENCE

Engineers and philosophers widely subscribe to the commonplace that life and consciousness are so-called emergent properties, arising in the course of evolution at a certain level of physical complexity. The notion of emergence is rooted in ancient ideas; for example, in the Aristotelian and, later, alchemical notion that "mixed" substances have properties not traceable to any one of their elements, and in arguments for the irreducibility of natural phenomena that likewise go back to Aristotle and beyond.[15] But in its modern form, the idea of emergence has its roots in Charles Darwin's theory of evolution by natural selection, and began to figure in philosophical treatments of the origins of life and consciousness during the decades following the publication of On the Origin of Species (1859).[16] Thomas Henry Huxley, Britain's leading Darwinist—who, upon first reading the Origin of Species, exclaimed, "How extremely stupid of me not to have thought of that"[17]—decided that mind, like other aspects of animal life, was the direct consequence of animal machinery, and that animals and humans were essentially "conscious, sensitive, automata"[18] (as discussed in M. Norton Wise's chapter in this volume). This conviction was founded in Darwin's theory of evolution, which gave Huxley a new way of connecting mind with body. Following the implications of Darwinism where they led him, Huxley decided consciousness was neither utterly separate from bodily machinery nor reducible to it, but was its "collateral product." Mind was to body, Huxley wrote, as the sound of a locomotive's steam whistle to its engine, as the ring of a clock's bell to its works.[19]

Huxley's idea that mind was a collateral product of bodily structure grew out of his commitment to Darwinian gradualism, the view that living beings had developed their present features by degrees. Gradualism led Huxley to conclude that there must be a continuum stretching from the higher mental powers all the way down to the rudiments of sensation and reflexive reaction, capacities that could be explained far more easily than self-conscious thought in purely physiological terms.[20] Darwin himself had argued that the "interval in mental power" between the lowest animals and man, be it ever so "immense," was all "filled up by numberless gradations."[21] Drawing upon Darwin's notion of the evolutionary emergence of consciousness, Huxley framed a new variety of the mechanist account of mind. He translated evolutionary emergence, which occurred in time, into structural emergence, which happened at a certain level of complexity. Accordingly, mind and consciousness were neither contained in a particular substance nor reducible to a given mechanism. Instead, they were features of a physical system that emerged when the system was sufficiently complex. As the system's intricacy reached the requisite level, consciousness simply began to sound its tone.

The idea of emergence in various renditions has been an organizing theme of modern projects in artificial intelligence from the beginning. John von Neumann, a leading figure in the development of the digital computer and in cybernetics, wrote that "below a certain minimum level of complexity, you cannot do a certain thing, but above this minimum level of complexity, you can do it," and so the salient difference between a brain and a computer was that "the human nervous system is roughly a million times more complicated."[22] Three decades later, Hofstadter proposed that viewing the "mind/brain" on a "high level" might involve "concepts which do not appear on lower levels," giving such a "high-level" view an "explanatory power that does not exist—not even in principle—on lower levels. It would mean that some facts could be explained on the high level quite easily, but not on the lower levels *at all*. No matter how long and cumbersome a low-level statement were made, it would not explain the phenomena in question."[23]

Emergence remains prominent in attempts to understand the origins of life and mind by reenacting them. Rodney Brooks, for example, has adopted an "incremental" approach to building intelligent systems in which he adds intelligence to the system "layer" by layer. Working in this way, he aims not so much to design an artificial intelligence as to bring one about by emergence.[24] His version of emergence, in keeping with most modern uses of the concept, connects intelligence with *structural* complexity, a feature of the system's organization, rather than with properties intrinsic to its *stuff*. But recently, as Bernadette Bensaude-Vincent's essay in this volume shows, some emergent theories of life and mind have returned to another idea with an ancient pedigree and have begun to invoke an "activeness" on the part of matter itself: an assumed capacity of organic matter to change spontaneously, assemble itself into patterns, and maintain and evolve these over time. Aliveness and sentience then emerge from complex arrangements of parts that are themselves not inert, but imbued with lifelike abilities and tendencies.

INTERACTION

"Interaction" designates the second current convention regarding the nature of life and mind. Closely connected with the principle of emergence, and similarly widespread among philosophers and engineers, is the conviction that the ability to interact with an environment, including exchanges with other creatures, represents the defining feature of living beings. The interactive principle, together with the concept of emergence, guides Brooks's incremental approach to designing an intelligent machine. The intelligence of his incrementally engineered system would ultimately be founded, he has said, in its "interactions with the world." Accordingly, such a system must be embodied, situated in the world, and in constant, physical engagement with it. Brooks builds robots able to move and sense.[25]

He and other designers of artificial life consider the crucial, interactive capacity to embrace psychological and social as well as biological responsiveness. Indeed, many current practitioners of artificial life research and their respondents place particular emphasis on the ability to have and express emotions as an important frontier in the design of lifelike machines.[26] Cynthia Breazeal, one of Brooks's former students and a roboticist at MIT's Media Lab, has argued that cognitive processes are inseparable from emotional ones in learning and intelligent decision-making. Evelyn Fox Keller discusses in her essay in this volume how Breazeal therefore designs robots able to interact emotionally—for example, by means of facial expressions. Whether the robots' emotions resemble human emotions on a subjective level, Breazeal argues, is irrelevant. A practitioner in the emerging field of robotic psychology, sometimes also known as developmental or humanoid robotics, Breazeal is a behaviorist. That is, she is interested only in the external manifestations of robotic emotions and in their effects, both on the robots' interlocutors and on the robots' own development.[27]

Like the concept of emergence in its modern form, the current emphasis on interaction also has a long lineage. This lineage includes eighteenth-century sensationist philosophy, whose promoters embraced the maxim that all thought originated in bodily sensations. Sensationist philosophers treated the senses, in turn, as the medium of interaction between any creature and the world outside. To think was to interact. The interactive view of life and mind in its modern form, again like the notion of emergence, has important sources in the middle of the nineteenth century. Darwinism made adaptation to external conditions and competition with other creatures fundamentally constitutive of the life of an organism. Nineteenth-century discussions of energy conservation recast animals in terms of what they took in and gave out, as fuel-consuming, work-producing machines.

Most directly in line with the interactive approach to artificial life, the nineteenth-century French physiologist Claude Bernard argued that responsiveness was the very condition of life. He took life to consist essentially of "a continuous and delicate compensation" enacted by a living organism in response to its external environment. Bernard therefore defined life as "a conflict between the external world and the organism," a conflict that, however, was "by no means a discord but a harmony."[28] This view of life as a continual, interactive balance between organism and environment became increasingly important during the last decades of the nineteenth century and the first decades of the twentieth, receiving the name "homeostasis" in 1932 from the Harvard physiologist Walter Cannon, whose work was foundational to cybernetics.[29] Cannon and his cyberneticist associates treated animal bodies as self-regulating devices on the model of the steam engine governor.[30]

Bernard had assumed that the ability to interact and compensate distinguished living creatures from machines, but the cyberneticists argued that, on the contrary, homeostasis constituted a link between animal and artificial machinery. Automata, "whether in the metal or in the flesh," needed to enact a continual exchange with the outside world.[31] The field of behaviorism in psychology likewise grew out of a rejection of Bernard's earlier view that "responsive" and "mechanical" were mutually antithetical. Pavlov and other founding behaviorists described an animal's ability to respond to its environments as "machine-like."[32] Early behaviorists argued that if psychology were to become properly scientific, psychologists must stop trying to study the ineffable, namely, consciousness, an invisible and nonmechanical phenomenon. Instead, they should set their sights on behavior, which, the behaviorists argued, was both observable and mechanistic. They thereby prepared the ground for cyberneticists to import behaviorism into their own projects, aiming to produce machines that responded to one another and to their environments, and evaluating their success on the basis of their machines' "behaviors."

Thus Breazeal's behaviorism, and that of her colleagues in the emerging field of robotic psychology, is part of a tradition. Again like the idea of emergence, behaviorism has run deep in artificial life and intelligence since the middle of the twentieth century. Cyberneticists took an explicitly behaviorist approach, designing their machines from within, but evaluating their performance from without, like zoologists observing the habits of a new kind of animal. Because a cybernetic automaton's defining feature was its ability to interact, its crucial parts, distinguishing it from the "blind, deaf and dumb" clockwork automata of the seventeenth and eighteenth centuries, were its sense organs. These might be photoelectric cells, able to tell light from dark, or instruments for measuring tension in a wire or changes in temperature. Indeed, any scientific instrument for measuring any quantity whatsoever was potentially a sense organ.[33] Wiener emphasized that automata with sense organs were no pipe dream: they already existed. They included thermostats, automatic door-openers, controlled missiles, proximity fuses, gyrocompass ship-steering systems, and anti-aircraft fire-control systems.[34] These had all been designed with immediate, practical aims in mind.

But Wiener and others also built sensing automata for purely experimental purposes. The Moth and the Bedbug, for example, were little three-wheeled carts equipped with propulsive motors, rudders, and photocells.

The Moth was designed to seek light, and the Bedbug to avoid it.[35] Two mechanical Tortoises, Elmer and Elsie, devised by a Cambridge University neurologist and engineer named Grey Walter, were more sophisticated, having two senses rather than one: they included touch sensors as well as photocells. The Tortoises explored and responded to a simple environment consisting of lightbulbs and obstacles.[36]

Their most impressive achievement, however, was entirely unexpected. Being themselves equipped with lightbulbs, they responded not only to the light sources and obstacles in their environment, but also to their own images in a mirror, and to one another, in a way that startled and delighted their inventor. They seemed to him to exhibit self-recognition as well as "mutual reactions which would be interpreted by most animal psychologists as social behavior if they were found encased in flesh and blood instead of brass and steel."[37]

Along with the Moth, the Bedbug, and the Tortoises, the genus of cybernetic automata included another species, the Homeostat, built by Ross Ashby, a British psychiatrist and cyberneticist. The Homeostat was a self-regulating electromechanical device with four pivoting magnets that, when disrupted, found a way back to its equilibrium state, "like a cat that turns over and goes back to sleep when it is disturbed."[38] Because it received information from the environment and acted accordingly to maintain a particular state, the Homeostat simulated deliberate behavior, giving "the appearance of purposefulness."[39]

And appearance was arguably everything. If responsiveness to its surroundings at a certain level of complexity was the criterion that made a living creature alive, then its vitality could only be judged from outside. The fact that Elmer and Elsie, Walter's Tortoises, *appeared* to recognize themselves and one another, and to dance little jigs of mutual greeting, was significant to their designer and his colleagues, despite the fact that they all knew the simple, electromotive causes of these phenomena. Taking a deliberately behaviorist approach to life and consciousness, cyberneticists set aside questions of the internal mechanisms or subjective experiences of the automata, and regarded them instead exclusively in terms of their outward manifestations. Walter's unexpected observation of his mechanical creatures' apparent social life was, Wiener predicted, "the beginning of a new science of mechanical behavior."[40]

From cybernetics to robotic psychology, a behaviorist approach to artificial life has characterized research programs whose goal is an embodied artificial intelligence. But behaviorism has not been the sole preserve of such research programs. A similarly externalist impulse was at work in one of the earliest moments of a sharply contrasting tradition of artificial life, classic Artificial Intelligence. In his 1950 essay "Computing Machinery and Intelligence," the inaugural paper of the discipline, the British mathematician Alan Turing famously proposed what has since come to be known as the "Turing Test" for artificial intelligence.

Rather than asking "can machines think," Turing proposed, we should instead ask whether they could pass a particular test. This test was modeled on a game, the "imitation game," in which a man and a woman were concealed from an interrogator, who tried to determine which was which by asking them questions. Both man and woman were constrained to be truthful in their answers, but the man was to try to mislead the interrogator, and the woman, on the contrary, was to try to help him or her. Now suppose, Turing proposed, that a computer were to take the place of the man in this game, and any human the place of the woman. The interrogator would then try to determine which was the human and which the machine. The human would try to help, while the computer would do its level best to persuade the interrogator that it was the human being. Any computer that could win this game as often as the man in the older version would be considered intelligent.[41]

In other words, for a machine to be intelligent, it must simply *appear* intelligent in conversation. To be intelligent was to seem intelligent, and to seem intelligent was to be intelligent. Turing anticipated an objection that he called "*the Argument from Consciousness*," namely, that intelligence was something that could be proven only from within, from the subjective experience of "feel[ing] oneself thinking." But, putting aside the fact that to accept this principle would be to doubt the intelligence of any other being, artificial or human, Turing argued that even the most rigorous skeptic would have to be persuaded by the following dialogue:

> Interrogator: In the first line of your sonnet which reads "Shall I compare thee to a summer's day," would not "a spring day" do as well or better?
> Witness: It wouldn't scan.
> Interrogator: How about "a winter's day." That would scan alright.
> Witness: Yes, but nobody wants to be compared to a winter's day.
> Interrogator: Would you say Mr. Pickwick reminded you of Christmas?
> Witness: In a way.
> Interrogator: Yet Christmas is a winter's day, and I do not think Mr. Pickwick would mind the comparison.

Witness: I don't think you're serious. By a winter's day one means a typical winter's day, rather than a special one like Christmas.[42]

Trying to conceive of a state of affairs that could conclusively demonstrate a computer's intelligence, Turing did not imagine, say, a physical theory of the mechanism of thought, or any other internal account of the nature of intelligence. Instead, he pictured chatting with a machine and figured that if it could hold up its end of the conversation, we should grant it the same courtesy—that of assuming it could think—that we would grant to any fellow human.[43] Like Grey Walter stepping back from his Tortoises' hardware in order to watch them interact, Turing imposed a screen between the internal mechanism of his hypothetical machine and its external (conversational) behavior.

The governing conventions of emergence and interaction are currently so ingrained that they can seem to be naturalizing, disappearing into the landscape. But I hope it is becoming abundantly clear that these are ideas with long and complicated lines of descent, during which they have taken many different and sometimes mutually contradictory forms. One cannot properly evaluate them or other current hypotheses without seeing them in historical terms, as generated over time and by an assortment of factors—philosophical, scientific, technical, social, and cultural.

All of the ideas canvassed so far—emergence, active matter, the various forms of holism, sensationism, the view of animal bodies as self-regulating mechanisms—have also developed partly as reactions to countering visions. It is hard, I have been suggesting, to find subscribers to the radically mechanist-materialist view that living creatures are fully reducible to inert bits of matter in motion. It is much easier to find philosophers and engineers struggling to compose explanations out of many elements, including, but not limited to, mechanical arrangements. But the scarcity of promoters of the extreme mechanist position has not detracted from its power as a foil against which more moderate theorists have continually defined their own hypotheses.[44] At the opposite extreme, proponents of the idea that the essences of life and mind are utterly immaterial—religious spiritualists and philosophical idealists of many sorts—have similarly provided a reference point for those who, by measuring life against machinery, have hoped to discover a middle ground.

Finally, these contending views of life and thought, as the essays in this book reveal, have been entangled everywhere with social and cultural concerns of every kind. To take just a few examples from this volume: the political economy of industrialization and Victorian ideas about gender and race figure prominently in M. Norton Wise's analysis; early modern views of sex differences are at work in William Newman's essay; and theological questions regarding God's role in nature constitute the substance of Elliott Sober's analysis.

The Information Revolution, like revolutions generally, tends to erase history, to wipe the slate clean and begin anew. But artificial life is not coeval with electronic computing. On the contrary, the project of reinventing life and mind is several centuries older and, by some reckoning, ancient. This book is an attempt to restore to the project of simulating life its full, complicated history.

Let us now turn to the essays by way of the synthetic creatures themselves: Hercules slaying a serpent with a bow and arrow; a Franciscan monk of iron and linden wood who paces, beats his chest, and kisses a cross; a she-devil who flaps her wings and spits fire; a "homunculus," a microscopic human-like being grown in a flask from a seed, having extraordinary powers of intellect and prescience; a lady harpsichordist who plays several tunes on a real instrument, her eyes raptly following her fingers, her chest rising and falling with her breath; a mechanical draftswoman, painstakingly sketching—over and over—an Eastern-looking temple surrounded by palm trees; robots that mutate and evolve; virtual organisms that mutate and evolve; virtual fishes that swim, perceive, exhibit behaviors, and learn from their "environment"; microscopic, self-replicating machines; virtual people who communicate by gestures and facial expressions; actual robots that communicate by gestures and facial expressions. These are this book's protagonists. Their line stretches back to antiquity and disappears over the present's forward horizon. It encompasses creatures made of wood, metal, stone, glass, papier-mâché, electronic circuitry, silicon, and information.

Some of the artificial beings populating these pages have been built; others have been spectacularly faked; and some have driven great controversies without ever assuming physical embodiment, purely as possibilities. They are accompanied by other entities, not precisely manufactured, but similarly testing the boundaries between vital and inert, intelligent and rote: Hermione, masquerading as a moving statue in the final scene of The Winter's Tale; human beings themselves, viewed as organic machines designed and built by an intelligent deity. In each case, the informing question is the same: can we arrive at an understanding of life and mind by measuring these mysteries against something we fully comprehend, namely, mechanical and material entities of our own devising?

The volume is organized according to a compromise between thematic and chronological considerations. It is divided into three parts, each defined by an organizing theme, while the essays within each part are arranged (roughly) chronologically. The essays in part 1, entitled "Connections," examine efforts to connect intellect with matter, souls with mechanisms, magic with engineering. The connections in question are not reductions to clockwork mechanism: intellect, soul, and magic remain whole and intact, but they assume an active partnership with bodily things. Part 1 opens with Sylvia Berryman's "The Imitation of Life in Ancient Greek Philosophy," in which Berryman examines some ancient automata representing certain principal functions of soul as described by Aristotle: the nutritive capacity, common to plants, animals, and people; the capacities for locomotion, perception, and desire, shared by animals and humans; and, finally, the capacity for thought, specific to humans.

For example, Berryman describes two ancient, hydraulic automata by Hero of Alexandria (ca. 10–75 CE) that seem to mimic acts of perception. One shows a figure of Heracles with a bow and arrow. An apple in the ensemble, when lifted, triggers Heracles to shoot his arrow at a serpent in a tree, whereupon the serpent hisses. In a similar arrangement, birds in a tree whistle as an owl pivots toward them. When the owl faces the birds, they fall silent as though intimidated. Finally, as the owl pivots away, the birds begin to sing again. From these and other instances, Berryman suggests that ancient natural philosophers drew analogies between technological arrangements and the workings of living beings. Tertullian (ca. 155–230 CE), for example, compared the soul to Archimedes' water organ.

The harmonious coexistence of mechanism and mechanisms with ideas about soul and spirit is again evident in Anthony Grafton's essay, "The Devil as Automaton." This essay's protagonist, the fifteenth-century engineer Giovanni Fontana, was largely disdainful of traditional magic and used his automaton devils to deflate stories told by magicians and theologians. At the same time, however, Fontana listed demonic and angelic inspiration and dreams as important sources of knowledge. Indeed, he reported having consulted with a spirit himself. He was keenly interested in certain forms of magic, including astrology and talismanic magic, and believed that statues could become "ensouled": inhabited by spirits. Grafton argues that Fontana and his contemporaries took automata to epitomize a new kind of magical practice whose powers did not arise from demonic intervention but had an equally metaphysical source. This new magical practice, which could produce mechanical beings that were "almost alive," originated in the God-given force of the human intellect.

If mechanical wonders arise from a divine gift in Grafton's story, the reverse—a reliance of transcendence upon mechanism—is the subject of Scott Maisano's "Infinite Gesture." Maisano draws a line from Descartes back to Shakespeare, who, he suggests, was already a psychophysical dualist. Shakespeare used the idea of the human body as a machine to invent for his characters, by contrast with their bodily machines, unprecedentedly real, crucially nonmechanical selves. The very finiteness and inadequacy of the bodily machine—for example, in The Winter's Tale, Polixenes' inability to give thanks in perpetuity, or Leontes' powerlessness to do eternal penance or to express infinite sorrow for the actions that led to the death of Hermione—are proof of the existence of a transcendent mind or soul able to grasp the infinite. In the same way, Maisano points out, Descartes would later cite his recognition of the existence of infinity, despite being unable to count up to it, as evidence of a mind furnished by God with clear and distinct ideas.

Thus, precisely by referring to his body as "this machine," Hamlet acquires an ineffable inner self. It is no coincidence, Maisano argues, that Shakespeare was creating characters with modern selves at the same time that engineers were filling Europe's courts with automata. Hermione's appearance as an ostensible automaton in The Winter's Tale is a clear reference to these actual machines. Only automata, weeping unending tears from eyes of stone, could embody the idea of "perpetual emotion" that the characters continually invoke. (Here Maisano's discussion resonates with Elizabeth King's account in this volume of a sixteenth-century mechanical monk as an expression of "perpetual devotion.") And Hermione herself can show the boundlessness of her love only by means of the machine that she is not.

While the idea of the bodily machine crucially helped to constitute the emerging notion of the modern, immaterial self, early modern mechanist physiology, in return, derived its contours from ideas about soul(s) and their province. Dennis Des Chene's "Abstracting from the Soul" shows how mechanist physiologists such as Giovanni Borelli and Charles Perrault, who followed René Descartes in taking the machine as the model of intelligibility, nevertheless assumed the existence of an animal soul to supply to the bodily machine its motive force. Borelli and Perrault placed the nature of this animal soul beyond the bounds of their science. This exclusion, Des Chene writes, was in keeping with the traditional division between mechanics, which dealt with the mathematics of forces, and physics, which explained their origins. A physiologist who adopted a mechanist approach could accordingly get on with the business of studying animal machinery not because he denied the existence of an animal soul, but precisely by assuming its existence.

In keeping with the recognition, implicit in both Maisano's and Des Chene's chapters, that the idea of the animal machine was double-edged, Joan Landes retraces the equation between creatures and machinery in reverse to arrive at a resulting "anatomical view of the machine." Her subject is eighteenth-century anatomical wax models and écorchés, models made from preserved cadavers, which she considers alongside automata in "The Anatomy of Artificial Life." In the models and cadavers, as in the automata, designers were looking for the point where life connected with matter. The results were sometimes intensely dramatic. The natural anatomist Honoré Fragonard presented fully flayed humans bearing expressions of extreme emotion and/or engaged in incongruous activities. The disturbing power of these models resides in their simultaneous embodiment of two apparently antithetical truths: creatures are made of stuff; stuff does not make a creature.

The papers in part 1 show a nonreductive impulse at work throughout the history of artificial life. The designers of synthetic beings featured here did not strive to puncture the magic of mind and soul; rather, they aimed to show how these metaphysical entities worked in conjunction with mechanism. Just what were these nonmechanical aspects of life and mind? The philosophers and engineers discussed in part 1 and, indeed, throughout this book have offered no unified account of the metaphysical entities whose existence they frequently assumed. Builders of artificial creatures have had many things in mind to supplement mechanism—consciousness, active matter, emergent properties—and, sometimes, no clear idea at all. One of this volume's chief purposes is to lay bare the constant shifting of the terms in which experimenters have envisioned this "something else."

While part 1 shows a nonreductive tendency operating at key moments throughout the history of artificial life, an impulse to *connect* machinery with soul and intellect rather than to *reduce* life to mechanism, parts 2 and 3 examine the history of attempts to explain the nature of this connection. Specifically, the essays in these parts analyze the origins, development, and ramifications of the two current conventions described above under "Emergence" and "Interactions." Interactive and emergent views of life are antireductive in different ways, although, to be sure, the two generally work in conjunction in current theories. Those who emphasize interaction assume that the key to understanding living creatures is to focus, not on individuals in themselves, but on their intricate web of exchanges with one another and with their environments. Those who invoke the concept of emergence claim that phenomena such as aliveness and sentience cannot be analyzed into simple components, but exist only at the level of composition. These emergent phenomena are features of the overall system, not traceable to its parts.

Part 2, entitled "Emergence," opens with an early modern instance of the idea of emergence at work. William Newman, in "The Homunculus and the Mandrake," examines a tradition of chemical rather than mechanical attempts to create life artificially. Followers of the sixteenth-century alchemist Paracelsus von Hohenheim described procedures for growing various living creatures in flasks, including the homunculus, a morally and intellectually superior artificial person. Paracelsus and his followers claimed to produce homunculi by the putrefaction of semen, sidestepping what they took to be the usual, contaminating role of menstrual blood in reproduction. Alongside the homunculus, Newman considers another tradition regarding small people grown atypically: the legend that the root of the mandrake grew in the shape of a man, would utter a fatal cry when pulled from the ground, and would thereafter, if treated properly, bring luck to its keeper. Paracelsus attacked those who carved mandrake roots in the shape of men to profit from this legend. In Paracelsus's attack, Newman identifies a more general objection on the part of alchemists to the procedures of visual artists, which, the alchemists claimed, imposed merely external, accidental changes on matter rather than shaping it from within. To create a true homunculus, according to Paracelsus and his followers, one needed not only matter in the right configuration, but a process of generation that would transform the essence of the matter from within.

The alchemical notion that Newman describes, that in properly mixed substances there could arise new properties not explicable in terms of the constituents or their physical arrangement, represented a form of the concept of emergence. A new and quite different form of this concept arose, as discussed above, with Charles Darwin's theory of evolution by natural selection, which has been foundational to modern theories of life and mind as so-called emergent phenomena. Yet the evolutionary version of the idea of emergence retained certain elements of the earlier, alchemical version. In particular, evolutionary emergence kept the core assumption—with attendant moral implications—that life, whether natural or artificial, required a process not just of construction but of development.

Elliott Sober, in "Sex Ratio Theory," shows how Darwin's theory inaugurated a steady progression toward an ever-sharper distinction between living creatures and artificial devices. His essay complements the others in this volume by discussing how natural life became nonartificial. Early modern "intelligent design" theorists, appealing to such aspects of nature as the balance of male and female births, viewed natural life as a work of artifice. Just as a watch implies a watchmaker, they argued, the intricate mechanisms of living creatures imply an intelligent designer. After probability ideas took hold in the seventeenth and eighteenth centuries, people began to think that a mindless, chance process might, in fact, have produced life after all. Darwinism added a third alternative: neither intelligent design nor a mindless, random process, but instead a mindless, nonrandom process—evolution by natural selection. After Darwin, living structures such as the eye and inanimate mechanisms such as the watch seemed increasingly to have resulted from fundamentally different sorts of processes.

It was a difference, moreover, with formidable consequences. Whereas a process of intelligent design can be evaluated in moral terms, evolutionary emergence cannot. Even if one assumed that the evolutionary process had been launched by an intelligent being, who could be a subject of moral evaluation, the process works according to rules that render any such assessment incoherent. For example, Sober argues, selection can operate at different levels of organization: at the level of individuals and at the level of populations. In cases of conflicts of interest between individual and group selection, what would constitute benevolence?

A further, very familiar moral consequence of Darwinian evolution is that it suggested new ways of evaluating living beings. Evolution by natural selection implied that one could arrange the attributes and capacities of "conscious automata"—Huxley's phrase for animals and humans—diachronically, according to which ones emerged earlier and which later in the process of evolution. A ramification of this idea is the subject of M. Norton Wise's essay, "The Gender of Automata in Victorian Britain," which again looks at a problem regarding sex ratios, in this case, those of automata. Wise analyzes the implications of Darwinian evolution for Victorian ideas about women, as instantiated in the proliferation of female automata. Darwin's theory included the claim that males, being in constant competition to win female mates, evolved more rapidly than females. Women accordingly occupied a spot on the evolutionary scale somewhere between animals and men. According to Darwin, a capacity to recognize and respond to rhythms, which he believed to have appeared low on the evolutionary ladder and to be the basis of imitative behavior, directed women's sentiments and behaviors.

The idea that women, being more primitive in evolutionary terms, were more fully governed by a primeval sense of rhythm had roots, Wise suggests, in a classification of machinery that had been widespread in engineering textbooks and discussions of factory economy since the 1820s: the distinction between engines, which produced power, and mechanisms, which merely transmitted force to execute work. This classification had spread to encompass factory workers as well, with men as the engines and women as the mechanisms, preparing the way for Darwin to construe "primitive" as "mechanical." The same idea was at work in experimental physiology, whose practitioners proposed that involuntary motions such as the beat of the heart originated in a primeval, protoplasmic throbbing. Beginning with the proliferation of female musician-automata in the Victorian period, Wise thus explores a tapestry of associations throughout British culture and society, from engineering to the political economy of the factory system, evolutionary theories in natural history, and experimental physiology. This tapestry is the complicated background to Huxley's and others' interpretation of evolution to mean that human consciousness had arisen at a stage in the elaboration of living automata.

The post-Darwinian idea that living beings were evolving machines suggested that artificial machines might also be capable of evolving. Samuel Butler laid out the possibility that steam engines and other machines could evolve to the point of acquiring consciousness in his 1872 novel Erewhon, which included a deeply ambivalent spoof of Darwin's theory.[45] The idea of evolving artificial machinery became a familiar motif in philosophical and literary conversation during the second half of the nineteenth century. This possibility has since spread from philosophy and science fiction to orient research across the sciences. Timothy Lenoir's essay, "Techno-Humanism," examines how "the digital and the real" have been fused in recent developments in artificial intelligence. Lenoir takes as an example the field of evolutionary robotics, exemplified in Brooks's "incremental" method of design described above, in which researchers apply models of biological evolution to the development of technical systems. These researchers assume that the best way to synthesize thinking beings is to reproduce the process of evolution, setting up the conditions for intelligence to emerge once again.

In light of such ideas and developments, Lenoir suggests, the cyborg, an image invented by the cultural critic Donna Haraway in 1985, has lost its relevance.[46] The cyborg, a static fusion of human and machine, served to focus critical studies of an older model of the relations between machines and living creatures. But now, Lenoir argues, we need something else: a critical stance that responds to the evolutionary assumptions, ambitions, and, Lenoir believes, the successes of current researchers. He sees recent developments as having put human beings, social and cultural systems, and computing technology into a state of inextricable coevolution. The cyborg, which oriented critical analyses of the conjunctions of humans and machines at a particular moment, must therefore give way to what Lenoir calls "techno-humanism," a critical program whose ambitions are on a par with those of the people at whom it is directed: both parties, that is, aim to shape the course of evolution.

Bernadette Bensaude-Vincent's essay, "Nanobots and Nanotubes," examines another instance in which researchers have been blurring the line between the synthetic and the organic. Bensaude-Vincent argues that materials scientists working in the hot, new area of nanotechnology (engineering carried out at the level of the atom) exhibit an approach to the project of artificial life that contrasts sharply with the traditional approach of computer scientists. Whereas computer scientists dream of "nanobots," microscopic automata made of passive and uniform matter and informed by a program, materials scientists instead design structures such as "nanotubes," which make use of processes exemplified in living creatures, such as self-assembly. Nanotubes are made of carbon C_{60} atoms, which have a spontaneous, inherent disposition to make balls and tubes. The designers of nanotubes exploit what they see as the essentially active nature of organic matter, relying on its properties of self-assembly, spontaneity, and complexity. And these designers, like the researchers discussed in Lenoir's essay, also model their methods of design upon what they take to be the organic process of emergence by which living and intelligent systems come into existence.

The essays of part 2 reveal that the idea of emergence currently generates much discussion among philosophers, historians, roboticists, and engineers alike. Yet this idea has such a mystery at its core that it can seem to be begging the question. Often, indeed, "emergent" corresponds with "beyond understanding," as in the prediction by Danny Hillis, designer of the Connection Machine discussed in Lenoir's essay, that bringing about an emergent artificial intelligence might mean being able "to create more than we can understand." How do life and mind emerge from a proliferation of physical parts in increasingly complex arrangements? Just what happens at the level of emergence to turn an inert and mindless system into a living, thinking being? How can the whole be both greater than and equal to the sum of its parts?

The concept of emergence has been subject to competing interpretations, some ontological and the others purely epistemological. According to the first sort of interpretation, some new entity comes into existence at the level of emergence, an entity not reducible to component parts. According to the second sort, no new entity arises: life is ontologically reducible to its parts but cannot be understood in reductive terms. But attempts to define a purely epistemological form of emergence are afflicted by a core problem: how, ultimately, can we distinguish epistemological from ontological emergence? That is, given an explanation of a living creature, whether natural or artificial, that relies upon the notion of emergence, how can one ever rule out the existence of an emergent soul? Thus, emergence returns us to the gap with which we began—between God and Adam, spirit and body, life and matter—but construes it in a new way. With the notion that life and mind are emergent properties of physical complexity, the Sistine gap becomes an unfathomable fissure between parts and whole, between the individual mechanisms of life and their additive results.

David Bates, in "Creating Insight," questions the underlying assumptions of both classic Artificial Intelligence, founded in the analogy between the brain and the digital computer, and the newer tradition that construes the mind as an emergent property of interacting, distributed, parallel processes. Rather than seeing the brain as analyzable into sets of interacting rules or as the synthetic result of aggregates of simple systems, Bates proposes a third possibility, taken from the shared history of computing, psychology, and Artificial Intelligence. In particular, he examines Gestalt psychology and its brief engagement with cybernetics to suggest that here was perhaps a missed opportunity.

Gestalt psychologists such as Wolfgang Köhler set out to demonstrate the inadequacy of the deterministic, behavioral models of cognition that underlay cyberneticists' attempts to liken the brain to a digital computer. These models of cognition, according to Köhler and other Gestalt theorists, could not account for insight: the creative ways in which living creatures respond to their world, the sudden restructurings of thought that allow them to grasp new situations and to solve problems. Moreover, Gestaltists believed that the physical systems involved in thought (or, for that matter, in other bodily processes such as embryonic development) were also not strictly deterministic. They postulated an irreducible, intrinsic unity or, rather, two unities, isomorphic to one another, one somatic and the other psychological.

Cyberneticists such as Norbert Wiener were hostile to what they saw as the organicism and mysticism of Gestalt psychology. But Bates suggests that von Neumann shared significant common ground with the Gestaltists. Von Neumann, too, saw processes such as visual perception as resting on a perhaps un-analyzable recognition of the "wholeness" of an object. He also proposed an isomorphic relation between such intuitive forms of thought and a corresponding unity in the physical structure of the nervous system, a unity he took to be far beyond present understanding. Such a unity would be very different from an emergent property, Bates writes, because it would be "*present* from the start," organizing the system rather than produced by it. Artificial Intelligence researchers and cognitive scientists, even as they transform their approach in a holist direction, continue to neglect this older idea, Bates argues, which might bring important insights (so to speak) into the nature of minds, brains, and their relations to one another.

The holists of part 2 emphasized the importance of interactions within living beings, among the distributed parts of the complex systems that comprise them. Part 3, entitled "Interactions," focuses on the exchanges between creatures and the outside world, and on attempts to understand life in terms of these exchanges, which include not only physiological and sensory engagements, but equally, as discussed above, social and emotional ones.

The section opens with Elizabeth King, who portrays the power of an artificial being to provoke a response in "Perpetual Devotion." Confronted by the sixteenth-century mechanical monk she describes, four centuries of viewers have felt their hair stand on end. The monk—wearing a tunic, cowl, and sandals, its mechanism hidden beneath its habit—is a fully self-contained device, sixteen inches high. In its left hand, it clutches a rosary and cross. As it paces, the monk lowers and turns its head, moves its mouth and eyes, beats its chest, brings the cross to its lips. Its undeniable aura, King suggests, relies crucially on its size: a small, moving thing, it compels a primal, animal reaction, a shudder of recognition. The monk's smallness, too, serves the same function, perhaps, as the remoteness of the virtual aquatic creatures in Stefan Helmreich's essay (see below): the distance it imposes helps to vivify the moving entity at the other end. According to legend, King Philip II of Spain asked his engineer, Juanelo Turriano, to build the monk after King Philip's son, Don Carlos, made a miraculous recovery following a head injury. A fifteenth-century Franciscan monk, Diego de Alcalá, whose remains were brought to the prince's bed at the moment of crisis, received credit for the cure. And so the king, who in prayer had promised a miracle if only his son were spared, asked Turriano to build the automaton monk.

What did this object mean to its makers—for, King shows, it must have been the collaborative work of several people—and to its original viewers?

Her exploration suggests it had a variety of meanings, and it is their conjunction in one little machine that ultimately gives her tale its lesson. To begin with, the monk embodied the body-machine metaphor of contemporary physiology, which corresponded in turn to the clockwork universe of early modern cosmology. At the same time, the automaton's juxtaposition with the corpse of Diego de Alcalá, with the miraculous recovery of Don Carlos, and—like the eternally weeping machines in Maisano's reading of The Winter's Tale—with a perpetual act of prayer seems to express an "impulse to reach beyond death." The automaton monk was also a devotional object.

Such images, King suggests, drawing on the work of the art historian David Freedberg, are traditionally seen in a particular way, one that collapses the distinction between representation and presence: the saint inheres in the icon. At the same time, King writes, the monk exemplified a shift in the way such images were seen—the same shift chronicled in Grafton's story of the automaton devils of Giovanni Fontana—in which human agency was gradually replacing divinity as the source of the spiritual or lively presence within. Even through this shift, the Christian mysteries of resurrection and transubstantiation continued to vivify—or "ghost"—the automaton-icon. Its potent combination of clockwork, artistry, iconic imagery, and Christian legend brings King to a conclusion whose surprise is a shock of recognition: the monk epitomizes "the birth of the machine in the Age of Faith" and the continued burden of devotional belief with which he and his mechanical kind were invested.

Like King's essay, Adelheid Voskuhl's "Motions and Passions" is about the relations between physical and emotional dynamics. And, like Wise's chapter, Voskuhl's examines the cultural and social resonances of female musician-automata. In this case, however, the context is not Victorian Britain but the German late Enlightenment. Voskuhl juxtaposes eighteenth-century music-playing automata, notably the 1772 Musicienne, built by the Swiss clock-maker Pierre Jaquet-Droz and his son Henri-Louis, with contemporary texts on musical performance. In descriptions of the Musicienne's performances, witnesses marveled at the evocativeness of her gestures: her head turned (and still does, for she remains in working order in Neuchâtel, Switzerland), her eyes follow her fingers as they play, she breathes (by means of an internal bellows), and her breast seems to heave with feeling. At the same time, pedagogical works such as Johann Joachim Quantz's flute tutor and Carl Philipp Emanuel Bach's instructional text on piano-playing urged students not to play "mechanically" or sit like a "piece of carved wood" but to capture their listeners' hearts through bodily motions and gestures. The two subjects—the evocativeness of lady-musician-automata and the advice to students of music—came together in a satirical story entitled "Humans Are Machines of the Angels" (1785) by the writer Jean Paul, who produced his work from notebooks filled with excerpts taken out of books, journals, and ephemera of all kinds. This procedure made his writings a kind of distillation of contemporary interests and preoccupations.

In Jean Paul's story, humans are machines constructed by angels who in turn construct their own automata of "earth." These machines of earth, Jean Paul writes in the final sentence of his story, are mere inferior copies of "those female machines [of the angels, i.e., humans] who . . . accompany the music with bodily motions, which obviously seem to betray affective involvement." Here Voskuhl identifies a send-up both of the advice given in contemporary performance manuals and of the popular comparison of humans to machines. Jean Paul's spoof, Voskuhl suggests, indicates a fault line in Enlightenment thinking about the relations between mechanism and sensibility.

Writers such as Quantz and C. P. E. Bach described these relations in ways that were not only complicated but downright contradictory. On the one hand, they contrasted the merely mechanical with the emotional, assigning feeling, by implication, to the realm of the spiritual. On the other hand, eighteenth-century writers took bodily motions to be the very medium of emotions, the means by which feelings were both generated and transmitted, so that mechanical and emotional movements were ultimately inseparable. Texts such as Quantz's and C. P. E. Bach's present a kind of emotional kinematics, a physical theory of emotional interaction, in which bodily motions transmit emotional forces from the performer to the audience.

The idea that feelings and the other defining interactions of living creatures require a medium of transmission has been crucial to experiments in artificial life. Stefan Helmreich, in "An Archaeology of Artificial Life, Underwater," describes how current designers of artificial creatures—in this case, of virtual ones—place their creations in a medium in order to make them seem lifelike. The computer monitor has been essential, Helmreich suggests, to sustaining the impression that computers can contain worlds populated by living beings. In the Artificial Life simulations Helmreich studies here—Demetri Terzopoulos's comically lifelike Artificial Fishes and Karl Sims's boxy, genetic algorithm creatures that adorably tumble, play, and learn to swim onscreen—the computer monitor appears as a "kind of fish tank into which users can peer to see artificial life forms swimming about."

Life, Helmreich suggests, is lifelike only with respect to something external, something it can breathe, eat, consume, move through. By appearing to interact with a medium, aquatic virtual creatures come to seem thoughtful, or cute, or sentient. Moreover, the apparent medium also serves to distance the virtual creatures from the viewer, recalling the deeply familiar experience of the problem of other minds: we sense the sentience of other beings, even natural ones, but through a glass darkly. In the same way, Helmreich suggests, in John Lilly's 1960s work on dolphin intelligence, in which he claimed his dolphins began to repeat English phrases, the "feeling of weirdness" that Lilly reported—"We felt we were in the presence of Something, or Someone, who was on the other side of a transparent barrier"—was crucially sustained by the watery distance between Lilly and his subjects.

In the current field of Artificial Life, in keeping with the long-standing behaviorist tradition I have described, the fact that virtual creatures look alive takes on its own significance. In keeping with the idea that life is defined by interaction, one possible criterion of aliveness is simply to appear alive: an entity is alive if it provokes living beings to respond to it as such. In evolutionary or developmental robotics, as in Artificial Life, researchers aim to produce particular responses in human spectators. Some, indeed, make these responses the ultimate arbiter of aliveness and sentience: a robot or program will be alive when it makes people treat it as alive. Evelyn Fox Keller, in "Booting Up Baby," as I have mentioned, analyzes the implications of a generation of robots whose builders, including Rodney Brooks, Cynthia Breazeal, and Brian Scassellati, have made the reactions their robots elicit from human viewers—including themselves—their paramount consideration.

The robots built by these practitioners of developmental robotics are not just able to sense—with sensors allowing them to respond to light, sound, and motion—on the theory that cognition requires embodiment. This new generation of robots is also, according to its creators, capable of engaging in social behavior. The underlying idea is that human learning and cognition are not only embodied but fundamentally social. Breazeal, Scassellati, and other developmental roboticists aim to build a robot with the sensory and social capabilities of an infant and then allow it to learn, through its relations and exchanges with its "caretakers," to the point of having the cognitive abilities of a toddler.

Cog, the first "humanoid robot" with social capacities, identifies faces in order to make eye contact. It can also track the visual gaze of its interlocutors and direct their attention to a third object or person in the room. Kismet, a more advanced model, has cartoon-like facial expressions corresponding with "sadness," "happiness," and "fear," among other affective "values." Kismet also has "needs"—the primary ones being for social engagement and rest—and uses its facial expressions to motivate its "carers" to satisfy them. These features are not merely meant to give an impression of sentience, but actually to bring about a process of cognitive development. Leonardo, who followed Kismet, can add to its repertoire of facial expressions through imitation. Breazeal's behaviorist approach means that she focuses exclusively on the functional aspect of robotic emotions: these must simply serve the developmental functions of human emotions, evoking the responses in human trainers that enable cognitive development.

The theories informing the design of these robots are taken from cognitive psychology, including studies of autism and infant development. In return, many researchers in humanoid robotics are now proposing that their robots could be used to test these same theories. This apparently circular trajectory, Keller writes, is already common in applications of computer simulation to the physical and biological sciences, though it is new in developmental psychology. Her essay ends with a question: does this use of simulations in fact follow a circular trajectory, a closed loop in which the theory and the model can do naught but confirm one another, or is it rather a "spiral" having "forward momentum" despite—even due to—the "back-and-forth traffic between model and modeled"? Do people respond to these robots as lifelike because they are designed to embody what are currently the strongest convictions about what is lifelike? Or do people respond to the robots as lifelike because they *are* lifelike? Do the robots become more lifelike because of the way people respond to them? Or do the responses instead enable the robots to get better at *seeming* lifelike, transforming from something like a cartoon into something more like a realist painting?

The question of what a socially adept artificial being can teach its creators about human social and mental life is at the heart of the final essay in this collection. The artificial creatures in question, once again, are designed according to a single goal: to make humans react to them the way humans react to each other. But in this case, the greater purpose is not to create or bring about an artificial sentient being, but simply to understand better certain aspects of sentience in natural humans. Justine Cassell, in "Body Language," describes her primary research tools, virtual humans called "embodied conversational agents" (ECAs), as mechanisms that make people attribute humanness and aliveness to them. Although they have a purely virtual existence, Cassell's ECAs are expressions of her conviction that language takes place not just in the head but "spread throughout the whole body—the hands, the torso, the eyes." The design of these virtual people is also founded in the closely related principle that language can only be understood through a study of its social (relational, interactive) as well as its semantic aspects, and of the relations between them.

Cassell's virtual humans are designed to provoke people to exhibit fundamental human, social traits in response to them. They do so by incorporating the bodily and/or propositionally empty but socially crucial aspects of language into their speech: gesturing, nodding, moving their eyes, saying "uh-huh, uh-huh" to indicate they are listening. One of the ECAs, named NUMACK, offers directions to a spot on the Northwestern University campus; another, REA, acts as a real estate agent, giving virtual tours of houses. Cassell and her colleagues study people as they interact with NUMACK, REA, and their ilk. Both in their successes—when they elicit natural responses—and in their (sometimes comical) failures to do so, the virtual people have revealed a range of information, from the very specific to the quite general, about the social activity of language and about the relations between its social and semantic aspects. At the specific end of the range, for example, is the fact that a person giving directions uses the same hand to gesture to the left and right rather than using his or her left hand to gesture left and right hand to gesture right. Not only is this true, but it turns out to be important in the sense that violating the rule made NUMACK seem glaringly fake. At the general end of the range of knowledge generated by ECAs is the realization that "the phenomena of hand gesture, intonation, and facial expression [are] not separate systems, nor [is] one a 'translation' of the others, but instead [they have] to be derived from one common set of goals." An early ECA whose gestures, tones, and expressions were designed by three different researchers, each seeking to convey the same thing through his or her assigned medium (gesture, tone, or expression), inspired hilarity by seeming to be addressing either a very small child or else someone who did not speak his own language, or perhaps simply someone very stupid.

Cassell performs experiments on the essence of humanness and aliveness by studying the interactions between actual and virtual humans, using the design of artificial interlocutors as a way to identify the crucial features of natural ones. Her work nicely exemplifies the theme of part 3, the assumption that life and sentience reside in the interactions of living and sentient creatures with their surroundings. At the same time, Cassell conducts her work at a critical distance from what we might call the alchemical form of the interaction thesis. She takes interactions between simulated and real people as a way to understand humans rather than as a method for re-creating them by other means.

The many ways in which philosophers and engineers have construed and tried to cross the chasm between body and soul, matter and consciousness by producing (or by imagining the production of) living beings out of parts constitute the substance of this book. The designers of artificial life discussed in these pages have been mostly interested in bridging the gap, not in eliminating it. Had they sought to eradicate the gap altogether, the story would have been a monotonous tale of a centuries-long quest to reduce life to clockwork. Struggling to bridge the Sistine gap, in contrast, has meant trying to bring the intelligibility of artificial mechanisms to the workings of living beings. This is a dynamic story full of changes and reversals, experimentation and ambivalence, for intelligibility can take myriad forms, and any given model of intelligibility is shaped by a multitude of forces.

Nevertheless, I can propose two generalizations across these instances drawn from many centuries and situations. First, the essays ahead show the history of artificial life as a quest more for comprehension than for clockwork. And second, while clockwork, to be sure, has undergone some transformations during the history covered here, the changes in ideas about what might constitute comprehension have been dramatically greater.

NOTES

1. Tom Standage, *The Turk: The Life and Times of the Famous Eighteenth-Century Chess-Playing Machine* (New York: Walker, 2002).

2. Gaby Wood, *Edison's Eve: A Magical History of the Quest for Mechanical Life* (New York: Knopf, 2002).

3. This is true of Standage's and Wood's books (nn. 1 and 2, above); another example is Raymond Kurzweil, *The Age of Intelligent Machines* (Cambridge, MA: MIT Press, 1992), part 1.

4. Aristotle, *De Anima* 2.1.412b17–19; 2.2.413a22–30. On Aristotle's theory of the nature of life, see G. E. R. Lloyd, *Aristotle: The Growth and Structure of His Thought* (London: Cambridge University Press, 1968), chap. 9; Allan Gotthelf, ed., *Aristotle on Nature and Living Things* (Pittsburgh: Mathesis Publications, 1985); and Allan Gotthelf and James Lennox, eds., *Philosophical Issues in Aristotle's Biology* (Cambridge: Cambridge University Press, 1987).

5. On the early modern division of labor in mechanist physiology, see Dennis Des Chene, *Spirits and Clocks: Machine and Organism in Descartes* (Ithaca, NY: Cornell University Press, 2001); Annie Bitbol-Hespéries, *Le Principe de vie chez Descartes* (Paris: J. Vrin, 1990); Thomas Fuchs, *Die Mechanisierung des Herzens: Harvey und Descartes, der vitale und der mechanische Aspekt des Kreislaufs*

(Frankfurt am Main: Suhrkamp, 1992); and Karl Rothschuh, *Physiologie. Der Wandel ihrer Konzepte, Probleme un Methoden vom 16. bis 19. Jahrhundert* (Freiburg/München: Karl Alber, 1968).

6. On vitalism in Enlightenment life sciences, see François Duchesneau, *La physiologie des Lumières. Empirisme, modèles, theories* (The Hague: Martinus Nijhoff, 1982); Jacques Roger, *Buffon: A Life in Natural History*, ed. L. Pearce Williams, trans. Sarah Lucille Bonnefoi (Ithaca, NY: Cornell University Press, 1997), and *The Life Sciences in Eighteenth-Century French Thought*, ed. Keith Benson, trans. Robert Ellrich (Stanford, CA: Stanford University Press, 1997); Robert Richards, *The Romantic Conception of Life: Science and Philosophy in the Age of Goethe* (Chicago: University of Chicago Press, 2002); and Shirley Roe, *Matter, Life, and Generation: Eighteenth-Century Embryology and the Haller-Wolff Debate* (Cambridge: Cambridge University Press, 1981).

7. On Darwin's theory and naturalists' transformed understanding of the nature of life and mind in the second half of the nineteenth century, see Robert Richards, *Darwin and the Emergence of Evolutionary Theories of Mind and Behavior* (Chicago: University of Chicago Press, 1987); and Alain Prochiantz, *Claude Bernard: La révolution physiologique* (Paris: PUF, 1990).

8. John Locke, *An Essay Concerning Human Understanding* (New York: Penguin, 1998), 210, 212–13. (Orig. pub. 1689.)

9. Norbert Weiner, *Cybernetics, or Control and Communication in the Animal and the Machine* (Cambridge, MA: MIT Press, 1948), 11; Wiener, *The Human Use of Human Beings* (New York: Da Capo, 1950), 26.

10. Wiener, *Cybernetics*, 132.

11. Wiener, *Human Use of Human Beings*, 17.

12. Douglas Hofstadter, *Gödel, Escher, Bach: An Eternal Golden Braid* (New York: Basic Books, 1979), 358.

13. Rodney A. Brooks, *Flesh and Machines: How Robots Will Change Us* (New York: Pantheon, 2002), 187–88.

14. Steve Grand, *Creation: Life and How to Make It* (Cambridge, MA: Harvard University Press, 2000), 4.

15. See Paul A. Bogaard, "Heaps or Wholes: Aristotle's Explanation of Compound Bodies," *Isis* 70 (1979): 11–29; Gad Freudenthal, *Aristotle's Theory of Material Substance: Heat and Pneuma, Form and Soul* (Oxford: Clarendon Press, 1995), esp. 11–12; Lloyd, *Aristotle*, 171–75.

16. For a history of the modern concept of emergence, see David Blitz, *Emergent Evolution: Qualitative Novelty and the Levels of Reality* (Dordrecht: Kluwer, 1992).

17. Thomas H. Huxley, "On the Reception," in *On the Origin of Species*, by Charles Darwin (Peterborough: Broadview, 2003), 627. (Orig. pub. 1887.)

18. Thomas H. Huxley, "On the Hypothesis That Animals Are Automata" (1874), in *Methods and Results: Essays* (New York: D. Applewood, 1899), 199–250, on 201, 238. (Orig. pub. 1874.)

19. Huxley, "On the Hypothesis That Animals Are Automata," 240, 242.

20. Ibid., 236.

21. Charles Darwin, *The Descent of Man, and Selection in Relation to Sex*, ed. Robert M. May, with an introduction by John Tyler Bonner (Princeton, NJ: Princeton University Press, 1981), 35. (Orig. pub. 1871.)

22. John von Neumann, "Role of High and Extremely High Complication," in *Theory of Self-Reproducing Automata*, ed. Arthur W. Burks (Urbana: University of Illinois Press, 1966), fourth lecture. (Orig. pub. 1887.) See also Burks, "Editor's Introduction," 20.

23. Hofstadter, *Gödel, Escher, Bach*, 708.

24. Rodney Brooks, "Intelligence without Reason," MIT AI Lab Memo 1293, April 1991, 3; see also his "Elephants Don't Play Chess," *Robotics and Autonomous Systems* 6 (1990): 3–15, and "Intelligence without Representation," *Artificial Intelligence Journal* 47 (1991): 139–59.

25. Brooks, "Intelligence without Reason," 3; see also "Elephants Don't Play Chess" and "Intelligence without Representation."

26. Rodney Brooks, "Prospects for Human Level Intelligence for Humanoid Robots," *Proceedings of the First International Symposium on Humanoid Robots* (HURO-96), Tokyo, Japan, October, 1996; Rodney A. Brooks, Cynthia Breazeal (Ferrell), Robert Irie, Charles C. Kemp, Matthew Marjanović, Brian Scassellati, and Matthew M. Williamson, "Alternate Essences of Intelligence," *Proceedings of the Fifteenth National Conference on Artificial Intelligence* (AAAI-98), Madison, WI, July 1998, 961–76.

27. Cynthia Breazeal, *Designing Sociable Robots* (Cambridge, MA: MIT Press, 2002), and "Emotion and Sociable Humanoid Robots," *International Journal of Human-Computer Studies*, 59 (2003): 119–55.

28. Claude Bernard, *Lectures on the Phenomena of Life Common to Plants and Animals*, trans. Hebbel E. Hoff, Roger Guillemin, and Lucienne Guillemin (Springfield, IL: Charles C. Thomas, 1974), 84, 251, 46. (Orig. pub. 1871.) See also his *An Introduction to the Study of Experimental Medicine* (New York: Dover, 1957), 59–99. (Orig. pub 1865.)

29. Walter Cannon, *The Wisdom of the Body* (New York: W. W. Norton, 1932). The term *homeostasis* is introduced and defined on p. 24.

30. Wiener, *Cybernetics*, 11–12.

31. Ibid., 42. Von Neumann, similarly, sought a theory of automata that would embrace both "artificial automata, specifically computing machines" and "natural automata," by which he meant primarily the human nervous system, but he also included biological mechanisms of reproduction, self-repair, evolution, and adaptation. All these phenomena, artificial and natural, crucially involved a single, lifelike feature: a system that shaped and maintained itself in response to its environment. See von Neumann, "High and Extremely High Complication," 64, 73; and Arthur W. Burks, "Editor's Introduction," in von Neumann, *Self-Reproducing Automata*, 18, 19, 21.

32. Ivan Pavlov, *Conditioned Reflexes: An Investigation of the Physiological Activity of the Cerebral Cortex*, trans. G. V. Anrep (London: Oxford University Press, 1927), 8, 14, 4.

33. Wiener, *Human Use of Human Beings*, 22–23, 157, and *Cybernetics*, 43.

34. Wiener, *Human Use of Human Beings*, 22–23, and *Cybernetics*, 43.

35. Wiener, *Human Use of Human Beings*, 165–66.

36. W. Grey Walter, "An Imitation of Life," *Scientific American* 182, no. 5 (May 1950): 42–45, and *The Living Brain* (New York: W. W. Norton, 1953), chaps. 5 and 7. See also Owen Holland, "Grey Walter: The Pioneer of Real Artificial Life," in *Proceedings of the Fifth International Workshop on Artificial Life*, ed. Christopher Langton (Cambridge, MA: MIT Press, 1997), 34–44.

37. The characterization of Walter's findings is given by Wiener in *Human Use of Human Beings*, 174.

38. George B. Dyson, *Darwin among the Machines: The Evolution of Global Intelligence* (Reading, MA: Perseus, 1997), 176.

39. Wiener, *Human Use of Human Beings*, 48.

40. Ibid., 174.

41. Alan Turing, "Computing Machinery and Intelligence," *Mind* 59 (1950): 433–34.

42. Ibid., 446.

43. Ibid.

44. Thomas Hobbes is the leading candidate. On Hobbes and materialism, see Daniel Garber, "Soul and Mind: Life and Thought in the Seventeenth Century," in *The Cambridge History of Seventeenth-Century Philosophy*, ed. Daniel Garber and Michael Ayers (Cambridge: Cambridge University Press, 1998), 1:23. See especially p. 775: "Hobbes's direct influence consisted more in providing a position against which to argue than a position to adopt." See also Tom Sorell, "Hobbes's Scheme of the Sciences," in *The Cambridge Companion to Hobbes*, ed. Tom Sorell (Cambridge: Cambridge University Press, 1996), chap. 2, esp. 58–60; Olivier Bloch, *Le matérialisme* (Paris: PUF, 1985); and John W. Yolton, *Thinking Matter: Materialism in Eighteenth-Century Britain* (Minneapolis: University of Minnesota Press, 1983).

45. Samuel Butler, *Erewhon*, ed. Peter Mudford (London: Penguin, 1985), chaps. 23–25. (Orig. pub. in 1872.) Soon after *Erewhon's* publication, Butler wrote to Darwin, "I am sincerely sorry that some of the critics should have thought that I was laughing at your theory, a thing which I never meant to do, and should be shocked at having done." Butler to Ch. Darwin, 11 May 1872, in *Samuel Butler: A Memoir (1835–1902)*, ed. Henry Festing Jones (London: Macmillan, 1919), 1:156–57.

46. Donna Haraway, "Manifesto for Cyborgs: Science, Technology, and Social Feminism in the 1980s," *Socialist Review* 80 (1985): 65–108.

PART ONE | # CONNECTIONS

2   The Imitation of Life in Ancient Greek Philosophy

SYLVIA BERRYMAN

If we are to consider seriously the question whether there is a history of the idea of imitating life in ancient Greek thought, it is important to be clear first about what might count as imitating life. What must ancient artifacts be, such that they might seem to be "like" living things? "Like" is of course an open-ended notion, and it is important to be clear in whose eyes the resemblance lies, and what features are the focus of interest. Those interested in art history, say, may be most taken by artifacts that outwardly mimic the appearance of organisms, crafted with care to fool the eye. But for those whose interests lie in the functioning of organisms and the contribution of mechanical technology to our understanding of these processes, a different kind of artifact might seem more central to the idea of imitating life.

Natural philosophers, I suggest, would be interested in artifacts that were perceived to resemble the specifically *living* character of certain things, that is, those that could imitate the kinds of functions thought to define life. If resemblances between artifacts and organisms are to transcend deceptive imitation and help us understand the functioning of living things, the relevant devices would not be those that merely *look like* organisms, but those that *work like* them. If we could build artifacts that do what organisms do, there would be less justification for claiming a distinction in kind between natural and artificial. Artifacts might show us how nature works or, at the least, undermine the view that some mysterious process—closed to investigation—is needed for the functioning of organisms. If we can build, say, self-moving devices "bottom up," by a structural arrangement of familiar materials, perhaps there is no need to invoke mysterious powers, hypothetical substances, or special transformations to study living things. Maybe organisms work "like that."

Here, I do not claim that ancient Greek technology went far toward erasing the distinction between living and nonliving, organism and artifact; rather, I consider what it would take to do so. Aristotle's natural philosophy serves as a useful point of departure, because he both formulates an influential distinction in kind between natural and artificial and tells us clearly what it takes to count as an organism. In On the Soul, Aristotle lists several capacities he takes to be definitive of psuchê, "soul," and says that having any one of these is enough to qualify something as living.[1] Plants have only the nutritive capacity, while animals add the capacities for desire, locomotion, and perception; human beings have in addition the capacity for thought.[2] For Aristotle, these functions are enough to mark out a distinction in kind between living things and artifacts. But increasingly complex devices were developed after Aristotle's time. The question I address is whether the devices found in ancient technology could be considered to imitate any of the capacities of living things.

Aristotle summarily dismisses the idea that artifacts could do what animals do. He draws a programmatic distinction between natural and artificial: natural things like plants and animals have an internal principle of change, whereas artifacts do not; this principle of change is closely tied to the organism's ability to develop and exercise its characteristic activities.[3] The notion that devices could perform a slave's work is compared to talk of the statues of Daedalus or the tripods of Hephaestus, stories from mythology.[4] Although mythology blurs the distinction between technology and divine power, my question here is whether actual technology available to ancient Greek natural philosophers might seem comparable to some of the capacities of organisms: whether ancient technology offered anything to challenge the distinction between living and nonliving as Aristotle formulates it.

Aristotle twice compares the capacities of animals to the operation of devices, "automatic puppets." Nonetheless, he does not think the similarity runs very deep. Aristotle distinguishes the ways organisms operate by attributing their functions to a special material with powers unlike those of the four elements from which the natural world is otherwise composed.[5] As the functions of artificial devices primarily depend on structural rearrangements rather than qualitative change,[6] this—to someone who accepts Aristotle's form-matter analysis—shows how far short they fall of being natural things. Animals are, of course, much more complex than ancient automata. Still, Aristotle to the contrary, one might think that the inventions of the later ancient world go some way toward bridging the divide and indicating the extent to which devices could be constructed to do *something like* at least some of the things that animals do. This in turn might seem to undermine Aristotle's reasoning about the distinction in kind between natural and artificial.

Take the nutritive faculty. This, for Aristotle, includes nutrition, decay, and growth, and also reproduction.[7] The capacity for nourishment is, for Aristotle, the lowest common denominator for living things, the one feature that all mortal beings must possess.[8] Nonetheless, he has difficulty defining this capacity without referring to other capacities involved in the concept of something with a nature. Mere addition of material is not enough to count as nutrition: Aristotle wants to deny that fire is nutritive, although it feeds on fuel and grows thereby.[9] What Aristotle takes to be different, in the case of organisms, is that their own internal nature limits their growth. A nutritive faculty, then, controls and directs the material it takes in. Aristotle hesitates to take as definitive the notion that food increases the bulk of what is fed, possibly to exclude cases where, say, water poured into a flexible container increases its size. However, it scarcely helps the project of defining natural things to say that it is only when the thing being fed is a natural substance that food really nourishes.[10] It is, Aristotle thinks, the capacity to maintain the organism *as the kind of thing it is* that is definitive of nutrition. The idea must be that the nutriment is assimilated so as to assist in the functioning of the whole; that is, the thing being nourished has a nature defined by its functions. In other words, there is no way to determine what counts as a case of true nutrition, independent of the other definitive capacities of organisms.

Could the capacity for nutrition alone be used to distinguish living from nonliving things? What of ancient devices in the later *Pneumatica* treatises that operate by taking in water? Some of these have little internal complexity or functioning: "drinking animals" are really only siphons rigged to drain a pool of water.[11] But other ancient devices use flowing water to operate complex, differentiated, and responsive internal mechanisms. The flow of the assimilated water enables them to do what they do.[12] There is surely potential here to link this use of intake to the idea of maintenance and nutrition. If Aristotle had insisted that nutrition involves qualitative transformation of the ingested matter, this would certainly have excluded many devices; others, however, turn water into steam.[13] Given Aristotle's failure to define nutrition independently of other capacities of the organism, it becomes difficult to maintain that the capacity to receive nourishment distinguishes living from nonliving things.

Some later ancient Greek thinkers apparently did take the technology of the day to offer an analogue to the growth of an organism, although not in the way suggested. Galen cites some unnamed doctors who think that the causal sequence in the development of an embryo is comparable to what happens in some theatrical devices.[14] The similarity they perceive is that, just as in embryos, a complex sequence of changes is predetermined by the structure of the device, so that one initial trigger instigates a progression of changes in a well-crafted series. If we think of growth as a process of ingesting and assimilating alien material into the form of the growing thing by qualitative transformation,[15] the devices available might not seem to offer much that could act as analogue. But the doctors Galen criticizes conceive of growth as a sequence of interconnected changes, each in turn triggering the next according to a predetermined pattern that is like the goal-directedness of art. They focus on the idea that growth requires an interconnected sequence of developmental changes, rather than on the material problem of assimilating new matter into the whole. The similarity of one process to another depends on the way the processes are conceived.

One of the cases where Aristotle explicitly compares animals to devices, only to distinguish them, concerns self-motion. Self-motion is one of the definitive ways to characterize animals as distinct both from simple bodies and from artifacts. He appears to derive this characterization from popular understanding, not an idiosyncratic or theoretically motivated classification. This notion of a self-mover in Aristotle is, I have argued elsewhere, much clearer if we keep in mind that the defining characteristic of animals is the initiation specifically of locomotion, rather than of every kind of change.[16] Aristotle's claim that animal motion is not uncaused is sometimes thought to be in tension with the notion that animals are self-movers.[17] In fact, there is no difficulty if we remember that what animals can do—and inanimate things cannot—is to instigate locomotion in a context where there are other environmental changes, but not some other body acting on them by locomotion. Stones roll when a child pushes them; carts move when the horse starts walking. But the child and the horse can do something rather different, which is to bring themselves into locomotion from a state of rest when some other kind of change—in Aristotle's account, a qualitative change involved in desire—triggers them to move.[18]

Few ancient devices, other than those worked by steam, involve qualitative change. But one might understand the term *self-mover* as referring straightforwardly to things that have the capacity to instigate and cease moving *locally*. Aristotle goes to considerable lengths to develop a philosophical account of the physiology of action and the ways in which desire produces local motion. But this is only one among many theories, and one can describe an animal as a "self-mover" without presupposing Aristotle's theory of qualitative change and its role in desire. The term could refer merely to something that can start or stop moving "of itself," that is, when it is not directly pushed or pulled by something else.

Some ancient devices might seem to have a claim to count as self-movers. Wind-up devices, which contain within themselves the ropes, axles, and falling weights that produce their motion, might seem to match the description of something that starts and stops moving "of itself" in the absence of anything pushing or pulling it.[19] It is true that these devices require an agent to wind up the rope, set the trapdoor, and release the trigger, but in a significant sense the devices contain within themselves the control of their own locomotion. Like animals, they require an instigating cause: a hand releases the trapdoor, and the automaton starts rolling. Where the child and the automaton differ from the stone is that the causes acting on them provide an occasion for moving, but the form of the locomotion as well as its extent and direction are determined by the constitution of the "self-mover" itself.

The ancients may have designed devices that could perform functions in some respects "like" animal capacities for self-motion and nutrition, but there are still other animal abilities remaining, including perception, desire, and thought. While ancient technology does not offer much in this regard, it is worth noting that some ancient devices respond to an environmental change and sometimes appear to mimic the animal ability to act in response. These include devices that might be thought to offer an "appropriate" response to a "perceived" signal or change in the environment, and those that act so as to restore themselves to a preexisting condition on "sensing" a change. Otto Mayr has noted the significance of the simple "feedback mechanisms" in the ancient pneumatics texts.[20] In simple water-level regulators, for example, a tank of water operated by a valve is set so as to maintain a constant level of water even when passersby drink from a vessel. The idea here is that a casual, unprogrammed trigger in the environment (someone takes a drink from a bowl) is sufficient to set off a reaction within the device (refilling the bowl) that restores it to (some aspect of) its former state.

Whether such a device bears any similarity to the idea of perceiving and desiring a certain end depends on what we take to be definitive of the latter.

An intriguing indication that some sources did recognize the potential of these float-level regulators as imitations of perception appears in the Arabic version of Philo of Byzantium's *Pneumatica*.[21] One passage dresses up such a self-regulator so that it has the appearance of a servant pouring drinks for guests.[22] This presentation of the device at least implies that its designers were aware that the responsive capacity of the device suggests perception. The device plays with the illusion that it *recognizes* that the vessel is empty and reacts appropriately.

Two kinds of devices described by Hero could perhaps have been the basis for the idea that devices can react in ways that provide an analogue to perception. They are certainly presented in ways that suggest this to the observer. One kind involves a device that operates when an agent releases a trigger, causing what follows. The stage setup presented to the audience consists of a figure of Heracles with a bow, a serpent, and an apple (see figure 2.1).[23] The mechanisms are hidden, as usual, and the observer's focus is on the apparent interactions between the figures. The trigger consists in the operator lifting the apple: the apple is connected by a rope through the bottom of the stage, out of sight, that pulls a lever, so that lifting the apple triggers the release of the arrow from Heracles' bow. Simultaneously, the rope attached to the apple also pulls a plug, causing water from a hidden tank to fall into a lower tank. The trapped air in the lower tank rushes out with a hissing sound through a pipe hidden in the serpent. The positioning of the visible figures and the timing of events suggests that Heracles is shooting his arrow in *response to* a signal—the lifting of the apple—and also that the serpent hisses *because* it was struck by the arrow. Heracles and the serpent seem to react because they *perceive* events in their environment, the above-stage world visible to the audience. Because the other causal processes are hidden, we readily fill in a narrative whereby perception supplies the missing link.

This may seem to be an illusion of, rather than an analogue of, perception. The device does not *really* operate when Heracles and the serpent *see* the apple move or *feel* the arrow strike: the causal chains do not operate via the specific sense organs thought to be involved in vision or touch. Nonetheless, the reactions are caused by the motion of the apple: in the case of Heracles, this is at least the appropriate trigger, although in the case of the serpent, it is not.[24] The later history of mechanical models echoes such devices: when Descartes conceives of devices that work like animals, his suggestion for a model of action relies on a device responding to an external signal. He talks of a statue of Diana in the pleasure gardens that hides in the reeds when a

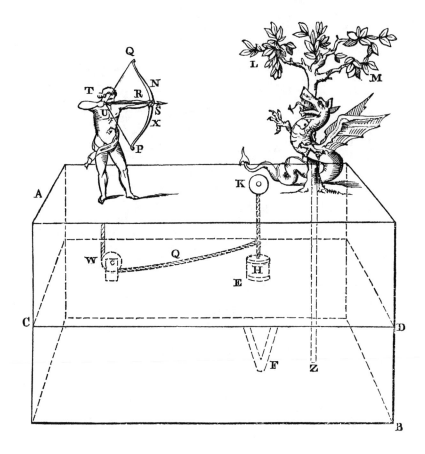

FIGURE 2.1 | Heracles with bow and arrow, apple, and serpent, from *The Pneumatics of Hero of Alexandria*, ed. and trans. Bennett Woodcroft (London: Taylor Walton and Maberly, 1851). The author thanks Irving K. Barber Learning Centre, University of British Columbia, for permission to reproduce this illustration.

passerby steps on a certain flagstone, or the sounding of an organ when the organist presses a given key.[25]

As a more convincing analogue to animal action, we might want a device that demonstrates differential responses to different "signals." Hero's owl-and-birds device shows figures responding alternately to two different situations (see figure 2.2).[26] It involves a flow of water into a tank and a self-starting siphon designed to start and stop the outflow at regular intervals.

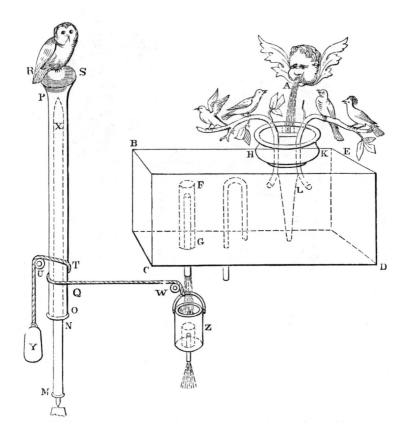

FIGURE 2.2 | Owl with birds, from *The Pneumatics of Hero of Alexandria*, ed. and trans. Bennett Woodcroft (London: Taylor Walton and Maberly, 1851). The author thanks Irving K. Barber Learning Centre, University of British Columbia, for permission to reproduce this illustration.

The flow into the tank is steady; the outflow is varying, started and stopped by a mechanism the observer cannot see. During periods when the tank is filling, the building pressure causes a whistling sound to emerge through a pipe. When the tank is emptying, the whistling ceases and a different mechanism operated by the outflowing water causes the owl to turn and face the birds. The device thus goes through two different phases. In phase one—while the tank is filling—the owl is looking away from the birds, and the birds are heard to whistle. In phase two—when the tank begins to empty—the owl turns to glare at the birds, and they fall silent. After a while, the owl turns away and the birds begin singing again.[27]

The coordination between the position of the owl and the whistling sound suggests to the observer that the birds perceive the menacing gaze of the owl and cease singing *because of this*. Here, there is no external agent reaching in to provide the trigger. Rather, the device is constructed so that two processes that appear to be caused by interactions between one another in fact have a common cause that produces both. If this were to be the "mechanical" equivalent of perception, it would suggest that the causal structure of perception is radically different from what we think it is. This is, of course, a possible result of building mechanical models of living things.

Hydraulic devices seem to have a particular appeal as models for the internal mechanisms governing animal action. Descartes takes the fountains and water-pressure devices of his day as a pivotal point of analogy to the organism, as he conceives it, which works by "animal spirit" in the nerves, connecting brain to muscles and tendons.[28] This aspect of his physiological theory echoes a view that became common in Hellenistic thought after the doctors Erasistratus and Herophilus linked the functions of perception and locomotion to the *neura*, or "nerves," and a "fluid"—*pneuma*—moving through them. Evidence indicates that Hellenistic doctors were looking to pneumatic technology to help understand the motion of other fluids like blood throughout the body.[29] It would not have been too far-fetched, then, to see the idea of a single fluid moving through various channels, operating by varying pressures, as a point of analogy between the ancient pneumatic devices and the systems governing perception and motor functions in animals. One text even suggests this comparison, albeit sketchily. Tertullian describes a model of soul he ascribes to three earlier philosophers: he likens that model of soul to Archimedes' water-organ, a device wherein a complex series of effects are caused by a single fluid going out through a variety of passageways.[30]

There is not much evidence, of course, that the technology of the ancient world inspired ancient natural philosophers to investigate the natural world in the ways that seventeenth-century mechanical philosophers did. But there are some scattered indications that a few ancient Greek thinkers did look to their technology in order to understand how the functions of organisms might be realized. For those who think that natural things are different in kind or work by qualitative transformations that human art cannot duplicate, the parallels would hardly be compelling. But for those who look to technological innovations to show what structural arrangements alone can achieve, the mechanics of the ancient world might suggest a direction for investigating the functions of natural things.[31] We have reason to think that this was a line of inquiry that at least some ancient Greek thinkers chose to pursue.[32]

NOTES

1. Aristotle On the Soul 2.2.413a21–25, in The Complete Works of Aristotle: The Revised Oxford Translation, ed. Jonathan Barnes, 2 vols. (Princeton, NJ: Princeton University Press, 1984).

2. Ibid., 2.3.414a29–415a12.

3. Aristotle Physics 2.1.

4. Aristotle Politics 1.4.1253b33.

5. Aristotle Movement of Animals 7.701b1–15, 10.703a9ff; Generation of Animals 2.1.734b11, 2.5.741b8. See also Sylvia Berryman, "Aristotle on Pneuma and Animal Self-Motion," Oxford Studies in Ancient Philosophy 23 (2002): 65–97; "Ancient Automata and Mechanical Explanation," Phronesis 48 (2003): 344–69.

6. The few exceptions are devices that employ fire to produce steam from water. For simplicity, I exclude this subclass of devices.

7. Aristotle On the Soul 2.4.416a19.

8. Ibid., 2.2.413a32.

9. Ibid., 2.4.416a10–12.

10. Ibid., 2.4.416a12–14.

11. Hero Pneumatica 30.

12. Ibid., 15. It does not cause growth, but then Aristotle recognizes that relying on the idea of growth rather than the maintenance of functions is problematic to defining nutrition.

13. Only those using fire to produce steam satisfy this demand, and even in these cases, the qualitative transformation does not result in the assimilation of matter.

14. Galen De Foetuum Formatione 4.688–9. For discussion of this passage, see Sylvia Berryman, "Galen and the Mechanical Philosophy," Apeiron: A Journal for Ancient Philosophy and Science 35 (2002): 235–53.

15. Aristotle On the Soul 2.4; On Generation and Corruption 1.5.

16. Aristotle On the Soul 2.4.415b22; Phys. 8.2.252b17ff; Berryman, "Aristotle on Pneuma."

17. Mary Louise Gill and James G. Lennox, eds., Self-Motion: From Aristotle to Newton (Princeton, NJ: Princeton University Press, 1994), xv.

18. I argue this in detail in Berryman, "Aristotle on Pneuma." For different accounts, see Gill and Lennox, Self-Motion.

19. Hero Automatopoetica 4.

20. Otto Mayr, The Origins of Feedback Control (Cambridge, MA: MIT Press, 1971), 2.

21. The surviving Arabic version contains material not found in the treatise that has come down to us in Latin: it is not always clear how much of this derives from an original or is a later addition to the compilation. See Carra de Vaux, "Les Pneumatiques de Philon de Byzance," Notes et extraits des MSS. de la Bibliothèque Nationale 39 (1903): 127; Frank David Prager, Pneumatica. The First Treatise on Experimental Physics: Western Version and Eastern Version (Wiesbaden: Reichert, 1974).

22. Philo Pneum. 36 (Prager).

23. Hero Pneum. 40.

24. The latter, after all, ought to be reacting merely because he is struck by the arrow, whether or not this was signaled by an apple.

25. René Descartes, The Philosophical Writings of Descartes, vol. 1, trans. John Cottingham, Robert Stoothoff, and Dugald Murdock (Cambridge: Cambridge University Press, 1985), 101.

26. Hero Pneum. 15.

27. Cf. Michel de Montaigne, The Diary of Montaigne's Journey to Italy in 1580 and 1581, trans. E. J. Trechmann (New York: L. and Virginia Woolf, 1929), 165.

28. For the details of his account, see Dennis Des Chene, Spirits and Clocks: Machine and Organism in Descartes (Ithaca, NY: Cornell University Press, 2001).

29. I. M. Lonie, "The Paradoxical Text 'On the Heart,'" Medical History 17 (1973): 1–15, 136–53; J. T. Vallance, The Lost Theory of Asclepiades of Bithynia (Oxford: Clarendon Press, 1990), 71; J. M. Longrigg, Greek Rational Medicine: Philosophy and Medicine from Alcmaeon to the Alexandrians (London: Gerald Duckworth, 1993), 207–9; and Heinrich von Staden, "Body and Machine: Interactions between Medicine, Mechanics, and Philosophy in Early Alexandria," in Alexandria and Alexandrianism, ed. John Walsh and Thomas F. Reese (Malibu, CA: J. Paul Getty Museum, 1996), 85–106; "Teleology and Mechanism: Aristotelian Biology and Early Hellenistic Medicine," in Aristotelische Biologie: Intentionen, Methoden, Ergebnisse, ed. Wolfgang Kullmann and Sabine Föllinger (Stuttgart: F. Steiner Verlag, 1997), 183–208; and "Andréas de Caryste et Philon de Byzance: Médecine et mécanique à Alexandrie," in Sciences exactes et sciences appliquées à Alexandrie, ed. Gilbert Argoud and Jean-Yves Guillaumin (Saint-Étienne: Publications de l'Université de Saint-Étienne, 1998), 147–72.

30. Tertullian De anima 14; see R. W. Sharples, "The Peripatetic School," in Routledge History of Philosophy, vol. 2, From Aristotle to Augustine (London: Routledge, 1999), 163; Berryman, "Ancient Automata."

31. On the idea that mechanism is a method, see Des Chene, Spirits and Clocks, 14.

32. For more on this, see Berryman, "Ancient Automata."

3   The Devil as Automaton: Giovanni Fontana and the Meanings of a Fifteenth-Century Machine

ANTHONY GRAFTON

No engineer of the fifteenth century thought harder about automata or devised more ingenious specimens of the genus than Giovanni Fontana. His designs included a bird that flew, a fire-farting rabbit, and a mechanical devil, brilliantly designed to move its facial features, arms, and wings. Fontana departed from engineering custom, moreover, when he had his devices illustrated both in real time, as they would look while in operation, and in diagrammatic form, to make clear exactly how they functioned.[1] In his drawing of a mechanical she-devil, for example, Fontana offered two different visions of his automaton on a single page. In the foreground, the artist showed the she-devil moving rapidly forward, flapping her wings and breathing fire. Behind it, he laid bare the structures that made the illusion work. With one very striking image, half diagram and half representation, Fontana's illustrator revealed the armature on which the devil's costume rested, the ropes that connected her head to her shoulders and her tail to make them move in synchronization, the basket-shaped structure that protected her moving parts, and the assembly of pipes inside her head that enabled her to shoot fire from her ears and mouth—a striking mechanism depicted in an even more striking image, oddly foreshadowing Picasso or Max Ernst.[2] Here Fontana made clear that he could produce a figure that moved and spat fire—a figure exactly like the devils that beat and spat fire at Santa Francesca Romana in the fifteenth-century frescoes of her miracles that line the upper chamber of the Tor de' Specchi convent in Rome.

FIGURE 3.1 | Fontana she-devil.

In the early modern period, as Horst Bredekamp argued in *The Lure of Antiquity and the Cult of the Machine*, material objects and images of them could make some complex arguments for which full verbal resources were hard to find. In his view, the Renaissance Kunstkammer, or museum, in all its magnificently weird, apparently random assemblage of natural objects and the products of human craft, embodied an argument: that the natural world itself had a history, one imposed on it by the human art and industry that had reshaped it.[3] Like the shells and weapons, narwhal horns and canoes that whirled giddily across the walls and ceilings of Renaissance museums, Fontana's automata were not only machines that moved, but also "things that talked."[4] But what arguments were these creations of an inventor with a scholastic background meant to make? By decoding these—and by using the verbal testimonies that he and other engineers left, as well as some of the objects he devised—we can tease out some of the cultural roles that automata played in early Renaissance courts and cities, before the scholar-engineers of the later sixteenth century revived the Hellenistic technology of Hero, and before the landscape architects of the same period created grottoes and gardens in which their musical abilities could astonish, and their hidden pipes soak, innocent visitors.[5]

A product of the University of Padua, where he studied with Biagio Pelacani, Fontana wrote fluent if unclassical Latin and studied the Aristotelian texts on statics and mechanics that circulated so widely in learned circles.[6] His works include a full-scale encyclopedia in Latin on the natural world as well as the funkily illustrated works on machines and cryptography that have made him notorious in modern scholarship. Though Eugenio Battisti and many others have reconstructed his life and work, they have not yet placed his automata in their full historical context—much less worked out exactly what these artificial creatures, and his designs for them, might have meant to him and his customers.

In fifteenth-century Italy, automata usually made their appearances in public and theatrical contexts: as part of the apparatus of temporary machines created to stage public ceremonies of welcome for rulers, as well as in pageants of other kinds and in plays. Theatrical machines—machines that worked, in public, with no visible human or animal effort to power them—became a node where engineering and the fine arts intersected, visibly and effectively. Lords and subjects, artists and craftsmen collaborated, devising machines that could inspire wonder in a sophisticated audience. In late May 1475, for example, Costanzo Sforza, lord of Pesaro, married Camilla of Aragon, natural daughter of Ferrante of Aragon, the king of Naples. Costanzo, a highly cultured young ruler, loved both technology and the arts. He fortified his pleasant seaside city, in the most up-to-date way, to protect it against tyrants, and he adorned it with a fine ducal palace. When he married a great and learned lady, accordingly, he left nothing to chance. Camilla's entrance to the city became the occasion for everything from "a splendid joust with four prizes" to an hour-and-a-half-long speech in Latin by the humanist Pandolfo Colenuccio, in praise of marriage, which he recited so beautifully that his audience, *mirabile dictu*, wished it longer.[7] A Florentine humanist, Niccolò di Antonio degli Agli, recorded this and the other festivities in a set-piece text, which the scribe Lionardo da Colle copied on vellum in 1480 as a gift for another great and cultured warrior prince, Federigo da Montefeltre.[8] Giovanni Santi, father of the painter Raphael, illustrated the booklet—and very likely planned the technical events for the occasion, though Niccolò did not name him.[9]

As Niccolò told the story, neither the prowess of the jousting knights nor the eloquence of the erudite humanist amazed the crowds more than the "various and delightful spectacles" that Giovanni Santi "laid out with marvelous skill and subtle craft."[10] When Camilla and Costanzo approached the city, forty merchants and citizens came to meet them on a ship "moved by men, with the most extraordinary ease and marvelous ingenuity, on wheels," its full sails swelled by the wind.[11] On Sunday, the day set aside for the wedding, the "great room of the court" was decorated with splendid images of the planets and the signs of the zodiac. Through a round door in the middle of the zodiac, which gave glimpses of "a splendid paradise full of gold and silver," figures representing the sun and moon descended on clouds to recite poems, after which they ascended again "with marvelous speed and skill."[12] On Monday, in the same room, the lords and their subjects gaped as a mountain of wood—covered with trees and populated by hares, goats, deer, bears, wild boars, and other animals—moved into the same room. It disgorged a lion and a wild man, whose antics amazed all beholders.[13] Later still, the Jewish community of Pesaro arrived, accompanied by elephants, one of them bearing the Queen of Sheba, and another mountain.[14] After them came floats bearing the gods of the planets, as well as a still more splendid one representing the triumph of love, with twelve rotating "spiritelli" that were presumably working automata.[15]

Even after the joust, a final float arrived in the Borgo. This represented the triumph of fame and bore a beautiful woman, three men "very richly armed in the ancient style, one for Scipio, one for Alexander, and one for Caesar," and a "round ball, which was entirely blue, and the color of water, except that the part of the earth which is inhabited, and above the water, was laid out and depicted according to the true cosmography."[16] The content of this elaborate program evoked cosmic harmonies, favorable planetary influences, and other elements of the magical universe as it had taken shape in antiquity.[17] But the methods used to carry it out were anything but magical, and Niccolò took care to make that clear.

To judge from Niccolò's comments, every aspect of the festivities pleased the audience, from the rich colors and materials used in the decor to the sugared biscuits and other precious delicacies that they savored between acts. But nothing astonished them so much, or so consistently, as the ingenuity and skill that had gone into the creation of so many vehicles and beings that moved on their own. Again and again, with great precision, Niccolò recorded the cause of the wonder that Giovanni Santi's technical feats inspired.[18] The ship that met Camilla and Costanzo, he noted, moved so smoothly that one could not see who was pulling it or even tell it from a real ship.[19] Ms. Moon and Mr. Sun impressed the feasters above all by the speed and silence with which they came down from the zodiac and rose back into it. The elephants created by the Jewish community impressed viewers because "the legs of the men inside them were so well covered by those of the elephant that it would be impossible for anyone who had not seen this marvelous device to write of it."[20]And the big camel that turned up a little later, ridden by an Ethiopian, "was so well represented and with so much art that it seemed alive, and it was large, and opened its mouth and extended its neck and sank to the ground as real camels do."[21] Movement—organically correct movements carried out by replicas of animals, swift movements carried out by images of the celestial bodies, smooth movements executed by a large model of a ship—was at the core of what made the experience dramatically effective and thus enabled it to serve Costanzo's political ends. Automata that worked swiftly and smoothly, that looked convincingly natural, could captivate a sophisticated public.

Niccolò found these devices so impressive that he gave credit for them not to Giovanni Santi or any other professional showman, but to the lord who had paid the piper. Most of these inventions, he claimed, emerged from the "wit and intellect" ("ingegno & intellecto") of Costanzo Sforza himself. Through challenge after challenge, the lord of Pesaro "never tired of showing the diligence and the strangeness of his mind."[22] The terms that Niccolò used here were freighted with meaning. Renaissance engineers insisted that their craft had a powerful intellectual discipline at its core. Their very title, *ingegneri* or *ingeniatores*, designated them as men of the intellect, *ingenium*, as well as creators of clever new devices, *ingenia*.[23] These were the terms that Antonio Manetti, Lorenzo Valla, and other humanists consciously echoed in the praise of human inventiveness that became a motif in the public conversation of the Italian republic of letters from the 1430s onward, a moveable feast of words that changed places and participants more than once. As Niccolò described Costanzo's feats for Federigo, that great patron of both architecture and engineering, he constructed the lord of Pesaro as an engineer in his own right, a techie Prospero who not only paid for but staged his own masques. By doing so he made clear how brightly the arts of the engineer now glowed in the eyes of patrons and humanists. The automaton, like the other dazzling products of fifteenth-century engineering—Brunelleschi's dome, for example, and the new machines he used to construct it—offered a new kind of testimony to the powers of the human intellect.

The genre of the text in which Fontana included his automaton is also highly revealing. A number of fourteenth- and fifteenth-century engineers—Giovanni Dondi, Georg Kyeser, and Mariano Taccola, among others—produced similar texts.[24] Though these vary widely in style, content, and connection to technological reality, all of them resemble Leonardo's notebooks. Like Leonardo, the engineers systematically combined illustrations—many of them very precise diagrams—with text. Like Leonardo too, and some before him, they made bold claims for themselves and what they had to offer, insisting on their high status and high ideals. Taccola, for example, threw all his rhetorical tropes into one basket as he appealed simultaneously to the best-known ancient engineer and to the sputtering late medieval crusading enterprise: "I, Mariano Taccola the notary, also called the Archimedes of the great and powerful city of Siena, did not draw up these devices with my own hand to operate against Christians, but I devised, created and drew them so that they might come into play against infidels and barbarous races."[25]

Like other engineers, Taccola tried to use his pretty manuscripts to make his way. He thought they might clear his path to the ear trumpets of potential patrons, like the Emperor Sigismund, to whom he presented one of them. But Taccola did more than boast. He also provided accurate drawings of contemporary devices, like the armed wagons of the Hussites. Taccola boasted that his lifting machines could do more than the most powerful draft animals: "intelligence is stronger than oxen."[26] Coming from Siena, a city whose industries relied on an elaborate network of underground waterworks, Taccola knew a great deal about hydraulics.[27] He himself devised both an elaborate caisson for working underwater and a powerful siphon. His work changed shape and content, moreover, as he saw and evaluated others' designs and contemplated the havoc that his own devices might wreak. When he illustrated his siphon and showed how it could be used to flood an enemy's fields, he also warned against doing so: "Flooding cities, castles or fields does not form part of the custom of warfare." "Therefore," he insisted, a little paradoxically, "we should remain silent about this matter."[28] Here Taccola implicitly criticized Brunelleschi, who had tried and failed to flood the city of Lucca. The new, powerful engineering had already begun to develop a rudimentary rhetoric of social and environmental responsibility. When Fontana included his she-devil in a treatise of this kind, he made clear that the maker of automata was not engaged in a trivial pursuit. Automata and similar devices were wonderful toys—but they were also something more.

These engineers' manuscripts, in fact, amounted to something like the realization of Roger Bacon's thirteenth-century project for a "science of experience." In keeping with Bacon's predictions, they showed how to work wonders, using the manifest properties of matter. In keeping with Bacon's practice, they also deployed the techniques of high magic and took advantage of occult properties—to say nothing of the demons and angels some of them summoned.[29] Like Bacon, the engineers took a special interest in devices that showed the human ability not just to emulate but to outdo nature: wagons that moved without draft animals, powerful weapons, and the like. And like Bacon, the engineers took care to distinguish their practices from the evil (if effective) ones of diabolic magic. Fontana—an educated man as well as an engineer—was certainly conscious of these traditions when he designed mechanical devils.

Fontana, moreover, compiled an encyclopedia as well as engineering treatises, and this work made his larger intellectual allegiance clearer than the magnificent illustrated codex that portrayed his mechanical devices: "God has given men so much force of intellect," he wrote, "that they have devised many arts, and brought about many things that nature could not have done on her own." As examples of arts that did not rest on the direct imitation of nature, Fontana cited the things

> which the people call *ingenia*, many of which architects have attained with geometrical measures, physical arguments or arithmetical reasoning. For example, there is the screw, from which is derived the pump, which expels water in a spectacular way. And there is the multiple set of pulleys, an extraordinary device through which ropes run, and which can pull or suspend incredible weights with a light pull. . . . There is also the terrible machine which we call the bombard, which destroys every form of resistance, even marble towers.[30]

Like Fontana's devices, his words were effectively Baconian—a hymn of praise to the positive powers of the human "ingenium."

Yet the devilish appearance of Fontana's automata is as revealing as their mechanical interiors. For, in fifteenth-century Italy, the making of statues that moved could smack of more than technical ingenuity. From late antiquity onward, readers of the central Latin Hermetic text the *Asclepius* were astonished, horrified, and exalted by the text's claim that the ancient Egyptians "discovered the art of making gods"—that is, they "called up the souls of demons or angels and implanted them in likenesses through holy and divine mysteries, whence the idols could have the power to do good and evil."[31] Throughout the thirteenth and fourteenth centuries, this passage continued to excite and worry Christian readers—especially those engaged in the making of amulets and talismans.[32]

Fontana, moreover, took a serious interest in magical as well as mechanical contrivances. In his work on the art of memory, for example, he acknowledged that men could obtain true knowledge by demonic or angelic inspiration and by dreaming as well as by trial and error. He treated the *ars notoria*, astrology and talismanic magic, as valid, powerful disciplines.[33] He himself claimed to have spoken to a "spiritual substance," from whom (or which) he received very useful information.[34] And he was more than aware of the older tradition that treated ensouled statues as a preeminent work of demonic magic. For him, as for Aquinas, this subject naturally belonged to the larger discussion of astrological talismans. "Do we not know," Fontana asked, "of the spirits, that, once impressed on the idols of the pagans, they gave answers, and lured the pagans to idolatry? First the Chaldeans, and then the Romans, thought them to be gods, and worshipped them, and again nowadays many oriental peoples similarly pray to them, in their blindness and confusion."[35] Fontana believed that certain "magi and necromancers" used "holy characters and the names of good angels" to compel these spirits to enter statues—as well as animals and dead bodies. He described these spells as stronger than iron chains, and expounded the orthodox view that the demons entered the statues of their own volition, to fool their worshippers.[36]

For the most part, however, Fontana treated magical traditions with scathing disdain. He made fun of popular beliefs and practices, and the supposedly wonderful objects that sustained them—like the stones with two horns that women thought could protect them against lightning.[37] And in his encyclopedia he argued, drawing on his study of optics at Padua, that armies seen marching in the heavens—a fairly common occurrence in this period, and one usually greeted with consternation by the learned and the common people alike—were not omens revealed by divine providence but illusions created by atmospheric refraction:

> Biagio Pelacani, my former teacher, tells the story that in his time, in the year of grace 1403, horsemen and footmen armed with lances and swords appeared, attacking one another, in the clouds every day for three days, before the third hour, in Lombardy, near the town called Busseto. Those who saw this were terrified, not knowing the cause of this novelty. For it happened that at a distance from this place, in a certain plain, there were armed cavalry and foot soldiers, whom the Duke of Milan had assembled so that they could besiege Bologna. And since at that time there was a watery cloud in the sky that could accept a

likeness of those armed men in motion, men who were in nearby places, seeing the cloud, concluded that there were demons in the air, fighting, with the appearance of armed men, and that they had come there either voluntarily or by the art of magic. And he says again that once in his time in . . . the city of Milan several images like angels were seen. Some of them seemed to be trying to come down to earth, and some to climb to heaven. Some had trumpets and some had swords in their hands. Everyone was terrified by these visions, as if God wished to hasten the last judgment. But later it became known that these fantastic appearances were seen because there was an angel at the lofty, sharp peak of the tower of S. Gottardo [in Corte], made with a trumpet and sword in his hands, and he presented himself to the aforesaid watery cloud as to a mirror. The images were multiplied by the motion of the cloud, as if in moving water, and seemed to move in various ways.[38]

The sophisticated perspectival studies of Fontana's time not only inspired some of his devices, but also allowed him to naturalize events that his contemporaries normally took as preternatural or even as supernatural. Many of Fontana's contemporaries became unhinged with fear when they saw, or heard reports of, ships, castles, cities, and "impossibly big dragons and unheard of, bizarre monsters" in the air.[39] Pamphlets and rumors that bore news of this sort could have a powerful impact on politics and group behavior.[40] Unhesitatingly, Fontana proclaimed that a master of optics could duplicate all of these phenomena "by art, and not even a magical one, but by pure craft."[41] Ignorance—"ignorance of the cause," as Fontana explicitly said—inspired fear. Knowledge—knowledge of the properties of light and the media it passed through—dispelled it. When Fontana inserted an image of a magic lantern, the earliest one known, in his treatise on military instruments, he showed it projecting the image of a winged she-devil armed with a sphere "to inspire terror in those who see it"—clear evidence that he believed that he could replicate the sightings and visions of supernatural beings that terrified his contemporaries.[42]

The evidence suggests that Fontana meant his diabolic automata to be understood in a similar way—not only as evidence of his own powers of invention, but also as a deflation of the stories told by magi and theologians alike. Magicians invoked devils. Fontana showed how to create a mechanical male devil, with moveable eyes, tongue, horns, arms, and wings, as well as his fire-breathing she-devil.[43] Necromancers raised the dead. A wooden skeleton linked by cords to a wheel and an escapement, Fontana explained, would appear to move its members, "just as the figures in mechanical clocks do." Tongue firmly in cheek, he called this design, perhaps destined for a clock, "resurrectio mortuorum artificiosa"—"a crafty way to bring the dead back to life."[44] There is every reason to think that Fontana meant not only to advance the claims of engineering, but also to deflate those of magic and its critics when he illustrated his designs. At times, he explicitly acknowledged that he had cast his inventions in provocative forms. Certain men at Padua, though "most learned," falsely accused him of raising "infernal spirits" from Tartarus to help him, using a pentacle of seals and secret seals.[45] This was probably the very reaction that he hoped to provoke with his resurrection machines and fire-farting rockets in the shape of rabbits and birds.

In these cases, Fontana showed how to use mechanical devices, instead of the magician's circle and incense, to create the same psychologically effective illusions. Though the form of his finished works rarely followed their

FIGURE 3.2 | Fontana he-devil.

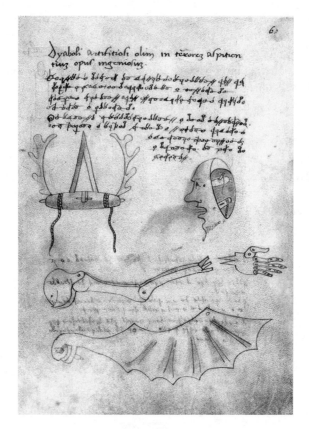

FIGURE 3.3 | Innards of he-devil.

function, it had a function of its own. Again and again, Fontana wrapped his purely mechanical devices in frightening exteriors that recalled the other great set of fifteenth-century claims to human power over nature, those of the magicians. By doing so, he suggested that engineers could claim the same immense powers as magi. By the same token, and more radically, he mechanized the apparently preternatural or supernatural—and by showing that he could do so, he exalted his own craft as no one before him had. Unlike later makers of automata, apparently, Fontana saw the fact that his figures required human motive force not as a form of cheating, but as a necessity, and did not distinguish in any clear way between automata that moved by automatic machinery and those that housed men pulling ropes.[46] In his readiness to surrender to spectacle, he was probably at one with his patrons and audiences—like those good citizens of Pesaro who took such delight in the fact that they could not see the men who moved Giovanni Santi's elephants and camels. Fontana's automata played to a period sense of what an automaton was—a sense soon to be transformed by the availability of new hydraulic and other technologies.

What Fontana argued in a series of partial ways, some visual and some verbal, the most erudite and articulate of contemporary engineers suggested in two works, both written around 1450 and both reflecting his experiences in the papal court in Rome and Florence. Leon Battista Alberti—who shared Fontana's interest in perspective, ciphers, and many other topics—was a proficient amateur at technological pursuits but highly skilled at the study and use of classical sources. Like Fontana, he knew the traditional stories about ensouled statues, and like Fontana, he made triumphant fun of them. In his elaborate satire of the rulers of the Roman church, the Momus, Alberti confronted this passage from the Asclepius directly, dramatically, and derisively. The gods, eager to observe human festivals in their honor, decide to descend into the theater. So they remove the statues of themselves and take their places. Alberti drew from Lucian's Zeus tragoedus the notion that one could imagine the gods as so many statues and make fun of their different qualities. But he took from his mother wit the carnivalesque and humiliating results that follow the gods' decision. A slave pisses on Jupiter. The philosopher Oenops, who drives the slave away, hurts the god Stupor by cleaning his face. Finally Aeolus accidentally releases the winds; they in turn knock down the great awnings that shade the theater; and the gods in their statues, helpless to defend themselves, are mutilated and broken. Jupiter falls and breaks his nose; Cupid breaks off Hope's wing and then is broken himself. Only Pluto fights back—by crushing the foot of one member of the gang of pimps who try to gouge out his glittering eyes, thinking they are jewels. In the end, the battered gods replace their statues, and return to heaven to lick their wounds, completely defeated.[47] The legend of the moving statue, for Alberti, was evidence of human superstition—nothing more. His ensouled statues could only stand and stare, helplessly, as the masterful humanist subjected them to imagined rituals of humiliation—a parody, perhaps, of the tradition of humiliating the statues and relics of the saints.

In another, quite different, context, Alberti readily acknowledged that the creation of moving machines formed a vital task for anyone interested in technology. In his massive work On the Art of Building, he continued and synthesized the work of the engineers, taking the ancient work of Vitruvius as his model and dealing in detail with the management of water and the construction of the basic machines. One passage is especially revealing.

Alberti compared the machines with which he and other builders lifted columns and other heavy objects to living beings: "But here," he wrote, "we need only consider the machine as something like extremely strong animals with hands, an animal that can move weights in almost the same way as we do ourselves. These machines must therefore have the same extensions of members and muscle that we use when pressing, pushing, pulling and carrying."[48] Evidently, Alberti saw the engineers' machines—of which, for Fontana, the automaton was one of the supreme specimens—as mechanical beings, almost alive: "velut animantia."

Analogies like this would play a vital role in the seventeenth century. For Descartes and others, they underpinned the notion that a single mechanical philosophy could account for the actions of animals and humans as well as material objects.[49] Alberti, however, had no clear textual precedent to cite. What he drew on, instead, was the work of the engineers—and the challenge they offered to an older way of understanding figures and machines that moved. The automaton was only the most prominent, and clearest, example of a larger argument. Though pictures like Fontana's do not appear in other engineers' manuscripts, it seems certain that similar considerations underlay—for example—Brunelleschi's efforts to understand the logic and play of mechanisms like cranes.[50]

It was no accident, then, that Manetti and Alberti, both professional masters of the word, celebrated a particular and somewhat peculiar set of skills as the hallmark of the modern mechanical arts—and as the evidence of human creativity. They emphasized the ability to move large objects on the one hand and to call up effective illusions on the other; they evoked the moderns' mastery of perspective, astrology, and the basic machines. By doing so, they both evoked an existing tradition and identified the Florentine artists they knew as the ones who had most fully realized its promise. Brunelleschi and the rest had altered the face of the world and the condition of mankind.

Alberti hailed the powers of architecture in a bravura paragraph. Architects engaged themselves heroically in "cutting through rocks, tunneling through mountains or filling in valleys, by restraining the waters of the sea and lakes, draining marshes, building ships, altering the course and dredging the mouths of rivers, building bridges" and by doing so "opened up new gateways to all the provinces of the world."[51] Manetti exclaimed in a memorable chapter of his treatise on the dignity of man that human industry had transformed the world into a replica of heaven: "They are ours, that is human, since they are all the products of human industry: all the houses, all the towns, all the cities, and finally all the buildings all over the world. They are so great that on account of their excellence, they ought properly to be evaluated as the work of angels rather than men."[52] Both men celebrated the feats of navigation that had created networks of trade and communications across the known world and beyond it. And both men made clear that machines embodied the creative energy of the human mind at its highest. The chief lesson that the humanists drew from the work of the engineers was clear: the advancing edge of human effort had transformed and was continuing to transform the natural world, both as it was and as it was perceived.

Long before the first Renaissance magi ranged the courts and squares of Europe, the artist-engineers and the literati who praised them developed a new vision of human power and a new sense of the creativity of modern culture. Even the most text-obsessed and speculative of the early magi, Marsilio Ficino, drew on the engineers' ideas and their work, as well as that of their observers, when he set out to define a new sort of magical practice and a new vision of the magus. And no wonder. The fifteenth-century engineer—as Fontana embodied and Alberti understood him—was ready to play Frankenstein. He could already build a partial replica of a human being, and by doing so, he showed not only, as Archimedes had said in another context, that he could move the world, but also that he could disenchant it. The automaton was one of the strongest pieces of evidence that he could wield in support of this bold and influential claim. No wonder that Fontana lavished so much ingenuity on his devils and fire-farting birds. They did important work—intellectual work that no ancient text known in the fifteenth century could do. And at the same time, they amazed the punters. No engineer could hope for more.

NOTES

1. For period conventions, see the rich case studies in Wolfgang Lefèvre, ed., *Picturing Machines, 1400–1700* (Cambridge, MA: MIT Press, 2004).

2. Bayerische Staatsbibliothek, MS Icon. 242, 63v.; published in Eugenio Battisti and Giuseppa Saccaro Battisti, *Le macchine cifrate di Giovanni Fontana* (Milan: Arcadia 2004).

3. Horst Bredekamp, *The Lure of Antiquity and the Cult of the Machine: The Kunstkammer and the Evolution of Nature, Art and Technology*, trans. Allison Brown (Princeton, NJ: Markus Wiener, 1995).

4. See in general Lorraine Daston, ed., *Things That Talk: Object Lessons from Art and Science* (New York: Zone, 2004); and Steven Shapin and Simon Shaffer, *Leviathan and the Air-Pump: Hobbes, Boyle, and the Experimental Life* (Princeton, NJ: Princeton University Press, 1985).

5. For a brief discussion, see Anthony Grafton, *Magic and Technology in Early Modern Europe*, Dibner Library Lecture, 15 October 2002 (Washington, DC: Smithsonian Institution Libraries, 2005). Alex Marr is preparing a full study of these later automata.

6. See especially Marshall Clagett, "The Life and Works of Giovanni Fontana," *Annali dell'Istituto e Museo di Storia della Scienza di Firenze* 1 (1976): 5–28; and Battisti and Saccaro Battisti, *Le macchine cifrate*, which provides a full bibliography. Fontana was *rector artistarum* at Padua in 1421.

7. Biblioteca Apostolica Vaticana, MS Urb. lat. 899, 1r.–v., not in the slightly different text of *Le nozze di Costanzo Sforza e Camilla d'Aragona celebrate a Pesaro nel Maggio 1475*, ed. Tammaro de Marinis (Florence: Vallecchi, 1946), 15r. (de Marinis 10). This text is cited hereafter as *Nozze*. For the oration and its larger setting in Renaissance wedding ritual and practice, see Anthony D'Elia, *The Renaissance of Marriage in Fifteenth-Century Italy* (Cambridge, MA: Harvard University Press, 2004).

8. Biblioteca Apostolica Vaticana, MS Urb. lat. 899, 124v. For the authorship of the text, see Claudia Cieri Via, "L'ordine delle nozze di Costanzo Sforza e Camilla d'Aragona del ms. Urb. Lat. 899," in *La città dei segreti. Magia, astrologia e cultura esoterica a Roma (XV–XVIII secolo)*, ed. Fabio Troncarelli (Milan: F. Angeli, 1985), 185–97. A good general discussion and full bibliography appear in Nicoletta Guidobadi, "Musique et danse dans une fête 'humaniste': Les noces de Costanzo Sforza et Camilla d'Aragona (Pesaro 1475)," in *Musique et humanisme à la Renaissance* (Paris: Presses de l'Ecole Normale Supérieure, 1993).

9. See Patrizia Castelli, "Cronache dei loro tempi II: Le 'allegrezze' degli Sforza di Pesaro, 1440–1512," in *Pesaro tra Medioevo e Rinascimento*, ed. Maria Rosaria Valazzi (Venice: Marsilio, 1989), 232–41; Maria Grazia Ciardi Dupré Dal Poggetto, "Un problema di coerenza: Memoria e realizzazione nella pittura di Giovanni Santi: Nuove proposte cronologiche e attributive," in *Giovanni Santi: Atti del convegno internazionale di studi, Urbino, Convento di Santa Chiara, 17/18/19 marzo 1995*, ed. Ranieri Varese (Milan: Electa, 1999), 109–10, 114 n. 18.

10. Biblioteca Apostolica Vaticana, MS Urb. lat. 899, 1r.; not in *Nozze*.

11. Ibid., 6r.–v.; *Nozze*, 4.

12. Ibid., 12v.–13r.; *Nozze*, 9; 52v.; *Nozze*, 13; 53v.; *Nozze*, 14; 66v.–67r.; *Nozze*, 23; 67v.; *Nozze*, 24.

13. Ibid., 82v.–83r.; *Nozze*, 33–34.

14. Ibid., 86r.–91r.

15. Ibid., 113r.–v.; *Nozze*, 54–55.

16. Ibid., 117r.–v.; *Nozze*, 57.

17. For Costanzo's interest in astrologers like Lorenzo Buonincontri and Camillo Leonardi, see Benedetto Soldati, *La poesia astrologica nel Quattrocento* (Florence: Sansoni, 1906), 127; and Castelli, "Cronache dei loro tempi."

18. On the meanings and experience of wonder in this period, see the now classical work of Lorraine Daston and Katharine Park, *Wonders and the Order of Nature, 1150–1750* (New York: Zone, 1998).

19. Biblioteca Apostolica Vaticana, MS Urb. lat. 899, 6v.; *Nozze*, 4–5.

20. Ibid., 86v.; *Nozze*, 35.

21. Ibid., 106r.; *Nozze*, 50.

22. Ibid., 124r.; *Nozze*, 61.

23. Christine Smith, *Architecture in the Culture of Early Humanism: Ethics, Aesthetics, and Eloquence, 1400–1470* (New York: Oxford University Press, 1992); Hélène Vérin, *La gloire des ingénieurs: L'intelligence technique du XVIe au XVIIIe siècle* (Paris: Albin Michel, 1993).

24. See in general William Eamon, *Science and the Secrets of Nature: Books of Secrets in Medieval and Early Modern Culture* (Princeton, NJ: Princeton University Press, 1994); Pamela Long, "Power, Patronage and the Authorship of 'Ars': From Mechanical Know-How to Mechanical Knowledge in the Last Scribal Age," *Isis* 88 (1998): 1–41, and *Openness, Secrecy, Authorship: Technical Arts and the Culture of Knowledge from Antiquity to the Renaissance* (Baltimore, MD: Johns Hopkins University Press, 2001).

25. Mariano di Jacopo detto il Taccola, *De machinis*, ed. Eberhard Knobloch (Baden-Baden: V. Koerner, 1984), 53; Frank Prager and Giustina Scaglia, *Mariano Taccola and His Book "De ingeneis"* (Cambridge, MA: MIT Press, 1972), 17.

26. Taccola, *De machinis*, 138–39.

27. On Taccola's career, see James Beck, "Introduzione," in *Liber tertius de ingeneis ac edifitiis non usitatis*, by Mariano di Jacopo detto il Taccola, ed. James Beck (Milan: Il Polifilo, 1969), 11–22; Prager and Scaglia, *Mariano Taccola*, 3–21; Bernhard Degenhart and Annegrit Schmitt, with Hans-Joachim Eberhardt, *Corpus der italienischen Zeichnungen*, 1300–1450, vol. 2, *Venedig. Addenda zu Süd- und Mittelitalien*, 4: *Katalog 717–719. Mariano Taccola.* (Berlin: Gebr. Mann, 1982). For the larger context in which he worked, see Paolo Galluzzi, ed., *Prima di Leonardo: Cultura delle macchine a Siena nel Rinascimento* (Milan: Electa, 1991), and the works of Long cited in note 24 above.

28. Taccola, *De machinis*, 1984, 96–97.

29. On Bacon, see the discussions in Eamon, *Science and the Secrets of Nature*, and in the two works by Long cited in note 24 above, as well as the recent collection of studies edited by Jeremiah Hackett, *Roger Bacon and the Sciences: Commemorative Essays, 1996* (Leiden: Brill, 1997).

30. Giovanni Fontana, *Liber Pompilii Azali Placentini de omnibus rebus naturalibus quae continentur in mundo videlicet coelestibus et terrestribus necnon mathematicis et de angelis motoribus quae [!] coelorum* (Venice: O. Scotus, 1544), 5.9, 110v.–11v. On this text and the ascription to Fontana, see in general Lynn Thorndike, "An Unidentified Work by Giovanni da' Fontana: *Liber de Omnibus Rebus Naturalibus*," *Isis* 15 (1931): 31–46.

31. *Asclepius* 37; Brian Copenhaver, ed., *Hermetica: The Greek Corpus Hermeticum and the Latin Asclepius in a New English Translation* (Cambridge: Cambridge University Press, 1992), 90.

32. See especially D. P. Walker, *Spiritual and Demonic Magic from Ficino to Campanella* (London: Warburg Institute, 1959).

33. Giovanni Fontana, *Secretum de Thesauro*, 1.3, in Battisti and Saccaro Battisti, *Le macchine cifrate*, 144; Fontana, *Liber Pompilii Azali*, 1.17, 18v.–19v.; 1.26–27, 26r.–28v.; 2.44 (misnumbered 46), 66r.–73r., e.g., 66v.

34. Fontana, *Liber Pompilii Azali*, 1.17, 19r.

35. Ibid., 1.17, 19r.

36. Ibid., 1.17, 19r.–v.

37. Ibid., 3.10, 73v. (misnumbered 72).

38. Ibid., 3.13 (misnumbered 14), 76r. (misnumbered 75).

39. Ibid., 3.13 (misnumbered 14), 76r. (misnumbered 75).

40. Ottavia Niccoli, *Prophecy and People in Renaissance Italy*, trans. Lydia Cochrane (Princeton, NJ: Princeton University Press, 1990).

41. Fontana, *Liber Pompilii Azali*, 3.13 (misnumbered 14), 76r. (misnumbered 75).

42. Bayerische Staatsbibliothek MS Icon. 242, 70r.; Battisti and Saccaro Battisti, *Le macchine cifrate*, 99–100, 140.

43. Bayerische Staatsbibliothek MS Icon. 242, 59v.–50r.; Battisti and Saccaro Battisti, *Le macchine cifrate*, 94, 134–35.

44. Bayerische Staatsbibliothek MS Icon. 242, 51r. For a helpful commentary, see Battisti and Saccaro Battisti, *Le macchine cifrate*, 88, 131.

45. See Lynn Thorndike, *History of Magic and Experimental Science*, 8 vols. (New York: Macmillan, 1923–58), 4:150–82.

46. For these developments, see Simon Schaffer, "Enlightened Automata," in *The Sciences in Enlightened Europe*, ed. William Clark, Jan Golinski, and Simon Schaffer (Chicago: University of Chicago Press, 1999), 126–65, and the essays collected in this volume.

47. Leon Battista Alberti, *Momus*, ed. Virginia Brown and Sarah Knight, trans. Sarah Knight (Cambridge, MA: Harvard University Press, 2003); cf. Lucian *Zeus tragoedus* 9–13.

48. Alberti *De re aedificatoria* 6.8, in Leon Battista Alberti, *L'architettura (De re aedificatoria)*, ed. and trans. Giovanni Orlandi, 2 vols. (Milan: Il Polifilo, 1966), 2:497; and *On the Art of Building in Ten Books*, trans. Joseph Rykwert et al. (Cambridge, MA: MIT Press, 1988), 175.

49. See Jurgis Baltrušaitis, *Anamorphic Art*, trans. W. J. Strachan (New York: Abrams, 1977).

50. Battisti and Saccaro Battisti, *Le macchine cifrate*, 97.

51. Alberti, *On the Art of Building*, 3; *L'architettura*, 1:9–11.

52. Giannozzo Manetti, *De dignitate et excellentia hominis*, ed. Elizabeth Leonard (Padua: Antenore, 1975), 77–78.

Infinite Gesture: Automata and the
         Emotions in Descartes and Shakespeare

SCOTT MAISANO

Let me see. [Takes the skull.] Alas, poor Yorick! I knew him,
Horatio, a fellow of infinite jest. | *Shakespeare,* Hamlet

At the dawn of the twenty-first century, as we have moved closer to
creating machines capable of demonstrating humanlike motion and emotion, a bit of folklore concerning the philosopher René Descartes has acquired new relevance and wider circulation.[1] Legend has it that, in his later
years, a lonely Descartes constructed a mechanical replica of his illegiti-
mate daughter, Francine, who had died a decade earlier at the age of five.
This automaton, it was said, served as the philosopher's constant, albeit secret, traveling companion; until, that is, the fateful evening of its discovery,
whereupon some deckhands (ignorantly fearing some demonic magic had
gotten aboard their ship) cast the mechanical child overboard to a watery
grave. This cautionary tale is situated in the opening pages of Gaby Wood's
recent account of the origins of artificial life and casts a rather morbid light
on Descartes, the mechanist philosopher, and on his apparent attempt at
reducing human existence to complex machinery. The same anecdote is also
rehearsed in the introduction to Stephen Gaukroger's recent "intellectual
biography" of Descartes. Gaukroger, however, cites the rumor in order to
reject it as anti-Cartesian "propaganda," a narrative that suspiciously recasts
the philosopher's psychophysical dualism in terms of a masculine agent of
reason and a feminine body of emotions. What this rumor falsely ascribes to
Descartes, according to Gaukroger, is the belief that women are emotional
creatures driven by their passions, with states of mind determined by the
internal fluctuations of bodily fluids: in short, the belief that women are not

fundamentally different from and may even be replaced by sophisticated hydraulic machines.

Whether the rumor is being promoted or disputed, this diabolical portrait of Descartes has taken center stage in recent writing on the seventeenth-century philosopher because it appeals to—and remains relatively easy to discuss within—currently popular perspectives on artificial life. My own attempt to explain the distance separating twenty-first-century theories of artificial life from seventeenth-century thinking about the relationship of bodies to machines also begins with Descartes contemplating the possibility of resurrecting a dead loved one, and a dead woman at that. But the conclusion I draw is quite different from Gaby Wood's anachronistically Frankensteinian view of Descartes. In place of the coldly rationalistic Descartes, who, according to Wood, first formulated "the notion, taken up by anatomists and philosophers alike, that man is a machine, and can only be understood as such," I substitute a deeply dualistic but nonetheless compassionate Descartes who agonizes over the fact that "man is *not* a machine, but can only be understood *by others* as such." With this revaluation of Descartes in mind, I hope to show that, even though Wood is in excellent company when she suggests that Descartes "laid the foundations" for the period's pervasive metaphor of the body as a machine, that dubious honor might just as easily be ascribed to another seventeenth-century philosopher, Shakespeare's Hamlet.

DESCARTES' PASSION PLAY

In his late work *The Passions of the Soul*, Descartes distinguishes "internal emotions," which originate in the mind itself, from ordinary emotions, which originate in the body and later influence the mind:

> When a husband mourns his dead wife, it sometimes happens that he would be sorry to see her brought to life again. It may be that his heart is torn by sadness aroused in him by the funeral display and the absence of a person to whose company he was accustomed. And it may be that some remnants of love or of pity occur in his imagination and draw genuine tears from his eyes. Nevertheless he feels at the same time a secret joy in his innermost soul, and the emotion of this joy has such power that the concomitant sadness and tears can do nothing to diminish its force.[2]

This hypothetical scenario causes Stephen Gaukroger, an otherwise able and ardent defender of Descartes, to concede that the "example suggests to me a low view of women on Descartes' part."[3] But *The Passions of the Soul* has its genesis in the long and often emotional correspondence between Descartes and Princess Elizabeth of Bohemia, to whom the book was dedicated.[4] Thus it seems unlikely that, if Descartes held a "low view of women," he would have chosen to display it in this particular work. Indeed, though Descartes himself never married, it is more likely that, in choosing the example of a husband mourning his dead wife, he purposely selected the loss he considered most profound, most likely to erode the faculty of reason in a sudden cataract of uncontrollable passion.

This interpretation of the passage also makes the most sense of the remainder of Descartes' explanation of these "secret joys" or "passions of the soul":

> When we read of strange adventures in a book, or see them acted out on the stage [Descartes continues, immediately following the example of the dead wife], this sometimes arouses sadness in us, sometimes joy or love, or hatred, and generally any of the passions, depending on the diversity of the objects which are presented to our imagination. But we also have a pleasure in feeling them aroused in us, and this pleasure is an intellectual joy which may as readily originate in sadness as in any of the other passions.[5]

Descartes here supplies a rationale for the paradoxical pleasures to be found in viewing a tragedy. Though the calamitous events "acted out on the stage" frequently reduce us to tears, we would be sorry, as in the earlier example of the dead wife, to forgo this experience of suffering and grief. Descartes, therefore, is not looking for away out of, or around, the experience of emotions. Instead he looks to take their full measure: "We have a pleasure," he insists, "in feeling [the emotions] aroused in us." Descartes simply wants his reader, say Princess Elizabeth, to recognize that her sense of self—her "secret joy"—is not being swept away by the exigencies of the moment, but is instead transcending them as the unshakable perspective on all such trials and ordeals.

The "passions of the soul," which can accompany both real and fictional tragedies, depend on the bodily passions as their impetus; but ultimately these "intellectual" emotions provide one's inmost self with a bulwark against the onslaught of emotions precipitated by external factors. "Such troubles will serve rather to increase its joy," Descartes continues, alluding to the soul, "for on seeing that it cannot be harmed by them, it becomes aware of its perfection."[6] The emphasis placed on the indestructible sense of identity, on "seeing that [the soul] cannot be harmed," reminds us that this late treatise is the culmination of years of informal "counseling" that Descartes had been offering to the melancholic Elizabeth, who suffered from fever and stomach-sickness, among other symptoms, prompted by the momentous ups and downs of her extended family's fortunes.

For Descartes, there is no question of removing the passions, only of mastering them: and mastering the passions requires "striving to separate within ourselves the movements of the blood and spirits from the thoughts to which they are usually joined."[7] This idea of "separating" thoughts from impulses ("the movements of the blood") may sound to us, today, like the kind of emotional abstraction we would associate with an intelligent machine, but *The Passions of the Soul* never makes its ideal state of abstraction dependent on the exclusion of bodily passions. To the contrary, Descartes concludes the entire work with the thought that "it is on the passions alone that all the good and evil of this life depends."[8] Far from discounting emotional reactions, therefore, Descartes' philosophy makes these passionate feelings the basis of—the necessary scaffolding for—the thinking self.

"WHAT A PIECE OF WORK IS MAN"

In act 2, scene 2, of Shakespeare's greatest tragedy, when Polonius concludes his reading of a letter that Hamlet has sent to his daughter, Ophelia, the signature at the end of the letter reads, "Thine evermore, most dear lady, whilst this machine is to him, Hamlet."[9] Most modern editions of the play inform readers, albeit without additional comment, that by "machine" Hamlet means "body."[10] But Hamlet's glib substitution of "machine" for "body" runs counter to the argument that it was Descartes who "laid the foundations for" the idea that "man is a machine, and can only be understood as such." Worse still, Hamlet not only refers to his body as a "machine," he even refers to himself in the third person: "whilst this machine is to him, Hamlet." And yet it is also Descartes, not Shakespeare, who has been faulted by John Cook and other analytical philosophers for "introducing an extraordinary use of the word 'body,' in which a 'distinction between himself and his body' makes sense."[11]

Though the received history of science and philosophy tells us that a great epistemic upheaval divides Shakespeare from Descartes, the careers of the dramatist and the philosopher peaked just a generation apart. Indeed, some of the same people, including Elizabeth, the Winter Queen of Bohemia (mother to Descartes' patron and correspondent of the same name), played host to both men.[12] These two authors, whose names have become synonymous with some of the greatest achievements of the literary Renaissance and the scientific revolution, respectively, are nonetheless routinely contrasted as if they had existed centuries apart from one another. To take just one brief example, Shakespearean scholar Gail Kern Paster, insisting on the importance of Galenic physiology and the role of internal bodily humors in determining the mental states of characters like Hamlet, points out that "Descartes succeeded in disturbing this continuum and beginning its slow demise," for it was "Descartes [who] set in motion the gradual process towards abstraction and dematerialization that, in the eighteenth century, overtook early modern discourses of mind and body."[13]

This Cartesian legacy, for Paster, acts as an irritating obstacle to readers' understanding of how Shakespeare and other early seventeenth-century writers conceived of the interaction between mind and body. "Only with great difficulty," Paster maintains, are literary critics able to appreciate the historically specific, pre-Cartesian discourse of feelings and emotions in "even the most familiar" of Shakespeare's texts. Paster reminds historically oriented audiences that Shakespeare wrote at a "moment in the history of bodies, minds, and souls when bodily fluids could still carry the full weight of a character's destiny, a moment when . . . to alter the character and quality of a body's fluids was to alter that body's passions and thus that body's state of mind or soul."[14] While it is important to point out that bodily fluids "could" transform Hamlet's passions, his state of mind, even his very soul, it is equally important to observe that Hamlet appears determined not to let that happen. Paster claims that Hamlet's rhetoric "bespeaks his natural embeddedness in the world," gives voice, that is, to a Galenic monism and holism still innocent of late seventeenth-century Cartesian "abstraction and dematerialization."[15] And yet, for all that, Hamlet refers to his body as a "machine."

Of course, Shakespeare's *Hamlet* predates the posthumous publication of Descartes' *Treatise on Man* by more than half a century. So, if Hamlet has not yet read Descartes while studying philosophy at Wittenberg, then what has he been reading that would lead him to refer to his body as an estranged "machine"? We can be fairly certain of one text, which likened human bodies to machines, that Shakespeare had at hand, or at least in mind, while composing *Hamlet*. In the same scene in which Hamlet refers to his body as a "machine," he also defends himself, to Rosencrantz and Guildenstern, against the charges of madness by saying: "I am but mad north-north-west; when the wind is southerly I know a hawk from a handsaw."[16] The source for this reference to "hawks" and "handsaws" is Timothy Bright's *A Treatise* of *Melancholie*, first published in 1586, in which the author suggests that all "actions are bodily performed of the soul, by employing that excellent, and catholic instrument of the spirit, to the mechanical works of the gross and earthly parts of our body."[17] In Bright's analysis, the hawk and the handsaw represent two distinct modes of instrumentality: the hawk "obtaineth power in itself, and requireth direction only"; the handsaw, by contrast, is "dead in itself, and destitute of all motion." The human spirit, according to Bright, is like the hawk: vital but requiring from the soul some channeling or guidance in the proper direction. The human body, on the other hand, is "slow, dull, and given to rest of itself . . . like flails, saws, or axes in the hand of him that uses them."[18] This proto-Cartesian body is inert, passive matter—pure extension in space—until it is put into motion by the spirit or the soul.

The idea that "man is a machine, and can only be understood as such" did not begin with Descartes. Instead, this analogy had materialist precedents as far back as Aristotle and Galen, particularly in the form of Galen's hydraulic model of fluid-based psychology, which historian Peter Brown has characterized as "a human Espresso machine."[19] As a theologian, Timothy Bright was not entirely comfortable with what he viewed as the reductive functionalism of Galenic humoral theory, which maintained that human beings are creatures whose actions might appear autonomous to an ignorant observer, but whose every movement—internal as well as external—is to some degree the predetermined effect of an unseen fluid dynamics.[20] Indeed, each of Bright's numerous mechanical analogies—he invokes everything from mythological automata, such as Archytas's mechanical flying doves and Vulcan's "walking stools," to more modern "automatical instruments," such as "clocks, watches, and larums [alarms]"—is used to illustrate a "top-down" model, a sort of psychosomatics in which the motive cause travels from the soul to the spirit to the body, that he hopes will complement and/or complicate the dominant, "bottom-up" paradigm of Galenic humoral theory.[21] Given Shakespeare's familiarity with Bright's *Treatise of Melancholie*, I suggest that the historical context for Hamlet's unique psychology is one in which the turbulent subjectivity of Galenic humoralism—what Gail Kern Paster calls the "swift metamorphoses of humoral corporeality"[22]—was slowly losing ground to a competing notion of subjectivity, later to be called Cartesianism, whose distinctive feature is precisely its irreducibility to such physical phenomena.

To suggest that Hamlet sets the stage for modern cognition is, of course, nothing new. In the opening pages of his best-seller, *Shakespeare: The Invention of the Human*, Harold Bloom calls Hamlet "the central hero of Western consciousness," adding that "something in him is determined not to be overdetermined."[23] For Bloom, Shakespeare has passed the theatrical Turing test: the thoughts, behaviors, actions, and speeches of Shakespeare's characters are more "lifelike" than many of Shakespeare's actual contemporaries. Hamlet, he contends, "seems no more fictive than Montaigne."[24] Nor is Bloom alone in this way of thinking. J. Leeds Barroll titled his 1972 study of the main characters in Shakespeare's tragedies *Artificial Persons*; and, a full century ago, A. C. Bradley was already writing of Hamlet, Othello, and Macbeth as if they were real people, with real psychological motivations and histories that extended beyond the words on the page.[25]

But Bloom also poses a number of questions that are—perhaps unwittingly—significant for the history of artificial life: "What made [Shakespeare's] art of characterization possible? How can you create beings who are 'free artists of themselves,' as Hegel called Shakespeare's personages?" And, more to the point, "Why do his personages seem so *real* to us, and how could he contrive that illusion so persuasively?"[26] It is no coincidence, I contend, that Shakespeare created the most powerful simulations of human psychology, in the form of his soul-searching characters, at precisely the same time that early modern engineers were creating legions of automata for display in courts across Europe. Automata posed a genuine crisis for human agency: this crisis, specifically, was how to prove that humanity itself was anything more than an elaborate machine or "a piece of work," as Hamlet says. Hamlet repeatedly runs up against this ambiguity between human and machine, the same ambiguity that would lead Descartes to ask in the *Meditations*: "If I look out of the window and see men crossing the square . . . do I see any more than hats and coats which could conceal automatons?"[27] For Hamlet, ironically, proof of one's humanity sometimes requires a *lack* of animation, an ability to do precisely *nothing* in response to one's turbulent inner passions. Here, for instance, is how Hamlet coaches an anonymous actor who wishes to make his performance more "lifelike":

> Do not saw the air too much with your hand thus, but use all gently, for in the very torrent, tempest, and as I may say, whirlwind of your passion, you must acquire and beget a temperance that may give it smoothness. . . . O, there be players that I have seen play . . . have so strutted and bellowed that I have thought some of nature's journeymen had made men, and not made them well, they imitated humanity so abominably.[28]

Although Hamlet's criticisms of the exaggerated gesturing and overloud speaking, which we all associate with amateur theatre, do not strike us as particularly novel, his choice of words does. Hamlet likens bad actors to poorly constructed automata, to men made by other men—by semiprofessional "journeymen" to be precise—who look and sound "abominably" to the eyes and ears of real human beings.[29]

Inferior actors and automata "imitate humanity so abominably," in Hamlet's words, because they place an undue emphasis on externalizing the passions, on making visible for all to see emotions that most of us would work equally hard to suppress or restrain. That is, these actors and automata—these "artificial people" in any case—make no distinction between inner passions and outward actions. The Prince of Denmark, contrariwise, counsels "temperance," "smoothness," and the ability to express oneself "gently" even under the most extreme circumstances. The murder of his father and the hasty remarriage of his mother notwithstanding, Hamlet famously resists acting on impulse; he strives, *pace* Descartes, to "separate within [himself] the movements of the blood and spirits from the thoughts to which they are usually joined." Hamlet, in fact, proves so successful in separating his thoughts from his passions that, as he confesses in soliloquy, he remains unable to act out his bloody impulses despite having "the motive and the cue," unlike the anonymous actor who is furnished only with fictitious reasons for an imagined revenge. At the same time, however, Hamlet seems to admire—indeed, almost to envy—the actor's ability to assume a fictional persona and to cry on command, even at a mere pretext for sorrow:

> Is it not monstrous that this player here,
> But in a fiction, in a dream of passion,
> Could force his soul so to his own conceit
> That from her working all his visage wanned;
> Tears in his eyes, distraction in his aspect,
> A broken voice, and his whole function suiting
> With forms to his conceit?[30]

And, still, Hamlet knows too well from observing the hypocrisy of King Claudius and Queen Gertrude, that crying on command is what we would expect from an actor, from a piece of ingenious garden statuary, or from what philosophers, such as Daniel Dennett, designate as a "zombie": "a human being who exhibits perfectly natural, alert, loquacious, vivacious behavior but is in fact not conscious at all, but rather some sort of automaton."[31] All expressions of emotion are subject to scrutiny because, from Hamlet's philosophical perspective, there can exist no "clear and distinct" signs of sincerity.

From his very first lines of dialogue in response to his mother, who is concerned that he seems to mourn his father excessively, Hamlet characteristically casts doubt on all "actions" as mere "mechanical" imitations or "rote" simulations of *real* feelings, the kind of "internal emotions" that he insists defy external representation altogether:

> Seems, madam! Nay it is, I know not "seems."
> 'Tis not alone my inky cloak, good mother,
> Nor customary suits of solemn black,
> Nor windy suspiration of forced breath,
> No, nor the fruitful river in the eye,
> Nor the dejected haviour of the visage,
> Together with all forms, modes, shapes of grief,
> That can denote me truly. These indeed seem,
> For they are actions that a man might play,
> But I have that within which passes show,
> These but the trappings and the suits of woe.[32]

Hamlet might saw the air with his hands and weep a tempest of tears, but doing so will fail to convince anyone, including himself, that his feelings are any more authentic than the melodramatic exhibitions of so many spectacularly second-rate actors and automata.

In his stubborn insistence on a nondemonstrable subjectivity, Hamlet adumbrates the philosophy of Descartes, who writes in the *Second Meditation*: "I am not that structure of limbs which is called the human body. I am not even some thin vapour which permeates the limbs—a wind, fire, air, breath, or whatever I depict in my imagination; for these are things which I have supposed to be nothing. Let this supposition stand; for all that I am still something."[33] This passage says quite the opposite of Wood's formulation "that man is a machine and can only be understood as such." Instead, it echoes Hamlet to the effect that man is not a machine—"that structure of limbs which is called the human body"—and yet can only be understood, known to others, as such. For Descartes, as for Hamlet before him, what seems doubtful—and what can, therefore, be doubted—are any "actions that a man might play." Humanity remains, thus, outwardly indiscernible from both actors and automata. Meanwhile, what "denotes [a person] truly" is not the body, not some humor or vapor, not even a breath, but only "that within which passes show," an invisible passion, inaccessible to the scrutiny of loved ones and physicians alike. What psychophysical dualism ultimately enables, Descartes' eminently rational methodology notwithstanding, is the ineffable inner experience—and, yes, the emotional extremes—of Romanticism, Expressionism, and Existentialism. Thus, even while writing a love letter—or, rather, *especially* while writing a love letter—Hamlet has the impression that he is a machine. While Descartes might never have been the poster boy for either the Romantic imagination or Expressionist angst, Hamlet has never *not* been the poster boy for both.

Thus what makes Hamlet "human" in Harold Bloom's sense of the word is neither what he says nor what he does; it is, to the contrary, something invisible and occluded, something pulsing within "this machine," something akin to what Gilbert Ryle called in his critique of Cartesianism the "ghost in the machine."[34]

THE GHOST *IN* THE MACHINE IS THE GHOST *OF* THE MACHINE

What was he that did make it? See, my lord,
Would you not deem it breathed? and that those veins
Did verily bear blood?

> *Leontes to Polixenes, upon seeing the "moving statue" of Hermione*
> *in Shakespeare's* The Winter's Tale

Prior to Descartes, Shakespeare, too, contemplated the reaction a husband might have to seeing his dead wife—in this case, dead for some sixteen years—suddenly revived.[35] In Shakespeare's late tragicomedy, *The Winter's Tale*, however, the resurrection actually does involve recourse to an automaton; or at least it seems so initially. In the play's famous final scene, King Leontes, who long ago precipitated the death of his wife, Hermione, through an Othello-like outburst of jealous rage, comes face-to-face with a "statue" commemorating his dearly departed spouse. Paulina, a friend of Hermione's, orchestrates the entire scene: after leading Leontes and several other characters to the statue of Hermione, Paulina claims, to the incredulous assembled party, that she can make the statue of Hermione move. And it does. Paulina then urges the statue to speak—and it does. Finally Paulina reveals, to the surprise of everyone in the theatre—both the characters onstage and the audience—that Hermione is not an ingenious automaton but merely herself, a real human woman. Here, in a stupendous "coup de theatre," Shakespeare succeeds where Descartes would later fail: he convinces his audience that the human body is, as Descartes argues in *The Treatise on Man*, little more than "a moving statue."[36] Not until the fascination for such mechanical exhibits as Jaquet-Droz's *La Musicienne* and Kempelen's "chess-playing Turk" would

" A WINTER'S TALE "
TWO HERMIONES

MARY ANDERSON MRS. CHARLES KEAN
From photograph by W. D. Downey *From an old print*

FIGURE 4.1 | Victorian Hermiones: Mary Anderson, from a photograph by W. D. Downey; and Ellen Kean, née Ellen Tree, from an old print.

spectators be so thoroughly taken in again. The crucial difference, however, is that, instead of making a moving statue look like a real human being, Shakespeare made a real human being look like a moving statue.

Although the ambiguity between human and machine is usually said to have begun with René Descartes, lifelike automata were quite prevalent, as the allusions to them in Timothy Bright's *Treatise* and Shakespeare's *Hamlet* suggest, even in the early seventeenth century. Indeed, in the *Treatise on Man*, where Descartes "supposes" the "body to be nothing but a statue or machine made of earth," he refers specifically to automata that had been in existence in Shakespeare's lifetime. More specifically, the remarkably animated statues designed by Tomasso Francini for "the grottos and fountains in the royal gardens" at Saint-Germain-en-Laye provide the basis for Descartes' provocative supposition. In these marvelous fountains, Descartes informs his readers, "the mere force with which the water is driven as it emerges from its source is sufficient to move various machines, and even to make them play certain instruments or utter certain words depending on the various arrangements of the pipes through which the water is conducted."[37]

Descartes, however, was not the first to visit or write about these automata: the French Huguenot engineer Salomon de Caus had previously provided illustrations and explanations of Francini's uncanny creatures in his textbook Les raisons des forces mouvantes. This book was dedicated in 1615 to de Caus's patron at the time, Princess Elizabeth, the daughter of James I.[38] Interestingly, two years prior to the dedication of de Caus's book, the same Princess Elizabeth had been honored—as part of the wedding festivities for herself and the Elector Palatine in 1613—with a performance of William Shakespeare's The Winter's Tale. Hermione's acting like an automaton—a simulation of a simulation—surely called to mind for some members of this royal audience the ingenious inventions of hydraulic engineers such as Francini and de Caus. The Stuart court gardens, after all, had been inspired by the late Prince Henry's fascination with automata like the ones housed in the Bohemian palaces of Rudolph II. Salomon de Caus's partner in garden design at the Stuart court, the "Italian Neoplatonic engineer" Constantino de Servi, had previously been employed by both Rudolph in the Bohemian court and by Francesco I at the Florentine court.[39] De Servi came to England as the direct result of a request made on behalf of Prince Henry for an engineer trained by Tomasso Francini, the fountain designer whose automata, a generation later, served to illustrate Descartes' theory of the human body as a machine.[40] Given that The Winter's Tale was written and performed at the same time that the Jacobean court had recruited engineers and artificers such as Constantino de Servi and Salomon de Caus to construct hydraulic automata—lifelike statuary—that looked and moved like real human beings for the gardens at Somerset House, it might not surprise us to see the English playwright attempting to one-up these newly arrived foreign manufacturers of mechanical marvels, to beat them at their own games of illusionism and trompe l'oeil.[41]

What is surprising, however, is that the figure of the automaton—this curiosity of court culture—functions in The Winter's Tale as something more than an amusing novelty. For, even though nothing physically changes from one moment, when we see the "statue" of Hermione, to the next, when we suddenly see Hermione herself, there is no denying that a definite change has taken place. The automaton in this late play is a fictional ruse, the audience's own unwitting supposition; as such, it is a window into the invisible, "intellectualized" passions of the human soul. Not even Descartes himself could have provided a more effective demonstration that what we "recognize" as Hermione is "not that structure of limbs which we call the human body." After all, audiences stare at the human actor for a prolonged period—upwards of fifteen minutes in most productions—and, all the while, mistake her for an ingeniously lifelike statue. Hermione is, therefore, simultaneously identified with—and yet proven irreducible to—mere bodily machinery. Just for a second, The Winter's Tale allows us to glimpse in the animated statue of Hermione the otherwise unobservable, invisible "anima" itself.[42]

Hermione, whose complete composure for some sixteen years provides a perfect foil to Leontes' "swift metamorphoses of humoral corporeality," is living proof, so to speak, that humans in some of their most passionate moments remain indistinguishable from inert machines. Indeed, it seems fair to say that the "psychological self-sameness" that Gail Kern Paster identifies as the fictional stuff of the Cartesian cogito is not merely a ghost "in" the machine, but a ghost "of" the machine. Both the Cartesian cogito and Hamlet's thing "within which passes show" are fantasy figures of consistency, competency, and self-control. So, too, are Renaissance automata. Elizabeth King's discussion in this volume of a sixteenth-century automatic monk who prays continuously, whether for himself or for others, is a perfect example of how a machine's endurance and efficiency can set the (impossible) standard for human conduct.

With this superhuman and quasi-spiritual dimension of Renaissance automata in mind, it is important to note not only that Hermione is believed to have been dead for some sixteen years prior to the final scene, in which she appears as "a pretend automaton," but also that the famous "statue scene" is not the first time that Hermione has made a visit from beyond the grave. Just after she has allegedly died, Antigonus, a lord in the court of Leontes, informs Hermione's infant daughter, Perdita, that he has seen her mother's ghost:

> I have heard, but not believed, the spirits o' th' dead
> May walk again. If such things be, thy mother
> Appeared to me last night, for ne'er was a dream
> So like a waking. To me comes a creature,
> Sometimes her head on one side, some another.
> I never saw a vessel of like sorrow,
> So filled and so becoming. In pure white robes,
> Like very sanctity, she did approach
> My cabin where I lay; thrice bowed before me,

And, gasping to begin some speech, her eyes
Became two spouts.[43]

At first glance this description sounds ghostly, purely spiritual, supernatural. But look closely at the language: the hallowed aura surrounding the "pure white robes, / Like very sanctity" can hardly conceal the embarrassment of references to Salomon de Caus's hydraulic automata. First, we have the eerily pivoting head—"sometimes . . . on one side, some another"—and the routinized bowing movements. Then there is the comparison of Hermione to a "vessel" that, like a hydraulically powered automaton, is always both "fill'd" and "becoming." Finally, we have the "gasping" speech produced by air being driven through pipes and the "eyes" that literally "became two spouts."

If this equation of the supernatural ghost and the perfunctory machine seems incongruous, it nonetheless reveals a truth about the proto-Cartesian "soul" or "self." The cogito, precisely because it is freed from the gross materiality of the body, is uniquely capable of what I call "perpetual emotion" (what we have already seen Descartes call "internal emotions," "intellectual joys," and "the passions of the soul"). Throughout *The Winter's Tale*, characters find themselves in situations where an automaton is required, situations where the proto-Cartesian logic—"a person is not a machine, but can only be understood by others as such"—becomes inescapable. In the opening exchange of *The Winter's Tale*, Polixenes, the King of Bohemia, wishes to express his gratitude to Leontes and Hermione, the King and Queen of Sicily, for their gracious hospitality. Immediately, Polixenes perceives that his body is not up to the task and therefore gives an exponentially enhanced form of thanks:

> Nine changes of the wat'ry-star hath been
> The shepherd's note, since we have left our throne,
> Without a burthen: Time as long again
> Would be fill'd up, my brother, with our thanks,
> And yet we should, for perpetuity,
> Go hence in debt. And, therefore, like a cipher,
> Yet standing in rich place, I multiply
> With one "We thank you" many thousands moe,
> That go before it.[44]

What exactly is Polixenes saying here? To begin with, he implies that his sense of indebtedness to this couple is so great that the process of repaying them, through curtsies and bows and all the other conventional courtly flourishes, exceeds his powers. Polixenes imagines himself saying thanks continually (first, for nine months and then still longer) and, faced with the prospect of such an infinite gesture, concludes that he cannot even imagine doing this, at least not physically. His solution, then, is to express his gratitude in such a way that he can avoid "going through the motions," the endless, monotonous motions that could conceivably continue "for perpetuity." Perhaps Polixenes, whose undertaking has a far greater affective dimension than mere counting, wishes not to appear like a slavish automaton, an unthinking machine whose repeated gestures of thanks would gradually be discounted due to their predictable, routine, and compulsory nature. Or, alternately, Polixenes may regret that he cannot, like an automaton, say thanks forever. And, "therefore, like a cipher"—that is, like a zero, an entity that registers or represents itself strictly as nothing—Polixenes imbues his parting compliment with a touch of genuine, heartfelt humanity. It would be a mistake to read Polixenes' choice of metaphor here as nothing more than the self-abnegating flattery appropriate to addressing royalty. To become "like a cipher" is to approximate nothingness, yes, but in such a peculiar way as to "multiply" by "many thousands" the significance of one's actions.

Is Polixenes here just another Hamlet, urging his auditors to trust that he has "that within which passes show," an inner self that feels more passion than outward semblances can reveal? Or is there something still more subtle about the efficiency of Polixenes' speech, about the arithmetical gymnastics that enable nothing—a "cipher," literally a zero—to stand in place of everything—a veritable "perpetuity" of affect? There is no clearer way to distinguish the mind, or soul, from the body than to engage in a "thought-experiment," as Polixenes has done, and to arrive thereby at a mental conclusion whose proof would physically require an eternity to demonstrate. That is, when we resort to demonstrating the solution to a problem by supposing a hypothetical actor in an imaginary scenario—because we know that we cannot physically perform the demonstration ourselves—we have to some degree isolated not only the modality of the possible from the actual, but also the mysterious Cartesian cogito from its biological substrate. Think for a minute of Descartes responding to the second set of "Objections" aimed at his *Meditations on First Philosophy*. Pressed by Mersenne on the issue of how he can be sure his idea of God is more than a product of his embodied human imagination, Descartes replies with the hypothetical example of someone attempting to count to infinity: "I notice," he says, assuming his position as the counter, "that, when I count, I cannot reach a largest number, and hence I recognize that there is something in the process of counting which exceeds my powers."[45] Ironically, the very recognition of this failure—or rather, incapacity—to perform "the infinite" serves as proof of its existence as a "clear and distinct" idea for the mind. In other words, Descartes reasons, I could never know for certain that my tireless enumeration had failed to reach the "largest number" without an innate "understanding," which transcends the embodied and thus finite "imagination," of the presence of still greater numbers that God alone could comprehend.[46]

Thought-experiments like the one Polixenes articulates at the start of the play reappear in every act, if not every scene, of *The Winter's Tale*, as characters consistently judge their somatically defined selves to be inadequate to the proper conveyance of their sentiments. Take, for example, the imaginary scenario proposed by Paulina at the moment when she finally convinces Leontes that he has been wrong in his persecution of Hermione, that his actions have led both to her death and to the death of their son, Mamillius:

> A thousand knees, [she estimates]
> Ten thousand years together, naked, fasting
> Upon a barren mountain, and still winter
> In storm perpetual, could not move the gods
> To look that way thou wert.[47]

The adjective "perpetual" here is a denomination of Polixenes' aforementioned "perpetuity"; and in both cases the respective gestures of gratitude and contrition are placed into a purely hypothetical state of perpetual motion.[48]

A reformed and contrite Leontes realizes, for his part, that while human beings may strive for perpetual emotion—that is, a sorrow too profoundly spiritual to be whisked aside by the "swift metamorphoses of humoral corporeality"—they nonetheless remain constitutionally incapable of perpetual motion itself. Thus, without availing himself of Polixenes' "cipher" or Paulina's "thousand knees," Leontes vows to mourn his wife, automaton-like, for the rest of his life:

> One grave shall be for both [Hermione and Mamillius]: upon them shall
> The causes of their death appear, unto
> Our shame perpetual. Once a day I'll visit
> The chapel where they lie, and tears shed there
> Shall be my recreation. So long as nature

Will bear up with this exercise, so long
I daily vow to use it.[49]

For the third time, that word—"perpetual"—leaps out and reminds us that all Leontes' vows and visits and tears, like all of Polixenes' "thank you's" at the start of the play, will never be enough to unburden him of his sentiment. But could any human be adequate to the task of performing, exhibiting, demonstrating, or otherwise proving such a "passion of the soul," this sense of "perpetual emotion"? If, as Gaby Wood suggests, "clockwork is the antithesis of our mortal selves," because we remain "subject to time, to its inevitable march toward death, whereas the clockwork automaton merely marks the time without falling prey to it," then only a machine—perhaps a piece of garden statuary that can really "turn on the waterworks" and spout endless tears from its granite eyes—will suffice to embody the idea of perpetual emotion.[50]

And that is precisely the miracle of Hermione. She is not a statue who, for love, becomes human; but a human who, for love, becomes a statue. Why "for love"? As Descartes writes in the concluding sentence of *The Passions of the Soul*, "The chief use of wisdom lies in its teaching us to be masters of our passions and to control them with such skill that the evils which they cause are quite bearable, and even become a source of joy."[51] That sentence, that sentiment, sums up (better than any Shakespearean critic to date) the emotional trajectory of *The Winter's Tale*, which begins with Leontes' murderous influx of choler and concludes with Hermione's artless composure. Descartes' "chief use of wisdom" is precisely the lesson that Leontes learns and that Hermione exemplifies at the end of the play.

For sixteen years Leontes has gone every day, like clockwork, to the site of his wife's grave, to weep and to pray. And, it seems, his prayers have been answered. The only thing that worries the ostensibly happy ending of *The Winter's Tale* is something that Leontes says in response to learning that Hermione is in fact alive: "I saw her, / As I thought, dead; and have, in vain, said many / A prayer upon her grave."[52] Have not Leontes' prayers, which he says were "in vain," just been answered? After all, Hermione is now alive. Perhaps Leontes is suggesting that the prayers were never necessary in the first place; but, then, what if the prayers themselves are what rescued her from death? Or perhaps Leontes, like Descartes' fictional husband in *The Passions of the Soul*, took some comfort in knowing that he had psychologically survived the devastation of his wife's death; had kept his promise to visit her grave and to say his prayers daily; had proven, alas, that his destiny would no longer be derailed or otherwise threatened by errant flows of bodily fluids. Leontes, too, was on his way to mastering his passions. And now he is concerned—anxious even—that he may not hold up as well a second time.

NOTES

1. For three recent works of scholarship that take this apocryphal anecdote as their starting point, see Stephen Gaukroger, *Descartes: An Intellectual Biography* (Oxford: Clarendon Press, 1995), 1–3; Susan Bordo, "Introduction," in *Feminist Interpretations of René Descartes* (University Park: Pennsylvania State University Press, 1999), 1–5; and Gaby Wood, *Edison's Eve: A Magical History of the Quest for Mechanical Life* (New York: Alfred A. Knopf, 2002), 3–8.

2. René Descartes, *The Philosophical Writings of Descartes*, trans. John Cottingham, Robert Stoothoff, and Dugald Murdoch (Cambridge: Cambridge University Press, 1984), 1:381.

3. Gaukroger, *Descartes*, 468, n. 93.

4. For English translations of the letters Princess Elizabeth wrote to Descartes, see Andrea Nye, *The Princess and the Philosopher: Letters of Elisabeth of the Palatine to René Descartes* (Lanham, MD: Rowman & Littlefield, 1999). For an insightful reading of Elizabeth's influence on Descartes as well as a sounding out of Elizabeth's own philosophy vis-à-vis Descartes, see Lisa Shapiro, "Princess Elizabeth and Descartes: The Union of Soul and Body and the Practice of Philosophy," *British Journal of the History of Philosophy* 7 (1999): 503–20.

5. Descartes, *Philosophical Writings*, 1:381.

6. Ibid., 382.

7. Ibid., 403.

8. Ibid., 404.

9. Shakespeare, *Hamlet*, 2.2.122–23. The quotation that forms the heading for this section appears in *Hamlet*, 2.2.296–97. Lineation for all quotes from *Hamlet* in this essay comes from *The Wadsworth Anthology of Drama*, ed. W. B. Worthen (Boston: Thomson/Wadsworth, 2004).

10. The edition used here, for example (Worthen, *Wadsworth Anthology of Drama*), provides the gloss "machine: body," 291, and the same simple note appears in many others. Richard Halpern identifies this line as one of the "hints and details" in *Hamlet* that "implicate the entire play in the traditions of puppet theater." Because he views Hamlet as a "puppet," however, Halpern repeatedly describes the prince as an inert mechanism, "a mere instrument in the hands of an animating spirit," "listless and apathetic," and characterized, strangely enough, by "an absence of passion." What Halpern does not consider—in his invocations of the early twentieth-century automata of Sigmund Freud, Edward Gordon Craig, Ivan Pavlov, and Frederick Winslow Taylor—is that hydraulic automata of the seventeenth century contained their "animating spirits" *inside* themselves and, thus, were not dependent on the external manipulations of a set of hands for their show of passions. Indeed, these mechanisms would not function at all were it not for some *inner*, occluded turbulence. For a reading of the play that

views Hamlet's mechanism as void of interiority, therefore, see Richard Halpern, *Shakespeare among the Moderns* (Ithaca, NY: Cornell University Press, 1997), 228 and 239, respectively.

11. John Cook, quoted in John Sutton, "Controlling the Passions: Passion, Memory, and the Moral Physiology of Self in Seventeenth-Century Neurophilosophy," in *The Soft Underbelly of Reason: The Passions in the Seventeenth Century*, ed. Stephen Gaukroger (London: Routledge, 1998), 121.

12. Gaukroger, *Descartes*, 293–94.

13. Gail Kern Paster, "The Body and Its Passions," *Shakespeare Studies* 29 (2001): 44.

14. Ibid., 46.

15. Ibid., 49.

16. Shakespeare, *Hamlet*, 2.2.364–65.

17. T. Bright, *A Treatise of Melancholie* [a facsimile text], *reproduced from the 1586 edition printed by Thomas Vautrollier*, with an introduction by Hardin Craig (New York: Columbia University Press, 1940), 48.

18. Ibid., 61.

19. Peter Brown, quoted in Michael Schoenfeldt, *Bodies and Selves in Early Modern England: Physiology and Inwardness in Spenser, Shakespeare, Herbert and Milton* (Cambridge: Cambridge University Press, 1999), 4.

20. See Hardin Craig, "Introduction," in Bright, *Treatise of Melancholie*, xii–xiii.

21. Bright, *Treatise of Melancholie*, 64–69.

22. Paster, "The Body and Its Passions," 50. Indeed, Paster's allusion to Descartes' "disembodied mind instrumentalizing a set of neutral objects" fits much better with Bright's conception of the body as a "handsaw" than a Galenic "co-conspiracy of like material agencies" does. For the preceding descriptions, see Paster, "The Body and Its Passions," 48.

23. Harold Bloom, *Shakespeare: The Invention of the Human* (New York: Riverhead Books, 1998), 418, 407. For another reference to Hamlet's proto-Cartesian "anachronistic inwardness," see Francis Barker, *The Tremulous Private Body: Essays on Subjection* (New York: Methuen, 1984), 23.

24. Bloom, *Shakespeare*, 419.

25. See J. Leeds Barroll, *Artificial Persons: The Formation of Character in the Tragedies of Shakespeare* (Columbia: University of South Carolina Press, 1974); and A. C. Bradley, *Shakespearean Tragedy: Lectures on Hamlet, Othello, King Lear and Macbeth* (London: Penguin, 1992). The latter is one of numerous reprints of the 1904 edition.

26. Bloom, *Shakespeare*, 6 (emphasis in the original).

27. Descartes, *Philosophical Writings*, 2:21.

28. Shakespeare, *Hamlet*, 3.2.1–32.

29. Although the artificers for hire to whom Hamlet alludes are said to be "nature's journeymen," and thus perhaps not a group of "rude mechanicals" like the ones featured in *A Midsummer Night's Dream*, the description remains a prima facie adumbration of Descartes' hypothetical scenario in his *Treatise on Man*, where the philosopher "supposes the body" of a fictional artificial man created by a deity "to be nothing but a statue or machine made of earth" (Descartes, *Philosophical Writings*, 1:99).

30. Shakespeare, *Hamlet*, 2.2.530–36.

31. Daniel C. Dennett, *Consciousness Explained* (Boston: Back Bay Books, 1991), 73.

32. Shakespeare, *Hamlet*, 1.2.75–86.

33. Descartes, *Philosophical Writings*, 2:18.

34. Gilbert Ryle, *The Concept of the Mind* (London: Hutchinson, 1949).

35. The epigraph for this section is from William Shakespeare, *The Winter's Tale* (5.3.63–65). Lineation for this play comes from *The Winter's Tale*, ed. John F. Andrews (London: Everyman, 1995).

36. The phrase "coup de theatre" is often used to describe "the statue scene" at the conclusion of the play. For several examples, see Bruce R. Smith, "Sermons in Stone: Shakespeare and Renaissance Sculpture," *Shakespeare Studies* 17 (1985): 1; Andrew Gurr, "The Bear, the Statue, and Hysteria in *The Winter's Tale*," *Shakespeare Quarterly* 34 (1983): 420–25; Bloom, *Shakespeare*, 661; and Jörg Hasler, "Romance in the Theater: The Stagecraft of the 'Statue Scene' in *The Winter's Tale*," in *Shakespeare, Man of the Theater: Proceedings of the Second Congress of the International Shakespeare Association*, ed. Kenneth Muir, Jay Halio, and D. J Palmer (Toronto: Associated University Presses, 1981), 205.

37. Descartes, *Philosophical Writings*, 1:99–100.

38. In his translation of Descartes' *Treatise of Man*, Thomas Steele Hall notes that "Descartes was especially impressed with the fountains at Saint-Germain-en-Laye, which he knew either personally or from the work of Salomon de Caus." See René Descartes, *Treatise of Man*, trans. Thomas Steele Hall (Amherst, NY: Prometheus Books, 2003), 4, n. 7. For more on de Caus and his relationship to the court of King James, see Luke Morgan, "Landscape Design in England circa 1610: The Contribution of Salomon de Caus," *Studies in the History of Gardens and Designed Landscapes: An International Quarterly* 23, no. 1 (January–March 2003): 1–21; John Dixon Hunt, *Garden and Grove: The Italian Renaissance Garden in the English Imagination, 1600–1750* (Princeton, NJ: Princeton University Press, 1986), esp. "Gardens and Theatre," 59–72, and "Jacobean Garden Mania," 119–26; Roy Strong, *The Renaissance Garden in England* (London: Thames & Hudson, 1979), 73–112; and Frances Yates, *The Rosicrucian Enlightenment* (London: Routledge & Kegan Paul, 1972), 11–13. For a thorough discussion of the garden design eventually undertaken for James's daughter, Elizabeth, the "Winter Queen of Bohemia," see Richard Patterson, "The 'Hortus Palatinus' at Heidelberg and the Reformation of the World, Part 1: The Iconography of the Garden," *Journal of Garden History* 1, no. 1 (1981): 67–104; and "Part 2: Culture as Science," *Journal of Garden History* 1, no. 2 (1981): 179–202. For further discussion of de Caus's contributions to the culture of curiosity collectors, see Horst Bredekamp, *The Lure of Antiquity and the Cult of the Machine: The Kunstkammer and the Evolution of Nature, Art, and Technology*, trans. Allison Brown (Princeton, NJ: Markus Wiener, 1995).

39. Letter from Ottaviano Lotti to Belisario Vinta, 15 June 1611, Archivio di Stato, Florence, 4189. The letter is translated by and quoted in Vaughan Hart, *Art and Magic in the Court of the Stuarts* (London: Routledge, 1994), 97, 224, n. 52. The description of de Servi as an "Italian Neoplatonic engineer" is on p. 97 of this work.

40. Morgan, "Landscape Design," 2.

41. In de Caus's own intricate gardens, including the fountain he built between 1609 and 1612 at Queen Anne's Somerset House, lifelike figures drawn from Ovidian mythology literally embodied the "happy rivalry and co-operation of art and nature" as they appeared to play various musical instruments and to produce the melodious sounds that surrounded visitors to the grottoes. The "happy rivalry" phrase—which nicely complements the debate between Perdita and Polixenes over the respective domains of Art and Nature in act 4 of The Winter's Tale—comes from Hunt, Garden and Grove, 122. De Caus also continued to draw up still more ambitious plans for the royal estates, including life-size automata modeled on those of Francini at Saint-Germain.

42. For a quite different, but altogether fascinating, reading of proto-Cartesianism in The Winter's Tale, focusing on Leontes' "radical doubt," see Stanley Cavell's "Recounting Gains, Showing Losses: Reading The Winter's Tale," in Disowning Knowledge in Seven Plays of Shakespeare (Cambridge: Cambridge University Press, 2003). See also Jennifer Knowles, "Your Actions Are My Dreams: Sleepy Minds in Shakespeare's Last Plays," Shakespeare Studies XXXI (London: Associated University Presses, 2003); and Lynn Enterline, "'You speak a language that I understand not': The Rhetoric of Animation in The Winter's Tale," Shakespeare Quarterly 48 (1997): 17–44. The "statue" of Hermione is also discussed briefly among the early modern "wonders of art and nature" in Lorraine Daston and Katharine Park's Wonders and the Order of Nature, 1150–1750 (New York: Zone Books, 1998), 261–65. As this book was going to press, Amy L. Tigner's essay "The Winter's Tale: Gardens and the Marvels of Transformation" appeared in English Literary Renaissance 36, no. 1 (2006): 114–34. Tigner also sees the "statue" of Hermione as a garden automaton of the sort constructed by de Caus. For more on the "gallery" in which this automaton appears, see Scott Maisano, "Shakespeare's Science Fictions: The Future History of the Late Romances" (PhD diss., Indiana University, 2004), 80–130. Finally, the forthcoming third edition of Arden Shakespeare's The Winter's Tale, edited by John Pitcher, will include in its annotation to the "statue scene" a substantial discussion of contemporary automata and some mention of Descartes.

43. Shakespeare, Winter's Tale, 3.3.15–25.

44. Ibid., 1.2.3–9.

45. Descartes, Philosophical Writings, 2:99–100.

46. Likewise, in the Sixth Meditation, Descartes distinguishes between the human faculties of imagination and understanding by comparing the triangle to a chiliagon, a thousand-sided geometric figure. Descartes claims that he not only understands triangles, but he can also "imagine" them, he can see them with his "mind's eye," as it were. With chiliagons, however, Descartes' understanding happens despite the fact that a thousand-sided figure remains technically "unimaginable" for the human brain, unrepresentable in any precise way before the mind's eye.

47. Shakespeare, Winter's Tale, 3.2.211–14.

48. For yet another example of the will toward perpetual motion, see Florizel's flattering remark to Perdita: "When you do dance, I wish you / A wave o'th'Sea, that you might ever do / Nothing but that: move still, still so, / And own no other function" (4.4.139–42).

49. Shakespeare, *Winter's Tale*, 3.2.237–43.

50. Wood, *Edison's Eve*, xvii. In the final act, when one anonymous gentleman recounts to another the scene of Leontes' reunion with his long-lost daughter, the aptly named Perdita, the old shepherd who had served for sixteen years as Perdita's adoptive father is described "[standing] by like a weather-beaten conduit of many Kings' reigns" (5.2.64–65). John Andrews annotates this line in the Everyman edition of the play by explaining that "to the Third Gentleman, the old Shepherd's weeping suggests a human-shaped fountain." Not just any human-shaped fountain either, but one, like de Caus's, specifically designed for a royal palace.

51. Descartes, *Philosophical Writings*, 1:404.

52. Shakespeare, *Winter's Tale*, 5.3.138–40.

5 Abstracting from the Soul: The Mechanics of Locomotion

DENNIS DES CHENE

Life was art before it was artificial. It was divine art, evidence of the infinite power and understanding of the Creator. More to the point, animal life, in its higher forms at least, was exclusively divine art. Only God, or something superior to human beings, could produce a soul. The very idea of artificial life—not just the simulation of it, and not just magical means to make existing forms, like those of demons, do one's bidding—required a certain amount of conceptual rearrangement. That rearrangement occurred in the seventeenth century with the rise of mechanistic natural philosophies, notably that of Descartes. What is striking in the later history of the mechanistic treatment of living things is that in some authors there is a kind of reversion to the pre-Cartesian conceptual situation. The examples here are Borelli and Perrault. That reversion, however, that did not inhibit the treatment of animals as machines. It would seem, rather, that the general direction of thought after (say) 1700 is to regard "matter" itself as already, in some low degree, living. Creating life would then be impossible, not because it was beyond our powers, but because it was already there.

The topic of this paper belongs more to the parahistory of artificial life than to its history. With the possible exception of Descartes, the natural philosophers I am looking at did not attempt to produce living things from nonliving materials. They studied living things *as* if they were artificial, that is, as machines; but since (after Descartes) the distinction between art and nature no longer set machines on the side of nature, it might be better to say simply that they studied machines, some of which were already made without human labor.

ARISTOTELIAN BACKGROUND

The "past," for natural philosophers of Descartes' generation, was the so-called philosophy of the Schools, or Aristotelianism, and in particular the Aristotelianism of the Jesuit commentators, whose monument is the series of commentaries that issued from the Jesuit college at Coimbra. Two standard questions in Aristotelian textbooks on physics, together with their offshoots, are of interest here.

The first is a question on "nature and art." Its usual form is: Does art imitate nature? The pretext for the question is a passage in Aristotle's *Physics* in which he says that art imitates or perfects nature.[1] Human art imitates nature, but only to a degree. In the arts of depiction, human art mimics the outward appearance of things; in the arts of production, it imitates those things that "ought to have pre-existed" and strives to fashion them as nature would have.[2] Nature did not give us claws or fur coats, but because we, unlike the animals, have a rational soul, we can make those things for ourselves, imitating their counterparts. Even so, what we imitate remains only the outward form of things, the sensible qualities by which we come to know their inward natures: "As Plotinus . . . and Averroes . . . observe, art concerns itself with the external surface. . . . But nature occupies itself in working with great force upon the hidden interior."[3] Those inward natures, or forms, we cannot create. The mark of our incapacity is that natural forms are "actuous and as if alive, but the forms of artifacts are as if inert and dead, and have no effective force."[4]

The second question is that of the origin of forms. Matter, though potentially all things, is entirely passive. The active powers of things, and especially those of living things, cannot come from matter, from the elements, or from mixtures. Questions on the "eduction" of form from matter generally conclude that the forms of living things, and especially those of higher animals, are bestowed on matter by celestial intelligences—the mover of the Sun's sphere, for example—or, in the case of humans, by God himself. Human industry is secondary and subordinate to nature and to God, and so it is incapable of introducing new forms, especially forms with active powers like those we find in animals, into matter. Natural forms can rightly be said to be educed from the *potentia* of matter, because it has a natural "order, propensity, or inclination" to natural forms that it does not have for artificial forms. A special *potentia*, the *potentia obedientialis*, is proposed to account for the possibility of giving artificial form to matter. Artificial forms are only "certain modes of quantity resulting from various ways of cutting [matter] up or of composing it."[5]

What then of the famous automata of antiquity—the statues of Daedalus, the dove of Archytas, the animated stools of Apollonius of Tyana? In every case, according to the Coimbrans, there is either fakery or else the redirection of natural forces. "Neither art nor artificial form by its own power is capable of the work of nature."[6] The works of alchemy, on the other hand, fare somewhat better. The Coimbrans, after giving a standard list of technological wonders (e.g., the compass, the printing press, flat glass), conclude that human art might be capable of understanding the conditions under which gold is produced in nature and of reproducing them. But that human art might go so far as to do the same for animals is impossible, because animal life requires more than the mere mixing of elements.

CARTESIANISM

To the School philosopher, the very idea of "artificial life" suggests an impossibility. Descartes thought otherwise. The fundamental principles of his physiology are evident to anyone who has studied what he has to say about physics and physiology in the *Discourse* and the *Principles*.[7] The active powers attributed to natural things, hence to their substantial forms, are neither needed nor wanted in natural philosophy: all of nature consists in extension variously configured, moving according to the laws of motion; the "actions" of natural things are but expressions of the divine act of conservation. What I want to emphasize here is the role of simulation and the machine in the elimination from Descartes' natural philosophy of the sensitive and vegetative souls, and thus of animal souls altogether.

The *Treatise on Man* is an exercise in simulation. The Simulator is God, not us. God has made soulless machines that resemble us in every way possible. Descartes shows that, using only the mechanistic principles put forward in *The World*, he can explain all those functions of the animal that in the Schools were explained by supposing a sensitive and vegetative soul. In fact, the *Treatise* does not explain generation. That lacuna was filled when Descartes returned to physiology near the end of his life. *The Description of the Human Body* includes an account of the formation of the fetus, beginning with the mingled male and female seeds in the womb.[8] That account is likewise presented as conforming to the principles of Cartesian physics—it mentions nothing but the sizes, shapes, and motions of various corpuscles, and (in principle) derives their movements and configurations in the developing animal from the laws of nature.

Setting aside the wild card Descartes deals himself—the already existing seed—the *Description*, unlike the *Treatise*, offers not a simulation of the human body and its processes but the thing itself. The pretense of the *Treatise* that the objects described are automata made in imitation of the body is dropped. The body is not like a machine—it is a machine. "It is true," Descartes says, "that one may find it difficult to believe that the disposition of the organs alone is sufficient to produce in us all the movements that are not determined by our thought"; he will therefore "explain the whole machine of our body in such a way that we will have no more ground for thinking that it is our soul that excites in it the movements we do not experience as being directed by our will than we have for judging that in clocks there is a soul which causes them to exhibit the hours."[9]

How then do animal-machines, which are of divine origin, differ from the machines we are capable of building? The only difference is that God is a far better engineer than we are: more knowledgeable, more delicate, no doubt more patient too: "The operations in the making of art, for the most part, are performed through instruments large enough to be easily perceived by the senses; this is required if they are to be fabricated by humans."[10] But human industry is no longer subordinate to that of nature or to divine industry. It is not limited to imitation, nor is it confined to reshaping extant natural materials. In principle we could configure matter down to whatever dimensions are required and give it the appropriate motions; the machines we made would be animals. The question of building animals is one of feasibility alone.

Descartes was reputed to have built or planned several automata, but details are lacking. I mention instead one example from the *Dioptrics*—a prosthetic addition to the human eye. Descartes is considering ways to improve vision. One way would be to move the point of convergence of light rays entering the eye as far forward from the retina as possible. To do this, Descartes imagines affixing to the front of the eye a tube of water whose outward end has the same shape as the cornea. "Vision will occur," Descartes writes, "in the same manner as if Nature had made the eye longer than it is"—as long as the tube, in fact. Moreover the natural pupil of the eye will become "not only useless but even deleterious, insofar as it excludes, by its smallness, rays that could otherwise proceed toward the edges of the back of the eye."[11] If the pupil were excised, we would have a hybrid, a fusion of machine and organ, superseding the eye God gave us, but no less "natural."

To go from this to the construction of an artificial eye might require feats of engineering. But the conceptual barrier has been removed.

BORELLI AND THE MECHANICS OF LOCOMOTION

Borelli's *De motu animalium* was published posthumously in 1680 and 1681. The preface to the work, which is dedicated to Queen Christina, sets forth the apologetic aim of exhibiting the "perfect Goodness" of God by way of eliciting our admiration for the "works and machines which Nature has elaborated with such great art."[12] The "idiom and characters" with which the Creator has inscribed his conceptions in the "Codex" of nature consist in "Geometric Configurations and Demonstrations." Because animals are bodies, and their operations either are motions or else have motion as a necessary condition, we can see God's geometry at work in their organs, using our own geometry to describe them and to measure their effects.

The first part of Borelli's work is devoted to locomotion on earth and in water and air: walking and crawling, swimming, flying. Borelli says that in his inquiry he will suppose certain things that are "shown by the evidence of the senses." The first of these is that the "principal and effective cause of the movement of animals is the soul." This everyone knows, because "once the animal has expired, that is, once the soul no longer is operating, the animal-machine is left entirely inert and immobile." The soul, moreover, does not move the body per se, but by way of an instrument, namely, the power or faculty of local motion, "which is commonly held to reside in the animal spirits." The animal spirits in turn move the body by acting on the muscles, communicating their active power to them by way of the nerves.[13]

In introducing his description of the muscles and their use, Borelli writes that "just as in other Physico-mathematical sciences it is customary so too we will attempt to set forth the science of the movement of animals starting from the phenomena, considered as foundations."[14] Hence, the first task is to describe the structure and operations of the muscles. What follows is a series of propositions, including various lemmas from mechanics, which are intended to show that the muscles can and do exert the force necessary to move the body.

After showing that the "action of the muscle is contraction" (see figure 5.1, nos. 1–4), Borelli presents what he calls a "new notion [discovered] in recent years of the true form of the muscles and their mechanical mode of operations," which, from his love of truth, he will explain.[15] The problem seems to be to explain how the muscle, in contracting, thickens without the addition of new matter.[16] Borelli's predecessors had explained this by supposing that the muscle is a rhomboid composed of fibers all running in the same direction (Prop. 5). Borelli argues that this arrangement is quite inept for the task of lifting any weight.[17] Given that force is exerted by the fibers through contraction along their long axis, the arrangement shown in figure 5.1, nos. 5 or 7, will not do the job, even though (as Borelli notes) the rhomboidal arrangement does explain what we see when a muscle grows longer or shorter: "What is most important in this affair is the mechanical reason by which the force of the muscle moves a resisting [body] by means of the organ [i.e., the muscle]";[18] the rhomboidal arrangement provides no such reason.

The "true figures" of muscles are seen in figure 5.1, nos. 9, 10, and 11. The difference between Borelli's figures and those illustrating the false opinion is that the line to which the ends of the fibers are attached is either perpendicu-

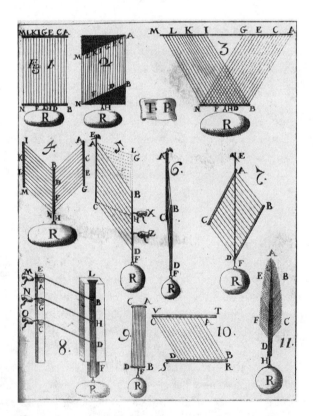

FIGURE 5.1 | Models of muscles. Giovanni Alfonso Borelli, *De motu animalium* (Naples: Bernardinus Gessarus, 1734), Tabula I.

lar to the force exerted or else to the line itself along which force is exerted. Mechanical reasons alone suffice to show that this must be the case.

I mention here one other proposition in which Borelli differs from his predecessors—not only Aristotle, Lucretius, and Galen, but also Gassendi. Those philosophers held that in animals a small force or power must be capable of moving a great mass. Borelli for his part holds that the force of the muscles can be "immense," and must be:

> I will show that machines are applied to the motions of animals, and that they are multiple and various; but that it is not true that a great weight is lifted by a small force, but rather a great force and effort of the animal faculty holds up a small weight; so that the motive force [of the muscles] exceeds by a hundredfold or a thousandfold the weight of the bones [. . .] and is never less than [their weight].[19]

What follows is a series of propositions on the shape and configuration of various muscles, and then a chapter consisting of lemmas concerning levers and weights. Borelli then shows what the force of various muscles must be: for example, the force exerted by the biceps when a weight of 28 pounds is being held by the hand with the arm extended horizontally is 560 pounds. The calculation is based on the known places of attachment of the muscles and the use of the forearm (in this case) as a lever, the fulcrum being the elbow. Much later, Borelli shows that in the flight of birds, the proportion of force to weight must be much greater; it is for this reason that humans will not be able to fly by their own power alone.[20]

The animal, or at least its muscular and skeletal structure, is a machine, the effects of which can be calculated on mechanical principles. What mechanics cannot explain is the source of the motive power by which the muscles exert themselves in lifting weights and locomotion. Borelli is here continuing a traditional division of labor between mechanics, which deals mathematically with the distribution of forces in machines, and physics, to which alone belong questions of the origin of force.[21] Descartes likewise distinguished mechanics, one of the three branches of the tree of knowledge, from physics proper, which is the trunk of the tree.[22] Among other things, mechanics takes weight for granted and measures or calculates its effects; the explanation of weight belongs to physics.

PERRAULT

Claude Perrault (1613–1688) is best known for his architectural work, which included the eastern façade of the Louvre, and for an edition of Vitruvius published in 1673 and 1684. He was trained as a physician and, after entering the Académie des Sciences at its founding in 1666, spent his last twenty years working on natural history and on the *Essais physiques*, the third volume of which is the *Mécanique des animaux* that I discuss here.[23] The *Essais*, as their title implies, do not purport to offer a systematic, orderly treatment of natural philosophy. Knowing that the taste for "philosophical physics" is rare, and yet hoping to satisfy the curiosity of "those who ordinarily have little of it" in these matters, Perrault not only writes in the vernacular but also promises to define all the terms of art he uses. Instead of the geometrical style of Borelli, we have the easy style of Descartes' *Discourse*.

Perrault is not a promoter of novelties. Of those which have been recently introduced into philosophy, he writes that they are mostly just the "explication of ancient opinions that modern authors have pushed a little further than their first authors did."[24] His own opinion, for example, concerning the movement of the muscles was "furnished" to him by Galen, and his preformationist theory of generation is credited to Hippocrates.

The *Mécanique des animaux* begins by disclaiming the implication of its ambiguous title. An animal, says Perrault, "is a being which has sensation and which is capable of exercising the functions of life by a principle called the Soul." The soul "makes use of the organs of the body, which are truly machines, by way of being the principal cause of the action of each piece of the machine." This even though "the disposition of those pieces with respect to one another" in the machine does almost nothing that it would not do "in pure machines."[25] For Perrault as for Borelli, mechanism in the study of the movements of animals does not preclude supposing them to have souls.

Despite the opposition to Cartesianism evident in the preface to the *Essais* and in the "Avertissement" to the *Mécanique*, Perrault's explanation of the action of the muscles proceeds in a manner not unlike that of Descartes in his *Treatise*. The chief difficulty in explaining their action is that the "fibers of the flesh of the muscle" are aligned transversely and thus do not seem to be responsible for the contraction of the muscle. To explain the action of the fibers, Perrault introduces a number of suppositions: (1) that the "fibers that compose the membrane that covers each muscle have a natural spring," and thus tend to return to their natural state after being stretched—thus do the claws of lions withdraw of themselves; (2) that the fibers are ordinarily stretched because each muscle has an antagonist; the "equilibrium" position of a limb is that in which the fibers of protagonist and antagonist are in equal tension; (3) that since both the protagonist and antagonist are in tension, the relaxation of one will lead to bending in the direction of the

other; (4) that the relaxation of the fibers occurs by way of the "introduction of the spirituous substance brought by the nerves from the brain,"
which "corrupts and relaxes" their spring.[26]

It follows (this is Perrault's principal claim) that the so-called animal
spirits operate to relax the muscles and not to tighten or shorten them. They
shorten of their own accord after being stretched. The opponent here is
Descartes, whose *Traité de l'homme* argues that the entry of the animal spirits
into a muscle shortens it.[27]

As with Borelli, we see a definite demarcation between the "mechanics"
of animals and what we might call their "energetics." The source of active
power is the soul, about which very little is said. What remains is to explain
the transmission and application of that power to the end of locomotion,
and that is a matter of applying mechanical knowledge. Perrault acknowledges that the animal-machine resembles "pure machines" in the manner
of its operation. But every machine requires a mover, and this the mechanism itself is incapable of supplying.

Rather than seeing Perrault's and Borelli's admission of animal souls as
a retreat from the more forthright and comprehensive mechanism of the
Cartesians, it might be fruitful to regard it instead as a division of labor
not unlike that proposed by Descartes himself in the *Dioptrics*, a division
taken by him from descriptions of the "mixed sciences" in his predecessors.
Descartes sets aside questions concerning the nature of light in favor of a
few assumptions about its action (e.g., that it travels in straight lines unless
interfered with) that allow him to get on with the business of applying
geometry to the description of its behavior when reflected or refracted. So,
too, in the study of animals, the vexing question of the soul, like analogous
questions about other causes and powers, can be set aside in favor of investigating the "instruments we can see and whose manner of acting we know
by experiences," in particular by the dissections that show all the pieces of
the animal-machine "distinctly and separately."[28]

There is a science of the movements of animals, of which Borelli's *De motu*
and the second part of Perrault's *Mécanique* are illustrations. There may also be
a science of the souls of animals, though it would seem that Perrault thinks
it is beyond our capacities. Only the latter can explain the source of animal
power, if it can be explained at all. The upshot is that the scientific treatment
of the animal-machine, considered as an object of mechanics, and that of
the animal-automaton, the self-mover, should part company. There will thus
be two sorts of "artificial life": that which simulates the operations of animals without going so far as to aim at self-motion, and that which does aim

at self-motion—which, without trying per se to make something that looks like an animal, is a self-mover.

To put the point another way: Descartes succeeded in introducing mechanism into the study of living things, or rather—more specifically—the new mechanism and the new mechanics put forward by Galileo, Descartes himself, and others in the first half of the seventeenth century. He and his contemporaries succeeded also in making the machine a "model of intelligibility." By that I mean that to understand something as a machine, or to understand that it is a machine, was to understand it sufficiently well—well enough that no further requirement of clarity or demonstrative certainty was in order. Borelli and Perrault do not dispute that. But Descartes sought to unify the science of life under the mechanistic model, and that included a reduction of the active powers of living things to mechanical forces. Borelli and Perrault reject that reduction. The result is a science of animal movement in which the soul is assumed only to be left behind, not because it can be eliminated but because it cannot be mechanized—nor is there any need to do so.

NOTES

1. Aristotle, Physics 2c2.199a18–20. All translations are mine unless otherwise specified.

2. Franciscus Toletus, Commentaria unà cum Quæstionibus in octo libros Aristotelis de Physica auscultatione, in Opera omnia Philosophica (Cologne: Birckmann, 1615–16; repr., Hildesheim: Olms, 1985).

3. Coimbra [Collegium Conimbricensis], Commentarii Collegii Conimbricensis . . . in octo libros physicorum Aristotelis, 2 vols. (Coimbra: Ioannes Baptista Buysson, 1594; repr., Hildesheim: Olms, 1984), 2c1q6a2, 1:217.

4. Ibid.

5. Coimbra, Comm. in phys. 1c9q12a7, 1:196.

6. Ibid., 2c1q7a2, 1:218.

7. René Descartes, Discours de la méthode (Leiden: Jean Maire, 1637), in Œuvres de Descartes, ed. Charles Adam and Paul Tannéry (Paris: Vrin, 1964–81), vol. 6; Traité de l'homme (Paris: Théodore Girard, 1664 [written in 1631–33]), in Œuvres, vol. 11; Principia philosophiæ (Amsterdam: Elzevir, 1664), in Œuvres, vol. 8, pt. 1. The Adam-Tannéry edition is subsequently cited as "AT."

8. René Descartes, Description du corps humain (Paris: Théodore Girard, 1664 [written in the late 1640s]), in AT, vol. 11.

9. Ibid., 226.

10. Descartes, Principia, AT 8: pt. 1, 326.

11. Descartes, *Dioptrique*, AT 6:156–57.

12. Giovanni Alfonso Borelli, *De motu animalium Giovanni Alphonsi Borelli* [...] (Rome: Angelo Bernarbo, 1680), a2r.

13. Ibid., 2, 3.

14. Ibid., 4.

15. Ibid., 2–3.

16. Ibid., 10.

17. Ibid., 12.

18. Ibid.

19. Ibid., 18 (prop. 8).

20. Ibid., 322–23.

21. Alan Gabbey, "Descartes' Physics and Descartes' Mechanics: Chicken and Egg?" in *Essays on the Philosophy and Science of René Descartes*, ed. Stephen Voss (Oxford: Oxford University Press, 1993), 311–23.

22. Ibid., 320.

23. Claude Perrault, *Essais de physique, ou Recueil de plusieurs traitez touchant les choses naturelles*, vol. 3 (Leiden: Peter vander Aa, 1720). Volume 3 includes the *Mécanique des animaux*, originally published in 1680–88.

24. Claude Perrault, *Essais de physique, ou Recueil de plusieurs traitez touchant les choses naturelles* (Paris: J.-B. Coignard, 1680–88), 1, "Preface," e ij.

25. Perrault, *Essais de physique*, 3:329 (Leiden edition).

26. Ibid., 362; see also plate 3, figs. 1 and 2, p. 365.

27. Descartes, *Traité de l'homme*, AT 11:135.

28. Perrault, *Essais de physique*, 3:333 (Leiden edition). In his "Avertissement," Perrault specifically rejects the Cartesian view that animals have no souls. But the only form of explanation he admits is the mechanical, even when discussing the operations of the brain in initiating action (ibid., 405–6).

6 The Anatomy of Artificial Life: An Eighteenth-Century Perspective

JOAN B. LANDES

After viewing a demonstration of Jacques Vaucanson's flute-playing automaton, the Abbé Desfontaines enthusiastically exclaimed, "It is doubtless the growth of human anatomy, and above all the anatomy of the nervous system, which guided the author in his mechanics."[1] A celebrity in his own day, the French engineer has lately attracted an audience in Artificial Life circles. The ("defecating") Duck, another of Vaucanson's automata, has even become a kind of logo for the new science of Artificial Life, and twentieth-century robotics is celebrated as the fulfillment of what could only have been a dream in the past: artificial life, defined as the mechanical simulation of life. However, Desfontaines' remark reveals an aspect of artificial life too often overlooked in such chronologies: anatomy, in its literal, tactile, or corporeal sense of cutting and dissecting, as well as its metaphoric sense of "dividing, separating, analyzing, fathoming permeated ways of thinking about, and representing, all branches of knowledge."[2] Although anatomy and dissection were "the eighteenth-century paradigms for any forced, artful, contrived, and violent study of depths," knowledge of the interior of things ultimately involved a *reassemblage* of the whole. Anatomy served Vaucanson in two overlapping respects: as a science of the body's structure and functions, and as a model for the artful construction of a material body. Moreover, despite the prioritization of movement in the case of machines, form and function in *écorchés*, both styles of simulated body aimed at an ever more genuine artificial representation of life.

In anatomy, the *écorché* is a flayed skeleton made either from real cadavers or other materials. In art, it is a two- (in the case of drawings) or three-dimensional (in the case of plaster or bronze casts) representation of the human body, in which the envelope of skin and fat has been removed. It is used to depict the surface muscles with anatomical correctness. Artists used *écorchés* as a check on their life drawings, and they offered surgeons a reminder of what they had dissected and guided future operations. *Écorché* figures were typically beautifully proportioned and usually shown in action, one leg before the other, one arm raised over the head.[3]

Eighteenth-century automata-makers and anatomists provide splendid examples of Vico's precept that we can only know what we make. Through human agency, both experimented with ways to enliven and ultimately bring dead matter to life. In an important sense, both risked usurping the place of God in the story of creation or of Prometheus in Greek myth. If mechanism is important in the history of artificial life, so, too, is the emphasis on the once-living human or animal body. By considering the cross-cutting influences of anatomy, mechanics, and art, we may be in a better position to comprehend not only mechanistic views of the body but also the anatomical view of the machine.

SIMULATION I: THE BODY IN VAUCANSON'S MECHANICAL FACSIMILES

What if we were to emphasize the second rather than the first term in the couplet *artificial life*, taking seriously the strong interest in anatomy exhibited by Jacques Vaucanson and many of his contemporaries and, above all, their ambition to manufacture anatomically correct simulations of life? In his correspondence with the Abbé Desfontaines, Vaucanson (1709–1782) commended his automata as appropriate instruments for instruction (see figure 6.1). He referred to the impression his three-dimensional mechanical objects could make on viewers, their anatomical accuracy, and their unique ability to demonstrate life processes (in real time), most spectacularly, in the case of the *Duck's* digestive system. He favored his mechanical constructions over a standard written account or what one would find in an anatomical treatise: "I don't believe the Anatomists can find any thing wanting in the Construction of its Wings," he wrote. "The Inspection of the Machine will better shew that Nature has been justly imitated, than a longer Detail, which wou'd only be an anatomical Description of a Wing."[4]

FIGURE 6.1 | *The Flute Player, the Duck, and the Pipe-and-Tabor Player*, in *An Account of the mechanism of an automaton, or image playing on the German-flute* [...] (London: T. Parker, 1742). Courtesy of the Brown University Library.

For Vaucanson, the conventional approach to anatomical knowledge—earned through careful empirical observation of cadavers and dissection—was by itself inadequate and ought to be supplemented by a visual demonstration of a working model or lifelike mechanical simulation. Motivated by a strong interest in anatomy, the engineer's inquiries were also intended as contributions to anatomy, designed to further the comparative science of animal and human bodily structure: "I believe that Persons of Skill and Attention, will see how difficult it has been to make so many different moving parts in this small Automaton; as for Example to make it rise upon its Legs, and throw its Neck to the Right and Left. . . . In short, I have endeavour'd to make it imitate all the Actions of the living Animal, which I have consider'd very attentively."[5]

Vaucanson's mechanical experiments borrowed directly from the materials, methods, and display techniques of the plastic and anatomical arts. His machines were meant to be didactic and scientifically accurate, as well as entertaining. They were designed for the attention of academicians as well as the more popular audiences for early modern scientific experiments and anatomical demonstrations. Noting this in his approval of Vaucanson's "Memoir" for publication, the Royal Censor compliments the engineer for appealing so directly to "curiosity" and public interest:

> Mr. Vaucanson explains in his Memoir those physical Principles that he has employed for the Invention and Execution of his Automaton, which is one of the most wonderful Productions of Art: It imitates a true Player on the Flute so perfectly, that the Publick continues to see and hear it with Admiration. Therefore we believe that the impression of Mr. Vaucanson's Memoir will be very useful to satisfy fully the Curiosity of the Publick.[6]

Aware of the risk that extreme favor with the populace could pose to professional reputation, the ambitious engineer and academic aspirant nevertheless took care to distinguish his works from the magic tricks or pleasing toys seen at boutiques or fairs. Speaking of the *Duck*, he announced, "The whole Mechanism of our artificial *Duck* is exposed to View; my Design being rather to demonstrate the Manner of the Actions, than to shew a Machine. . . . I have been desir'd to make every Thing visible; [so] I wou'd not be thought to impose upon the Spectators by any conceal'd or juggling Contrivance."[7]

Such protestations notwithstanding, Vaucanson's mechanical figures were nothing less than art, or even artifice, in a very basic sense: that is, three-dimensional, moving sculptures, or "images," as his English translator, the natural philosopher John Theophilus Desaguliers, preferred to call them.[8] They corresponded to the *écorché* paintings and statues adorning the studios of most artists, which served as anatomical guides to the accurate depiction of bones, muscles, and the outer surface of the body.[9] In addition, the life-size, seated *Flute Player* automaton was modeled after a sculpture in the royal gardens at the Tuileries, *The Flute-Playing Shepherd* (alternatively titled *Faun Playing the Flute*) (1709) by Antoine Coysevox, a favorite of Louis XIV.[10] The wooden android was even painted white to resemble the marble surface of Coysevox's *Faun*.

Vaucanson chose the German flute for his android because of the instrument's reputation for difficulty: the need to control the force of the breath, the shape of the lips, the position of the instrument, and the tongue's movements up and down, backward and forward. However, when faced with the challenge of finding an appropriate material for the android's mechanical fingers, since the effect of wooden fingers playing on a metal flute was nowhere as pleasing as that produced by human fingers, he substituted *peau* or, we assume, animal skin. As Gaby Wood remarks, "Pure mechanics were not enough, and Vaucanson had to import organic matter into his dead creation."[11]

MECHANICAL AND ORGANIC MATTER

The connection between the mechanical and the organic is not just a question of materials, but of outlook. To a practicing anatomist, Vaucanson's effort to comprehend the biomechanics of the body by making visible what is ordinarily hidden from sight would have been unsurprising. By complimenting Vaucanson on his anatomical learning, the Abbé Desfontaines acknowledges as much. Yet his remarks also call attention to anatomy's contribution to the period's much-debated question of how the senses influence the mind. By making visible the nervous system, anatomists had a central role to play in advancing empirical philosophy and medicine.[12] In his "Memoir," Vaucanson displays his anatomical credentials, aiming also to impress the academicians with the seriousness of his purpose. Speaking of the Duck, he writes:

> Not only every Bone has been imitated, but all the Apophyses or Eminences [protuberances] of each Bone. They are regularly observ'd as well as the different Joints: The bending the . . . Cavities [The cavities, the contours], and the three Bones of the Wing are very distinct. The first, which is the Humerus, has its Motion of Rotation every Way with the Bone that performs the Office of the Omoplat, Scapula, or Shoulder-Blade: The second Bone, which is the Cubitus of the Wing, has its Motion with the Humerus by a Joint which the Anatomists call Ginglymus; the third, which is the Radius, turns in a Cavity of the Humerus, and is fasten'd by its other Ends to the little End of the Wing, just as in the Animal. The Inspection of the Machine will better shew that Nature has been justly imitated than a longer Detail, which wou'd only be an anatomical Description of a Wing.[13]

Vaucanson was not a latecomer to anatomy. From an early age, he showed an interest in the subject, along with mechanics, aspiring "to create artificial beings, 'moving anatomies' . . . [to] reproduce as faithfully as possible the organs and functions of human beings or animals; their real role was not to divert and amuse but to teach, and so promote medical progress."[14] On his first sojourn in Paris between 1728 and 1731, he followed a course devoted to mechanics, physics, and anatomy. There were many such opportunities for the interested layperson, including the extremely popular courses in anatomy and medicine at the Jardin du Roi in Paris. The most celebrated courses— for example, those of Pierre Dionis (1643–1718) and Joseph-Guichard Duverney (1648–1730)—even became fashionable events. According to one description, "Not only did students press around the dissection table, but all of high society came to assist in the spectacle mounted like a play. And they commented on the anatomy lesson even in the salons."[15] Later in the century, the observer of eighteenth-century Paris Sébastien Mercier claimed that the "passion for anatomy" was "part of the indispensable equipment of every well-educated man."[16]

At age twenty-two, Vaucanson met the famous surgeon of the Hôtel Dieu Hospital in Rouen, Claude-Nicholas Le Cat.[17] An admirer of mechanist philosophy and physiology, and a corresponding member of the Academy of Surgery, Le Cat was also inspired by the challenge of building a human automaton. He is credited with authoring a lost mechanist treatise, *Déscription d'un homme automate*, which predated La Mettrie's *L'Homme Machine* of 1747.[18] Having attended anatomical courses and surgical demonstrations, Vaucanson would be no stranger to the practice of dissection, including the handling of specimens. He could have easily supplemented his knowledge of the body's parts and functions by further study of anatomical texts, images, and models.

In the ideal situation—that is, the dissection of a fresh or very recently deceased corpse—the audience to an anatomical demonstration witnessed the resistance of the flesh to the dissector's cutting and also saw the size and structure of actual organs. But anatomy involved more than the practice of dissection. As a descriptive science, "anatomy is an organized report, written or visual, recording what there is under the skin. It describes the structures of the body seen with the naked eye, structures that are uncovered in the experience of dissecting cadavers."[19] To record, document, and describe the results of their investigations, anatomists employed visual illustrations, often produced by an artist collaborator. Yet an illustration—whether in the form of a manuscript, printed text, or drawing—is at best an imperfect vehicle for the teaching of anatomy. As K. B. Roberts explains, this is largely because anatomy's ultimate reference point is necessarily "the structure of living human beings. The anatomical report therefore has to take notice of physiology—the way in which the body's structures work and function."[20]

For all these reasons, body models were introduced as early as the Renaissance, with wax being the preferred material; and anatomical artists were employed in medicine to fashion anatomical models for instruction and display.[21] Valued for their didactic potential, "artificial anatomies" were widely adopted as the principal alternative to the other major school of anatomy in the early modern period—"natural anatomy," or the use of the cadaver as the principal anatomical specimen.[22] Less authentic than the corpse, anatomical models had many didactic advantages for working anatomists. They were able to exaggerate or emphasize a detail for further study and also to overcome the limitations of a description limited to a single, specific cadaver.[23] Indeed, artificial bodies and body parts were intended to be more

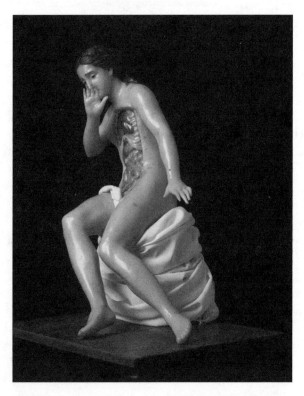

FIGURE 6.2 | André-Pierre Pinson, *Femme assise, anatomie de viscères* (Seated female figure, visceral anatomy). Courtesy of the Muséum National d'Histoire Naturelle.

general than specific, both as depictions of what amounted to a statistical mean and as reflections of the idealized, classicizing, aesthetical preferences of scientists and artists of the age (see figure 6.2).

In contrast, artificial models presented other limitations, whose parallel was noted in the prior account of Vaucanson's experiments in building automata. Solving the problem of manual dexterity was not enough to fashion properly the hands of his flute-playing automaton. To approximate both the feel and texture of organic matter, Vaucanson needed to find a material resembling skin. Whether employing the techniques of natural or artificial anatomy, early modern anatomists and artists, along with their assistants, faced multiple dilemmas. As Martin Kemp proposes, "All the anatomists found that achieving the accuracy they desired was not simply a matter of placing artists skilled in naturalistic representation in front of a specimen. Looking and representing are inevitably directed and selective processes, and the deputing of them to other eyes and hands caused inevitable problems."[24]

SIMULATION 2: CADAVERS AS ARTIFICIAL LIFE

And what about the need to reproduce the look of a living body, to enliven a dead corpse, to bring dead matter to life? As Deanna Petherbridge observes, one of the most persistent visual tropes in anatomy is "the picturing of skeletons, flayed cadavers and partially-dissected bodies as if they are alive."[25] Although both artificial and natural anatomists shared this goal, it is vital to consider the manner in which natural anatomists confronted the difficult challenge of depicting a body in a lifelike way while working with "natural" but potentially "abject" materials.[26] The act of dissection—the dismemberment of an already-decaying corpse, the opening up to the gaze of what is inside and concealed, the cutting that is no longer curative but inquisitive—always carried with it what Julia Kristeva refers to as the ambiguous effect of abjection. In discussing that which is expelled, cast out, and away, Kristeva draws upon Bakhtin's category of "the grotesque"—a body that is "blended with the world, with animals, with objects"; associated with the bodily stratum, degradation, filth, death, and rebirth; a "pregnant death, a death that gives birth"; open, secreting, multiple, irregular, and changing. In Kristeva's account, the self only becomes social by expunging the elements that society deems impure: a goal that she regards, however, as impossible to achieve. The abject, therefore, haunts the subject's identity, threatening the self with danger and dissolution.[27]

In surprisingly similar terms, the prominent anatomist Félix Vicq d'Azyr (1748–1794) remarked on the ambiguity at the heart of his enterprise, the study of life by way of decaying matter. In defense of anatomical practice against its detractors, he maintained:

> Anatomy is perhaps the science whose advantages are most celebrated and which has been the least favored. It is perhaps also the one whose study offers the greatest difficulty. Its investigations are not only devoid of the charm that attracts, but in addition it is accompanied by circumstances that repulse: Torn and bloody members, infectious and unhealthy odors, the ghastly machinery of death are the objects that it presents to those who cultivate it.[28]

No one could have put the matter more emphatically. Vicq d'Azyr's characterization captures directly the extent to which anatomical knowledge is tied to death, order to disorder, and the mind to the most abject, degraded elements of the body. Even without the smell and fluids of death, the very process of dissection would remain uncanny. As Barbara M. Stafford notes, "as the dissector dug deeper into the corpse, he exposed a mutable organism at once unified and fragmented, simple and complex, ideal and grotesque."[29]

After dissection, however, comes a moment of what Bakhtin might have called "rebirth" had he looked at anatomy through Vicq d'Azyr's lens: that is, the fashioning of a durable anatomical work, in particular, the écorché or flayed specimen, made from preserved human and animal cadavers. An excellent example of this version of artificial life is provided by the marvelous series of life-size écorchés produced by the natural anatomist Honoré Fragonard (1732–1799). Many of Fragonard's works are housed at the Veterinary School of Alfort in southeast Paris, created during the Revolution to succeed the Royal Veterinary School at Alfort, then on the city's outskirts.[30] As a practitioner of natural anatomy, Fragonard's techniques of dissecting, injecting, and preserving fresh cadavers resembled those of the eminent Dutch anatomist Frederik Ruysch.[31] Fragonard's procedure involved injecting the arteries with alcohol and aromatic spices in order to prevent putrefaction, following which the skin was removed, the body fixed in a frame in the desired position, and the dissection carried out, leaving the muscles and sometimes the nerves and the vessels, which were then injected with colored wax. Fragonard excelled in the handling of the whole body and, even more remarkably, of bodies in relation to one another.

Paradoxically, the natural anatomist had at least one advantage in the fashioning of a lifelike simulation because the natural preservation techniques of the period helped to produce a spirited appearance in the specimens.[32]

Ruysch exploited this "liveliness" of expression when he announced in 1696 the dissection of bodies "which appear still to be alive but which have been dead for about two years."[33] The French academician Fontenelle maintained that Peter the Great was so moved by one of Ruysch's preparations that "he tenderly kissed the body of a still lovable small child who seemed to smile at him."[34] Even the strange proportions of the dissected cadaver utilized in natural anatomy, in contrast to the more accurate proportions sought in artificial anatomy, could help achieve these effects. Indeed, Fragonard's "injection of the resin into the blood vessels unnaturally enlarged them, giving an effect of swelling that is not found in the wax models."[35]

Ruysch is perhaps most famous for his dramatic, teasingly lifelike presentation of anatomical subjects. From this Fragonard learned as well, though he did not go so far as Ruysch, who dolled up anatomized fetuses with scarves and embroidered baby hats.[36] By choosing to pose his figures in an animated fashion, Fragonard called attention to the liminal status of the anatomical subject, not alive but also not wholly dead. As if to confirm his intentions, the écorché in figure 6.3 is known today as *Fetus Dancing the Gigue*. As a result, the anatomized subject here exceeds its purpose as an object of demonstration in the context of veterinary or medical instruction.

This "surplus" effect, or artifice, in Fragonard's work produced in onlookers both admiration and discomfort. For some, the anatomist was a brilliant technician whose artistry preserved the cadaver against its inevitable decomposition. But for critics, far from being truthful, such affectations amounted to a trivial addition or a playful ruse, at odds with the developing protocols of serious science. What is at stake here is the place of the "natural" in the practice of anatomical modeling, as well as the physical status of dead matter.

One of Fragonard's most remarkable écorché, *Man with a Mandible* (figure 6.4) is a testament to his extraordinary skill. In *Man with a Mandible*, the full human figure is displayed, stripped of its outer covering, not just of the garments of culture but of the body's own protective sheath, the skin. What is revealed is the material inner substance of the body, very much in the manner suggested by Descartes—who spent much time observing the slaughterhouses and gallows in Amsterdam—when he said, "I consider myself first, as having a face, hands, arms, and the whole of this machine composed of bones and flesh, such as it appears in a cadaver, which I designate by the name of body."[37]

Using a routine technique of the natural anatomist, Fragonard replaced the eyeballs with glass replicas. This simple artifice breathes a sort of life—sensible, irritable life—into the dried, injected body. This sensibility accorded

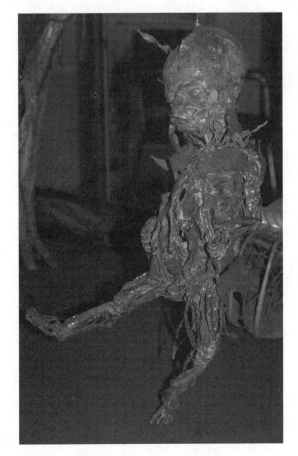

FIGURE 6.3 | Honoré Fragonard, *Fetus Dancing the Gigue.*
Photo credit: C. Degueurce, Musée Fragonard de l'Ecole
Vétérinaire d'Alfort (France).

well with the new, more activist conception, which by mid-century was
altering earlier materialist suppositions concerning inert matter. As Sylvie
Hugues explains, "The body becomes a closed field where vital forces and
forces of dissolution are opposed: 'life is the ensemble of functions which
resist death.'"[38] In this context, *Man with a Mandible* offers a solution (artistic
and scientific) to the problem within vitalist philosophy of how to represent the dead body. In any event, Fragonard has certainly introduced a kind
of turmoil in this figure—in the jumbled network of arteries and veins
that lie below the surface, his turned head and raised arm, which holds the

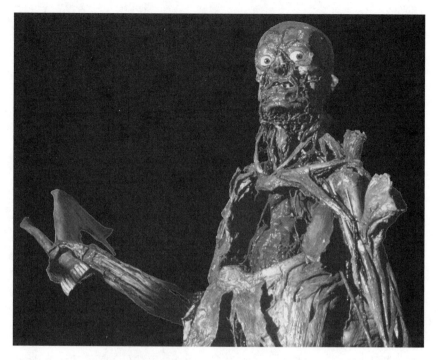

FIGURE 6.4 | Honoré Fragonard, *Man with a Mandible*. Photo credit: C. Degueurce, Musée Fragonard de l'Ecole Vétérinaire d'Alfort (France).

mandible as a potential weapon against a possible intruder. It is as if this man sees and hears something surprising and horrifying. Though dead, he appears to be stiffening against danger.[39] The injected veins and arteries on his skull and face further communicate horror.

Whereas *Fetus Dancing the Gigue* gently mocks the conventions established in the Vanitas tradition, *Man with a Mandible* is influenced by the baroque canon of anatomical illustration in which the *écorché* recalled Christ's torments. In these representations, the immaterial soul confronted the material body, with all its worthless flesh. Fragonard's *écorchés* are strong testimony to his skill as a natural anatomist. But they also have a spectacular quality, which simultaneously participates in and exceeds the demands of normal science. Thus, in addressing the impulse behind these works, one critic stresses, "Science furnished Fragonard the occasion for modeling sculptures in flesh . . . dissecting, injecting, anatomizing: manual activities that gave the tactile joy of materialized flesh and embodied matter, shared by the surgeon and the sculptor. By means of a biological diversion, a chemical disguise [travestissement] and a subtle surgery, he conferred on the cadavers the eternity of statues."[40] But this analogy fails to come to terms with the extent to which Fragonard's preparations are not frozen images, entrapped as if in a block of marble. To the contrary, they share a great deal with the animated statues so popular among Enlightenment artists and philosophes, such as Condillac, Buffon, and Bonnet, as well as with the mechanical automata built by Vaucanson. "Coming alive ex nihilo, like Locke's 'blank sheet,' the statue represented the logical point of departure for a psychology based on sequential development. As a means of visualizing the invisible, the figure also conjured up the technologies used to demonstrate hidden bodily functions, like the écorchés of human musculature in the Encyclopédie article 'anatomy,' the gynecological wax model (or femme invisible) of Marie-Catherine Bihéron, and Vaucanson's automated digesting duck."[41]

Like such figures, Fragonard's écorchés seemed to come to life, or better yet, to rise from the dead. And what is so astonishing is that he achieved this end without moving cams, wheels, gears, and levers. His écorchés were fashioned out of the same frustratingly intractable object about which Vicq d'Azyr despairingly remarked:

> A cold, inanimate body, deprived of life, offers only fibers without elasticity, slack and empty vessels. The art [of dissection], in truth, succeeded in filling them, but a foreign and coarse fluid leads the widest canals to become excessively distended. If, on the other hand, one uses a more refined fluid, it transpires through the pores like dewdrops, and teaches us nothing about the structures of the pathways through which it has passed. The nervous system that determined the strongest reactions, this pulp which was the focus of the most varied shaking, on which light itself imprinted images and left traces of its vibrations, is wholly insensible, wholly mute. The muscle no longer stiffens under the instrument that wounds it. The nerve is torn without excitement, trouble or sorrow. All connection, all sympathy is destroyed, and the bodies of animals in this state are a great enigma for those who dissect them.[42]

THE ARTIFICIAL AND THE NATURAL IN
EIGHTEENTH-CENTURY ARTIFICIAL LIFE

Most modern accounts of Artificial Life subscribe to a basic historical narrative, according to which mechanically driven automata of the early modern period are the antecedents of present-day robots, part of a long line of mechanical devices stretching back to ancient Egypt, Greece, Rome, and medieval Europe.[43] The cover art of the proceedings of the First International Conference on Simulation of Adaptive Behavior, *From Animals to Animats*, pictures four robotic ducks (see figure 6.5), each seemingly more completely robotic than its predecessor.[44] The first construction in the series still sports ducklike feet; the last has replaced them with metallic wheels. A closer inspection reveals that, far from being a fusion of the electromechanical and the "biological," all the ducks are simulations. Nevertheless, the first duck is perhaps more quaintly old-fashioned, reminiscent of a child's toy from pre-industrial times. Its brown color (as opposed to a companion's bright turquoise head) and the choice of material (suggesting hardwood) makes for the illusion of the "natural," especially because of the simulated effect of organic, traditional materials, which contrasts dramatically with the polished chrome exoskeletons and circuit-board innards of all the ducks. Still, as with the others, its "living" parts are but decoys for the eye. They are as "artificial" as the rest of its body. Subverting the viewer's impulse to detect

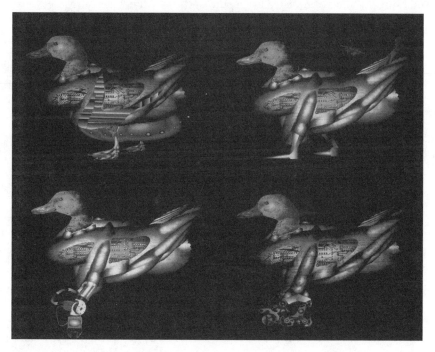

FIGURE 6.5 | Cover illustration for *From Animals to Animats: Proceedings of the First International Conference on Simulation of Adaptive Behavior*, ed. Jean-Arcady Meyer and Stewart W. Wilson (Cambridge, MA: MIT Press, 1991). Credit: Louis Bec, MIT.

the real (natural) animal in these machinic (animat) bodies, the artist also bestows a similar brown head on the last duck in the series, who, by all other definitions, is the most artificial of the group.

The artist has cleverly combined two themes of present-day thinking about Artificial Life: its early modern incarnation in evermore lifelike automata, for which Vaucanson's legendary, defecating Duck has become emblematic; and our own history of radical automation or progressive cyborgization in statu nascendi, a process that involves the progressive substitution of the brazenly artificial for the product of biological evolution. It is as if the early modern automata inhabited a lost arcadia of the artificial in contrast to today's post-animal, even post-human, utopian vision of a world that, finally, "works."[45]

The pastoral theme of the early modern Garden of Automata deserves exploration in its own right. In the present context, it serves as a pointer to the status of what we call "artificial life" in its eighteenth-century setting: less an usurpation of the natural by the artificial than a project that plays this substitution in reverse. That is, the "goal" (if there really is one) is to use mechanical technology to disguise its own existence, to produce the imitation of life in ever-increasing verisimilitude to its natural and human "originals." Recognition of this reverse-order artificial life involves a shift of attention from the clockmaker's workshop to the anatomist's theatre and studio. As the Abbé Desfontaines recognized upon seeing Vaucanson's Flute Player, it is there that the inspiration and the knowledge came to produce such a daring simulation. Like the anatomist, the mechanician was also in the business of vivifying dead matter.

Later in the century, we find other attempts to fashion artificial life without recourse to either of the two modes of simulation discussed here: on the one hand, producing the appearance of life through motion; and, on the other, enlivening dead matter by the techniques of natural anatomy. Perhaps the most significant figure in this development is the sculptor Jean-Antoine Houdon, who is widely known for his multiple écorché statues, which became prized objects for students of anatomy in medicine as well as the arts, and for his marvelous, lifelike sculptures of so many of the age's most renowned individuals. During his studies in Rome, as a winner of the Prix de Rome, Houdon attended dissections and studied anatomy, thereby supplementing the regular art curriculum. Throughout his career, he worked in the standard media of stone, plaster, and bronze. Nevertheless, he managed to solve the formerly intractable problem of how to capture the living eye's glistening and shifting appearance within a sculpture. Remarkably, he was able to reproduce the translucent effect of light shining on a living eye without recourse to the device of either a moving machine or the employment of dead (that is, formerly living) matter. Reversing Pygmalion's dream of bringing to life ("real" life?) a stone sculpture, Houdon instead animated the stone statue in such a way that he endowed it with life ("artificial" life?).

In all of these undertakings, one detects a strong ambition that surpasses the desire to acquire knowledge in its quotidian, technical, or scientific senses. Beyond the story of Pygmalion, beyond the contribution of ingenious automata to the staging of the pastoral myth of the aristocracy, there looms the original Myth of the Garden, in the context of which the anatomist, like the mechanician, risked infringing on the life-giving powers of the Creator.[46] Undergirding the human mechanical simulation of life is the creation of the latter by the Great Engineer, something that Descartes well appreciated when he wrote in his (suppressed) "Treatise on Man":

> I suppose the body to be nothing but a statue or machine made of earth, which God forms with the explicit intention of making it as much as possible like us. . . . We see clocks, artificial fountains, mills, and other such machines which, although only man-made, have the power to move of their own accord in many different ways. But I am supposing this machine to be made by the hands of God, and so I think you may reasonably think it capable of a greater variety of movements than I could possibly imagine in it, and of exhibiting more artistry than I could possibly ascribe to it.[47]

Little wonder, then, that a superior at the Order of the Minimes, where the young Vaucanson was a novice, objected to the androids that he had fabricated, supposedly to serve dinner and clear the tables. Declaring Vaucanson's tendencies "profane," he ordered the young man's workshop destroyed.

NOTES

1. "C'est sans doute la croissance de l'anatomie de l'homme, & surtout de la Névrologie, qui a guidé l'Auteur dans sa Mécanique." Lettre 180 (30 Mars 1738) in *Observations sur les écrits modernes, par MM. Desfontaines et Granet*, by M. l'abbé (Pierre-François Guyot) Desfontaines et Granet, vol. 12 (Paris: Chaubert, 1738), 337–42.

2. Barbara M. Stafford, *Body Criticism: Imaging the Unseen in Enlightenment Art and Medicine* (Cambridge, MA: MIT Press, 1991), 47; see 47–129, for Stafford's view of the place of dissection in enlightenment culture and medicine. Emily Jane Cohen also insightfully addresses the connections between anatomy, surgery, and enlightenment philosophy in "Enlightenment and the Dirty Philosopher," *Configurations* 5, no. 3 (1997): 369–429.

3. This description is drawn from John Cody, M.D., *Visualizing Muscles: A New Écorché Approach to Surface Anatomy* (Lawrence: University Press of Kansas, 1990), 1.

4. "Mr. Vaucanson's Letter to the Abbé De Fontaine," in *An Account of the mechanism of an automaton, or image playing on the German-flute*, trans. John Theophilus Desaguliers (London: T. Parker, 1742), reproduced in the facsimile edition of Vaucanson's original text (1738), along with the eighteenth-century English translation: Jacques Vaucanson, *Le Mécanisme du Fluteur Automate/ An Account of the Mechanism of an Automaton or Image Playing on the German-Flute* (Paris: Archives Contemporaines; Conservatoire National des Arts et Métiers, 1985), 22. As it turns out, Vaucanson's claim was misleading, at least as a description of the anatomical correctness of the whole machine. Finding it difficult to reproduce the digestive system's "Chymical Elaboratory," especially in the amount of time required for a live demonstration, he loaded a hidden chamber with a prepared mash, which made it appear that the duck had digested what it defecated. The corn that it ate was stored in a chamber below its throat. This discovery was made in 1783, some years after Vaucanson's death.

5. Vaucanson, *An Account*, 23.

6. H. Pitot, "The Approbation of the Royal Censor," in Vaucanson, *An Account*, 21. On popular science demonstrations, see Geoffrey Sutton, *Science for a Polite Society: Gender, Culture, and the Demonstration of Enlightenment* (Boulder, CO: Westview Press, 1995); Simon Schaffer, "Natural Philosophy and Public Spectacle in the Eighteenth Century," *History of Science* 21 (1983): 1–43, and "The Consuming Flame: Electrical Showmen and Tory Mystics in the World of Goods," in *Consumption and the World of Goods*, ed. John Brewer and Roy Porter (London: Routledge, 1993), 489–526; Larry Stewart, *The Rise of Public Science* (Cambridge: Cambridge University Press, 1995); Michael R. Lynn, "Enlightenment in the Public Sphere: The Musée de Monsieur and Scientific Culture in Late-Eighteenth-Century Paris," *Eighteenth-Century Studies* 32, no. 4 (Summer 1999): 463–76.

7. Vaucanson, *An Account*, 22–23.

8. See above, n. 4, where Desaguliers uses the word *image* in the English title for *Le Mécanisme du Fluteur Automate*. Along with the *Duck* and the *Flute Player*, Vaucanson demonstrated to the academicians, as he had previously to Parisian audiences for the price of entrance, a third figure that played simultaneously a (three-holed) pipe with his left hand and, with his right hand, a tabor, or small snare drum.

9. On the *écorché* and the commanding place of Jean-Antoine Houdon in this tradition, see Anne L. Poulet, *Jean-Antoine Houdon: Sculptor of the Enlightenment* (Washington, DC: National Gallery of Art; Chicago: University of Chicago Press, 2003); Louis Reau, *Houdon: Sa vie et son oeuvre* (Paris: F. de Nobele, 1964); Zofia Ameisenowa, *The Problem of the Écorché and the Three Anatomical Models in the Jagiellonian Library*, trans. Andrej Potocki (Warsaw: Polska Akademia Nauk, 1963).

10. The Louvre describes Coysevox's marble sculpture as a faun playing a flute, "part of a group devoted to the forest, with Hamadryad and Flora, placed in the Park of Marly at the 'Fer à cheval' (horseshoe) at the bottom of the 'river.' He wears the 'nebridos,' a faun skin. His head is crowned with oak leaves. The syrinx, flute with seven pipes, is at his feet. The pastoral staff is posed on the trunk on which he is seated. Behind him, a little faun puts his finger to his lips to demand silence. Commanded in 1707 and dated 1710,

the group was transported to the Tuileries Gardens as early as 1716." Available online at http://www.louvre.fr/anglais/collec/sculp/mr1820/txt1820.htm. Daniel Cottom compares Vaucanson's refusal of Frederick the Great's invitation to join his Prussian court with the celebrated sculptor's privileged place at the court of the Sun King. For Cottom, Vaucanson, "the engineer of the Enlightenment," appears to be the reincarnation of Coysevox, though presumably a "Coysevox" better suited to enlightened rather than court culture. See Daniel Cottom, *Cannibals and Philosophers: Bodies of Enlightenment* (Baltimore, MD: Johns Hopkins University Press, 2001), 70. Still the contrast only goes so far. As is widely acknowledged, Vaucanson also sought and received court attention, and his social rise certainly depended on the cultivation of such ties. On Cardinal Fleury's recommendation, and possibly to prevent the engineer's departure for Prussia, Vaucanson was appointed by the king in 1741 to the position of inspector of silk manufacture.

11. Gaby Wood, *Edison's Eve: A Magical History of the Quest for Mechanical Life* (New York: Anchor, 2002), 26.

12. There is a growing literature on the topic of eighteenth-century sensationalism; see, for example, John C. O'Neal, *The Authority of Experience: Sensationist Theory in the French Enlightenment* (University Park: Pennsylvania State University Press, 1996); Anne C. Vila, *Enlightenment and Pathology: Sensibility in the Literature and Medicine of Eighteenth-Century France* (Baltimore, MD: Johns Hopkins University Press, 1998); Jessica Riskin, *Science in the Age of Sensibility: The Sentimental Empiricists of the French Enlightenment* (Chicago: University of Chicago Press, 2002); *The Cambridge History of Science*, vol. 4, *Eighteenth-Century Science*, ed. Roy Porter (Cambridge: Cambridge University Press, 2003).

13. Vaucanson, *An Account*, 22. For greater clarity, I have inserted my own translations in brackets, alongside the version reproduced from Desaguliers's translation of Vaucanson, *Le Mécanisme* (see n. 4 above). The entire French passage reads as follows: "Je ne crois pas que les Anatomistes ayent rien à désirer sur la construction de ses aîles. On a imité, os par os, toutes les éminences qu'ils appellent apophyses. Elles y sont réguliérement observées comme les différentes charniéres [sic]: les cavitez, les courbes, les trois os qui composent l'aîle, y sont très-distincts. Le premièr qui est l'Homerus, a son mouvement de rotation en tout sens, avec l'os qui fait l'office d'omoplate; le second os qui est le Cubitus de l'aîle, a son mouvement avec l'Homerus, par une charniere [sic], que les Anatomistes appellent parginglime; le troisiéme, qui est le Radius, tourne dans une cavité de l'Homerus, & est attaché par ses autres bouts aux petits os bout de l'aîle, de même que dans animal. L'inspection de la machine sera mieux connoître l'imitation de la nature qu'un plus long détail, qui resembleroit trop à une explication anatomique" (Vaucanson, *Le Mécanisme*, 20).

14. Catherine Cardinal, "Preface," in Vaucanson, *Le Mécanisme*, xv–xvi.

15. Michel Lemire, *Artistes et mortels* (Paris: Chabaud, 1990), 4. Anita Guerrini discusses Duverney's preparations in "Duverney's Skeletons," *Isis* 94 (2003): 577–603. See also her "Anatomists and Entrepreneurs in Early Eighteenth-Century London," *Journal of the History of Medicine and Allied Sciences* 59, no. 2 (2004): 219–39.

16. Sébastien Mercier, *Tableau de Paris*, cited in Philippe Ariès, *The Hour of Our Death*, trans. Helen Weaver (New York: Knopf, 1981), 365, 368.

17. André Doyon and Lucien Liaigre, *Jacques Vaucanson, mécanicien de génie* (Paris: Presses Universitaires de France, 1967), 18–19.

18. For information on Vaucanson and Le Cat, see Doyon and Liagre, *Vaucanson*, 17ff. The full title of Le Cat's missing work is *Déscription d'un homme automate dans lequel on verra executer les principales functions de l'économie animale, la circulation, la respiration, les sécretions, & au moyen desquels on peut déterminer les effets méchaniques de la saignée, & soumettre au joug de l'expérience plusieurs phénomènes interessants qui n'en paroissent pas susceptibles* (Rouen, 1744).

19. K. B. Roberts, "The Contexts of Anatomical Illustration" in *The Ingenious Machine of Nature: Four Centuries of Art and Anatomy*, by Mimi Cazort, Monique Kornell, and K. B. Roberts (Ottawa: National Gallery of Canada, 1996), 71.

20. Roberts, "Contexts of Anatomical Illustration," 71.

21. Another tradition of anatomical models is provided by Chinese obstetrical dolls, which were introduced for diagnostic purposes so that the female patient, without being touched or viewed by the male physician, could nonetheless report on her symptoms and their location. In the West during the seventeenth and eighteenth centuries, ivory figurines were used as anatomical manikins by doctors for training barber surgeons and midwives, and also to instruct the lay public. For examples from China, France, Italy, and Germany, see the exhibition catalog by Julie V. Hansen and Suzanne Porter, *The Physician's Art: Representations of Art and Medicine* (Durham, NC: Duke University Medical Center Library and Duke University Museum of Art, 1999).

22. "Artificial" anatomy, the period's contrasting method, involved making a representation of the body from other materials: typically wax but also papier-mâché or, as in the work of Fragonard's contemporary Marie-Catherine Bihéron (1719–1786), silk, wool, threads, and wax-coated feathers.

23. José van Dijck speaks of the struggle to reconcile "the contradictory requirements of authenticity and didactical value in the teaching of medical knowledge"; see "Bodyworlds: The Art of Plastinated Corpses," *Configurations* 9, no. 1 (2001): 102. On the gender implications of such normative expression, not just in waxes but also in engraved images, see Londa Schiebinger, *Nature's Body: Gender in the Making of Modern Science* (Boston: Beacon Press, 1993).

24. Martin Kemp, "Foreword" to Hansen and Porter, *Physician's Art*, 14.

25. Deanna Petherbridge, "Art and Anatomy: The Meeting of Text and Image," in *The Quick and the Dead: Artists and Anatomy* (Berkeley: University of California Press, 1997), 27.

26. On the theme of abjection, see Julia Kristeva, *Powers of Horror: An Essay on Abjection*, trans. Leon S. Roudiez (New York: Columbia University Press, 1982). See also Mikhail Bakhtin, *Rabelais and His World*, trans. Helene Iswolsky (Bloomington: Indiana University Press, 1984); Mary Douglas, *Purity and Danger* (New York: Praeger, 1966); Sigmund Freud, *Civilization and Its Discontents*, trans. James Strachey (New York: W. W. Norton, 1989).

27. Bakhtin, *Rabelais*, 27, 20–21, 25–26; Kristeva, *Powers of Horror*, 9.

28. Félix Vicq d'Azyr, *Discours sur l'anatomie et de physiologie avec des planches coloriées, réprésentant au naturel les divers organs de l'homme et des animaux* [...] (Paris: l'Imprimerie de France, F. A. Didot l'Aîné, 1786), 1.

29. Barbara M. Stafford et al., "Depth Studies: Illustrated Anatomies," *CADUCEUS: A Humanities Journal for Medicine and the Health Sciences* 8, no. 2 (1992): 40.

30. Originally named the "École pour le traitement des maladies des bestiaux," the world's first veterinary school was founded in 1761 by Claude Bourgelat, friend of Diderot and contributor to the Encylopédie. Honoré Fragonard served as its director and chair of anatomy. In 1764, the school acquired its designation as the Royal Veterinary College. In 1765, Bourgelat was invited to found a royal veterinary school in Paris on the Lyon model, where Fragonard was again appointed director. See Alcide Louis Joseph Railliet and L. Moulé, Histoire de l'École d'Alfort (Paris: Asselin and Houzeau, 1908).

31. Ruysch served as praelector of anatomy for the surgeon's guild in Amsterdam from 1666 until his death in 1731. In 1679, he was appointed doctor of the court of justice, and he was appointed professor of botany at the Athenaeum Illustre in 1685 and thus became supervisor of the botanical garden.

32. Michael Hagner, "Enlightened Monsters," in The Sciences in Enlightened Europe, ed. William Clark, Jan Golinski, and Simon Schaffer (Chicago: University of Chicago Press, 1999), 180.

33. "Frederik Ruysch," available online at Who Named It? http://www.whonamedit .com/doctor.cfm/1142.html. For an entrance fee, Ruysch displayed his preparations, which were sometimes referred to as "the Eighth Wonder of the World." His cabinet attracted numerous foreign visitors, including Peter the Great of Russia, who purchased the collection from Ruysch in 1717 for thirty thousand guilders.

34. Bernard Le Bovier de Fontenelle, Éloges des académiciens (The Hague: Isaac van der Kloot, 1740).

35. Jonathan Simon, "The Theater of Anatomy," Eighteenth-Century Studies 36, no. 1 (2002): 70.

36. On Ruysch, see also Van Dijck, "Bodyworlds," 99–126; Julie V. Hansen, "Resurrecting Death: Anatomical Art in the Cabinet of Dr. Frederick Ruysch," Art Bulletin 78, no. 4 (1996): 663–79; and A. M. Luyendijk-Elshout, "Death Enlightened: A Study of Frederick Ruysch," JAMA 212, no. 1 (1970): 121–26.

37. René Descartes, Méditations métaphysiques (Paris: Presses Universitaires de France, 1970), 39.

38. Sylvie Hugues, "Esthétique et Anatomie: Science, Religion, Sensation," Dix-huitième siècle 31 (1999): 151–52; she quotes Bichat, Recherches physiologiques sur la vie et sur la mort (1800). Hugues argues that, given the refutation of materialism in new conceptions of irritability and sensibility, Fragonard is using an already outmoded language. She sees the dominance of aesthetic discourse in this work as troubling its scientific aspect (152).

39. Although not pictured in this particular photograph, the figure's attentiveness is mimicked by his penis's turgidity.

40. M. Ellenberger, L'Autre Fragonard: Essai biographique (Paris: Jupilles, 1981), 32.

41. Julia V. Douthwaite, The Wild Girl, Natural Man, and the Monster: Dangerous Experiments in the Age of Enlightenment (Chicago: University of Chicago Press, 2002), 70. See also J. L. Carr, "Pygmalion and the Philosophes: The Animated Statue in Eighteenth-Century France," Journal of the Warburg and Courtauld Institutes 23, nos. 3–4 (1960): 239–55.

42. Vicq d'Azyr, Discours sur l'anatomie, 2.

43. Another motif is that artificial life involves the influence of magic in ideas and myths about the imitation of life (e.g., the Jewish fable of the Golem). On the influence of

automata on political, philosophical, and medical thinking, see Otto Mayr, *Authority, Liberty and Automatic Machinery in Early Modern Europe* (Baltimore, MD: Johns Hopkins University Press, 1986). On automata, see also Alfred Chapuis and Edmond Droz, *Automata: A Historical and Technological Study* (Neuchâtel: Éditions du Griffon, 1958); Alfred Chapuis and Edouard Gélis, *Le Monde des Automates: Etude historique et technique*, 2 vols. (Paris, 1928); Roland Carrera, Dominique Loiseau, and Oliver Roux, *Androids: The Jacquet-Droz Automatons* (Lausanne: Scriptar and F. M. Ricci, 1979).

44. Jean-Arcady Meyer and Stewart W. Wilson, eds., *From Animals to Animats: Proceedings of the First International Conference on Simulation of Adaptive Behavior* (Cambridge, MA: MIT Press, 1991).

45. On these issues, compare Hans Moravec's *Mind Children: The Future of Robot and Human Intelligence* (Cambridge, MA: Harvard University Press, 1988) to MIT roboticist Rodney Brooks's *Cambrian Intelligence: The Early History of the New AI* (Cambridge, MA: MIT Press, 1999), where the human cognitive subject is set aside as a model for robotics. Just as the Roboducks in figure 6.5 lose their feet in favor of more efficient appendages, so Brooks's often insectoid robots shed both reason and representation. Alan Kay has written somewhere that "classical AI" had failed to construct even a form of intelligence worthy of termites. Brooks has labored to overcome this limitation.

46. On the Pygmalion craze in eighteenth-century culture, see Mary Sheriff, *Moved by Love: Inspired Artists and Deviant Women in Eighteenth-Century France* (Chicago: University of Chicago Press, 2003).

47. René Descartes, "Treatise on Man," in *The Philosophical Writings of Descartes*, trans. John Cottingham, Robert Stoothoff, and Dugald Murdoch, 2 vols. (Cambridge: Cambridge University Press, 1985), 1:99.

PART TWO

EMERGENCE

The Homunculus and the Mandrake:
Art Aiding Nature versus Art
Faking Nature WILLIAM R. NEWMAN

In the period from about 1200 to 1700, the human ability to re-create the products of nature emerges with increasing frequency in a wide variety of European sources. Just as we moderns experience conflicted feelings when we hear of fluorescent bunnies, strawberries with genes taken from fish, IBM computers that can outthink the grandest of the grandmasters, and genetically modified "Frankencorn" that may have escaped its human-imposed boundaries to hybridize with wild varieties of maize, so medievals and early moderns worried about the erosion of the boundaries between the artificial and the natural wrought by their real and perceived technological prowess.[1] The issue receives extensive treatment, for example, in scholastic theological commentaries; in sixteenth-century discussions of the liberal and mechanical arts, such as those of the humanist Benedetto Varchi and the ceramicist Bernard Palissy; and in seventeenth-century works of natural philosophy ranging from the university manualists like Daniel Sennert to nonacademic virtuosos such as Robert Boyle and Margaret Cavendish. Despite the heterogeneous character of these widely divergent texts, one additional element of commonality emerges.[2] Whenever the discussion turns to the issue of whether humans can really reproduce the products of nature, the example almost invariably used to illustrate and solve the problem is the alchemists' claim that they can make a gold equivalent in all respects to the natural metal. Just as Friedrich Wöhler's nineteenth-century chemical synthesis of urea would test the claim that nature could make things that man cannot, so the assertions of alchemists pushed much the same envelope for earlier generations.[3]

Medieval and early modern writers on alchemy therefore found themselves at the center of a controversy whose stakes extended much further than the mere issue of selling alchemical gold. Their art, unlike most others, made a striking and unequivocal claim. To alchemical practitioners of the High Middle Ages, alchemy occupied a privileged rank in claiming to alter the deep structure of matter in a way that was purely natural. This claim would form the focus of debate in subsequent centuries as a disputational literature gradually grew up with the aurific art at its center. Alchemy was a discipline that claimed not to imitate nature by deceiving the senses, but to replicate it in every detail. Just as modern chemists claim not only to make a likeness or similitude of a natural product, but to *synthesize* the very molecules out of which the natural thing is made, so alchemists claimed to utilize natural processes in order to fabricate products identical to those that nature itself makes. To opponents of alchemy, this claim seemed to make the alchemist a second deity, since he was creating new gold, precious stones, or minerals out of hitherto baser matter. The charge of "playing God," commonly leveled against the pioneers of genetic engineering today, was already raised against those medievals who would change the face of nature by replacing one species with another.

At the same time, alchemical writers from the Middle Ages onward consciously pitted themselves against practitioners of the visual arts. There is considerable irony in this, since alchemy itself was originally an offshoot of the decorative arts. It is commonly held that the earliest stage of alchemy is reflected in two late antique papyri now found in Leiden and Stockholm, which contain recipes for dyeing fabrics and simulating pigments and other materials, apparently for making temple artifacts and paste jewelry.[4] Yet as early as the fifth century, alchemy underwent a strange metamorphosis—its practitioners began to view their products in Pygmalion-like fashion as replications, rather than representations, of the natural world. Always aware of the potential charge that they, too, were engaged in a sort of *trompe-l'oeil* trickery, the medieval and early modern alchemists explicitly claimed that their discipline perfected nature rather than merely imitating it. This view built on the distinction that Aristotle draws in the *Physics* (2.8.199a, 15–17), where he states that "the arts either, on the basis of Nature, carry things further [*epitelei*] than Nature can, or they imitate [*mimeitai*] Nature."[5] To alchemical writers, this meant that most other fields leading to physical production, such as shipbuilding, fabric-making, or the visual arts, merely mimicked natural products—either in a loose and general sense, as in the old stories that based the invention of architecture on the observation of swallows' nests and weaving on the activity of spiders, or in the specific sense that pertained to painting, sculpture, and other representational arts. The supporters of alchemy, on the other hand, claimed that the aurific art actually duplicated natural products in ways that improved nature's own methods. When an alchemist purged a base metal and converted it to gold or silver, he was *perfecting* what nature had been unable to complete—he was not producing an *ersatz* replacement or copy of a natural product.

To facilitate this distinction between the perfective and the merely imitative, alchemical writers often used architecture, sculpture, and painting as paradigmatic counterexamples or foils to their art. The housebuilder or sculptor did not change the nature of matter itself, but merely imposed a superficial, accidental form upon it. Wood and stone remained wood and stone, even if they served as the components of joists, studs, and walls. The painter, likewise, merely manipulated the external accidents of matter when he depicted an image by shifting pigments from one place to another. Such illusory appearances were like the "sophistical transmutations" described by Geber in his famous *Summa perfectionis*, composed around the end of the thirteenth century.[6] They were impositions upon reality that could not withstand examination—like a spurious metal, they might trick us into accepting their veracity, but a closer inspection would reveal them as mere fraud. In effect, the alchemists relegated the illusionistic tricks employed by plastic and pictorial artists to the status of bad alchemy. As I argue in the rest of this paper, the aspersions that alchemists cast on the visual arts in comparing their genuine but artificial gold with the superficial changes wrought by painting and sculpture play out in different form when alchemical writers come to discuss the homunculus, or artificial test-tube baby. It is within this agonistic context, in which alchemists were trying to promote their art at the expense of other arts—and even in the face of nature itself—that we should view the early modern claim that alchemy can replicate human life in a flask. The alchemical homunculus involved a transmutation of substance, whereas the carved and painted effigies of mandrake roots were mere artistic representations produced by the manipulation of exterior accidents such as color and shape. Although this might seem obvious on one level, it represents a facet of the dispute between alchemists and visual artists about the nature of representation that has far-reaching ramifications. But let me first present the facts.

In 1572, the iatrochemical physician Adam von Bodenstein published a work supposedly written in 1537 by the famous medical reformer of alchemy,

Paracelsus von Hohenheim. This work, called *De natura rerum*, opens with a manifesto asserting the power of human art. There are two sorts of generation, the pseudo-Paracelsian author asserts—one that occurs purely naturally when one thing is corrupted and another is generated from it, and a second type that occurs artificially, as when transmutation occurs in a flask.[7] By means of such artificial generation, an alchemist can make birds in a flask, for example, by burning up a natural bird and converting it to a mucilaginous phlegm, which is then putrefied in a heated flask until the bird is regenerated. But the regenerate bird will be better than the natural one, since it will be "clarified"—its matter will have lost some of the turbidity that characterizes the terrestrial world.[8]

Pseudo-Paracelsus follows this rewriting of the Phoenix myth with the claim that other artificial creatures can also be manufactured *in vitro*. One can make a basilisk, for example, by sealing up the menstrual blood of a human female and heating it for an unspecified time. The resulting creature will have the traditional lethal ability of the basilisk to kill by means of its looks alone—and why not, since pseudo-Paracelsus accepts the long-standing view that catamenial women have the ability to damage a mirror by looking at it or to ruin good wine by breathing on it? In short, the basilisk is a concentrated form of femininity made by alchemical means in the same way that an alchemist might make brandy or aquavite by distilling out and concentrating its active ingredient.

The same principle applies to the male analogue of the basilisk, namely, the homunculus. The homunculus is produced out of sperm that is putrefied separately in a heated flask, just as the raw material of the basilisk was putrefied. Here the author of *De natura rerum* tweaks the traditional Aristotelian theory of generation, according to which human semen was almost entirely pure form, and menstrual blood supplied the matter to the human fetus. Pseudo-Paracelsus clearly thinks that he can obviate the need for menstrual blood and thereby create a homunculus that is almost entirely immaterial. No longer chained down by the poisonous matrix of its feminine materiality, the homunculus will be able to perform marvels, such as foretelling the future and winning great battles for its possessor. From the perspective of pseudo-Paracelsus, the homunculus is, as it were, the distilled essence of masculinity, concentrated and purified of its material dross.[9]

Now it is important to note that this strange eugenic thought-experiment is not found in the work of the genuine Paracelsus, but in a text of uncertain authorship. Nonetheless, *De natura rerum* does contain ideas drawn from the genuine Paracelsian *Nachlass*, and the same naturalistic impulse impels those comments on the homunculus that belong more definitely to the genuine Paracelsian corpus. In addition to an Arabic tradition of artificial generation transmitted by medieval alchemy, there is another source that Paracelsus seems to have used for his homuncular ruminations. I refer to the popular tradition of the mandragora, known even in Middle High German as *Alraun* or *Alraune*. The mandrake had a large and well-established mythology built up around it. Already in late antiquity the Jewish writer Josephus believed that the plant *Baaras*, or mandrake, would emit a deadly scream upon being uprooted, and that the only way to avoid death was to employ a dog for this job, by tying the unfortunate beast to the exposed part of the plant and then inducing it to pull.[10]

In his *Liber de imaginibus*, of uncertain date, Paracelsus attacks dishonest scoundrels who carve roots to look like a man, and sell them for *Alraun*. He denies categorically that any root shaped like a man really grows naturally:

> I reply and say that it is not true that the Alraun root has a shape like a man, but rather it is a cheat and *bescheisserei* of vagabonds, who *bescheissen* the people with more than this alone, for there is indeed no root that has the shape of a man, unless it is carved and shaped; none is formed thus by God or grows thus from nature. Hence there is no need to speak of it further.[11]

Cheating peddlers and mountebanks who carved false mandrake roots in the shape of men or women were a widespread topic of complaint in the sixteenth and seventeenth centuries. A near contemporary of Paracelsus, Otto Brunfels, bewails the fact that "false cheaters carve [*Alraun*] from the little root *Brionia*," in his herbal. Another German herbal writer of the same period, Leonhard Fuchs, complains of vagabonds who carve roots in a human shape complete with hair, beard, and other human features. Hieronymous Bock, a third writer in the same genre, says that the charlatans who carve *Alraun* place the completed figure in hot sand for a considerable period of time, in order to give it a properly shriveled appearance. Nor was this effusive lament against *Alraun*-carvers a mere literary *topos*. One Ambrosi Zender, a self-styled student, was apprehended in Lucerne in 1562 for various offenses. As it turned out, he had been selling counterfeit *Alraun* for the considerable sums of six and seven ducats. Zender had been supporting himself by the sale of such "mandrakes" for some time: he would carve them from the roots of the white lily and provide a testimonial letter to the buyer, describing the uses to which the purchased *Alraun* could be put. To his fellow prisoners, Zender confessed a bizarre final wish—that if he should be hanged on the Emmenbrücke bridge, his corpse should be turned toward the passersby, so that those who knew him could get a "lovely" view.[12]

FIGURE 7.1 | An artificially enhanced "mandrake," probably dating from the sixteenth century and kept at the Kunsthistorisches Museum in Vienna. Although this specimen is obviously Christlike, and hence presumably not intended to represent a plant grown from a decomposing body on a gallows, the manner of its production closely conforms to the usual one used in making artificial "mandrakes." Note particularly the beard, possibly consisting of rhizomes projecting from embedded seeds in conformity with the descriptions given by early modern herbalists.

We see, then, that Paracelsus was not hallucinating when he complained of cheating *Alraun*-carvers. But was he really as negative about the *Alraun* as his categorical comments in the *Liber de imaginibus* might lead us to believe? Despite his negative statements there, Paracelsus affirms in another passage that the mandrake can indeed be produced, even if the natural philosophers and physicians have enveloped it in error. In his work on the prolongation of life, *De vita longa* (1526/1527), after discussing the theory that pearls are generated from sperm, he says, "The homunculus, which the necromancers falsely call *alreona* and the natural philosophers *mandragora*, has become a topic of common error, on account of the chaos in which they have obscured its true use. Its origin is sperm, for through the great digestion which occurs in a *venter equinus*, the homunculus is generated, like [a man] in all things, body and blood, with principal and lesser members."[13]

Here Paracelsus argues that the mandrake incorrectly described by necromancers and philosophers is really a homunculus, which they have misidentified. Paracelsus is probably thinking here of the old German folk legend that the *Alraun* (his *alreona*) grew primarily beneath gallows, where it was generated from the sperm or urine of hanged criminals: in honor of its provenance, the *Alraun* was also called *Galgenmann* or *Galgenmännlein* "gallows-man."[14] This belief has been traced back to Avicenna and is mentioned by early modern German authors such as Brunfels. In the seventeenth century it was still believed in some quarters that such a gallows-man could also be produced by burying the sperm of a young man underground and periodically feeding the developing embryo with more of the same.[15] In order to understand Paracelsus's reasoning in the above passage, one must realize that he customarily employs the expression *venter equinus*, a technical term in alchemy for decaying dung used as a heat source, to mean any source of low, incubating heat. Thus it was easy for him to interpret the mandrake legend as a garbled recipe for the homunculus, where the earth beneath the gallows acted as a *venter equinus*, or incubator.

Although Paracelsus himself does not recommend that we dig up gallows-men or try to make them, the early modern folk sources that we have been considering put a high premium on their possession: as we have seen, Ambrosi Zender was able to sell such effigies for inflated sums in the sixteenth century. In part, this reflected the supposed danger that the seller incurred in extracting the plant from the ground. But this does not account for the market that existed for such items. The real value of the *Alraun* lay in the benefits that it was supposed to confer upon the possessor. A widespread folk belief can be traced back to early modern sources according to which the gallows-man

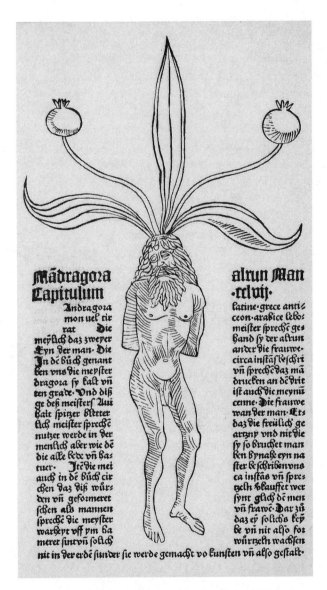

FIGURE 7.2 | Fanciful depiction of a "male mandrake" (*Mandragora officinalis*), from Johann Wonnecke von Cube, *Hortus sanitatis germanice* (Mainz: Peter Schoeffer, 1485). The hands tightly held behind the figure's back like the tied hands of a hanged prisoner indicate that this is probably supposed to represent a "gallows-man," a mandrake that has grown from the fluids dripping from a decomposing body left on the gallows.

brought good luck to his owner and would even "double his money." The *Alraun* was supposed to be kept in a little flask or coffinlike box—if a coin was inserted into its container at night, two coins would be found there the following day. But the owner had to follow an elaborate ritual: a Leipzig burgher in 1575 writes his brother that the *Alraun* he is sending him must be bathed in warm water and clothed in silk three days after its arrival. This bathing must be repeated four times per year. Other sources specify that this ritual bathing must be performed with wine, and that the "coffin" in which the gallows-man is kept must be ornately decorated.

A very strange description of such a "coffin" and its contents appears in a report dating from March 24, 1679, in Hamburg. On that day an impoverished old woman was buried in the graveyard of St. Catherine's Church; in accordance with the local regulations, her possessions were then auctioned. The *Kirchenvogt* responsible for her goods found a chest among her furnishings, and in one of its drawers there was a box, which in turn contained a smaller box. Within the smaller box, the official found a "dainty, little coffin." In the tiny coffin lay a tiny figure, wrapped like a mummy in white cloth, upon which a black cross had been painted. When the eager *Vogt* unwrapped this cloth, he found an *Alraun*, which looked like "an old, venerable, strange little man." This extraordinary being had a long beard and hair that almost reached its feet. In addition, it possessed a long nose, a mouth full of pointed teeth, and hands and feet complete with fingernails and toenails. The detail was so remarkable that one could even make out the skeleton beneath the shriveled skin of the *Alraun*. But the *Alraun* bore an additional characteristic that signified its violent death—its arms were fastened behind its back and its neck was broken like that of a hanged man. Clearly the *Alraun*-carvers had been busy to give every appearance of a gallows-man to this remarkable specimen.

Independent reports of the *Alraun's* container can be found in other seventeenth-century sources. The literary figure and alchemist Johann Rist (1607–1669) relates in his *Aller edelste Thorheit der ganzen Welt* (The noblest folly in the whole world) that the gallows-man is usually kept in a tiny coffin that should be painted red and equipped with colored blankets. On the inside of the coffin lid a cross is painted, and on the top, a gallows, from which a dead thief hangs. Other sources, however, maintain that the *Alraun* can be kept in a bottle or flask, as long as this is not exposed to public view. The image of a miniature human kept in a bottle is strikingly reminiscent of the homunculus described in the pseudo-Paracelsian *De natura rerum*.[16] If we consider the powers that were ascribed to the gallows-man—in addition to bringing good fortune and doubling one's money, it was sometimes said to be capable of prophesying the future and of bringing victory in battle—this similarity becomes even more pronounced. Perhaps the pseudo-Paracelsus, like his master, was also drinking from the dark and glimmering waters of German folk legend.

To return briefly to the theme with which I began, then, Paracelsus's diatribe against mandrake-cutters clearly follows the same pattern as the traditional attack that alchemists had been making on the visual arts since the High Middle Ages. By an extension of the scholastic viewpoint tradition-ally adopted in medieval alchemical texts, the mandrake-cutters impose a mere accidental change on matter. Carving and painting a preexisting root, they impose upon it the simulated form of a miniature human being. The Paracelsian alchemist, on the other hand, can create a genuine mandrake or *Alraun* in the form of the homunculus, by sealing up human semen for a proper period of time with the requisite application of heat. This process will lead nature to a goal that it could not otherwise attain, namely, the manufacture of a human being devoid of feminine materiality. Although the genuine Paracelsus does not recommend this process and even warns against the danger of making homunculi in other works, his contrasting of the two sorts of art—one purely mimetic and the other perfective—is a straightforward elaboration of the traditional distinction between arts that merely mimic nature and those that lead to an altered, but still natural, product. We moderns tend to elide the distinction between alchemists and visual artists when we speak of "the alchemy of the creative process" and argue that Renaissance artists saw themselves as something like alchemists. Although some artists may have seen themselves in this light, it is unlikely that many would have continued to do so if they were genuinely conversant with the literature penned by the alchemists themselves.

NOTES

The following paper is heavily dependent on my recent book, *Promethean Ambitions: Alchemy and the Quest to Perfect Nature* (Chicago: University of Chicago Press, 2004).

1. For the obtaining of antifreeze protein from winter flounder, see the Web site of the International Society for Horticultural Science, at http://www.actahort.org/books/484/484_99.htm. For Eduardo Kac and the bioluminescent rabbit, see his Web site at http://www.ekac.org/. For the argument that transgenic maize has made its way into the wild in Mexico, see David Quist and Ignacio H. Chapela, "Transgenic DNA Introgressed

into Traditional Maize Landraces in Oaxaca, Mexico, *Nature* 414 (2001): 541–43. On robotics and artificial intelligence, see Rodney A. Brooks, *Flesh and Machines: How Robots Will Change Us* (New York: Pantheon Books, 2002); and Hans Moravec, *Mind Children: the Future of Robot and Human Intelligence* (Cambridge, MA: Harvard University Press, 1988).

2. For documentation on these authors and their involvement in the art-nature debate, see Newman, *Promethean Ambitions*, 132–63, 271–89.

3. John Hedley Brooke, "Wöhler's Urea and Its Vital Force? A Verdict from the Chemists," *Ambix* 15 (1968): 84–115. See also Douglas McKie, "Wöhler's Synthetic Urea and the Rejection of Vitalism," *Nature* 153 (1944): 608–10.

4. Robert Halleux, *Les alchimistes grecs: Papyrus de Leyde, papyrus de Stockholm, recettes* (Paris: Belles lettres, 1981), 5–78.

5. Aristotle, *The Physics*, tr. Philip H. Wicksteed and Francis M. Cornford (London: Heinemann, 1929), 173.

6. For Geber's tripartite division of alchemy into three branches, of which the first is "sophistical," see William R. Newman, *The Summa perfectionis of pseudo-Geber* (Leiden: Brill, 1991), 752–69.

7. The distinction that pseudo-Paracelsus makes here between the powers of nature and those of art points to the elasticity inherent in the very concept of the artificial. From a modern vantage point, one could argue that generation in a flask would have been perfectly natural (if it could have been achieved by sixteenth-century methods), since natural ingredients such as semen and blood were involved. From the Aristotelian perspective employed by most premodern authors on the subject of the art-nature distinction, however, the *in vitro* generation described by pseudo-Paracelsus would have been either flagrantly unnatural (if one chose to take a conservative viewpoint) or a case of nature perfected by means of art (from a more liberal point of view). For more on the elasticity of the art-nature distinction and the view that art can perfect nature, see Newman, *Promethean Ambitions*.

8. [Pseudo-]Paracelsus, *De natura rerum*, in *Theophrastus von Hohenheim, genannt Paracelsus, Sämtliche Werke, I. Abteilung*, ed. Karl Sudhoff, 14 vols. (Munich: Oldenbourg, 1922–33), 11:312–13.

9. Ibid., 11:315–17.

10. Alfred Schlosser, *Die Sage vom Galgenmännlein im Volksglauben und in der Literatur* (Münster, Inaugural Dissertation, 1912), 23. See also A. R. von Perger, "Über den Alraun," *Schriften des Wiener-Alterthumsvereins*, special issue, 1862, 259–69, especially 260. Josephus does not use the term *mandragora*.

11. Paracelsus, *De imaginibus*, in *Theophrastus von Hohenheim*, 13:378: "dem geb ich zur antwort und sag, es sei nicht war, das alraun die wurzel menschen gestalt hab, sonder es ist ein betrogne arbeit und bescheisserei von den landfarern, die dan die leut mer denn mit disem alein bescheissen, dan es ist gar kein wurzel die menschen gestalt hat, sie werden dan also geschnizlet und geformirt; von got ist keine also geschaffen oder die von natur also wechst, darumb ist weiter darvon nit zu reden &c."

12. Schlosser, *Galgenmännlein*, 25–34. Schlosser cites several other authors who complain of carved *Alraunen*, for example, on pp. 27 and 29. See also Lynn Thorndike, *A History of Magic and Experimental Science*, 8 vols. (New York: Columbia University Press, 1923–58), 8:11–13.

13. Paracelsus, *De vita longa libri quinque*, in *Theophrastus von Hohenheim*, 3:274: "homunculus, quem necromantici alreonam, philosophi naturales mandragoram falso appellant, tamen non nisi in communem errorem abiit propter chaos illud, quo isti obfuscaverunt verum homunculi usum. origo quidem spermatis est; per maximam enim digestionem, que in ventre equino fit, generatur homunculus, similis ei per omnia, corpore et sanguine, principalibus et minus principalibus membris. . . ." A parallel passage is found in the German text of Paracelsus, *De vita longa*, 3:304.

14. Will-Erich Peuckert, *Handwörterbuch der Sage* (Goettingen: Vandenhoeck & Ruprecht, 1961), 1:406.

15. Anton Birlinger, *Aus Schwaben: Sagen, Legenden, Aberglauben, Sitten, Rechtsbräuche, Ortsneckereien, Lieder, Kinderreime* (1874; repr., Scientia Verlag, 1969), 1:157–71. See also Peuckert, *Handwörterbuch*, 1:406.

16. Schlosser, *Galgenmännlein*, 34–38, 40–41, 52–55.

8 Sex Ratio Theory, Ancient and Modern: An Eighteenth-Century Debate about Intelligent Design and the Development of Models in Evolutionary Biology ELLIOTT SOBER

Artificial life, in its modern meaning, contrasts with life found in nature. But in a different sense, and from the point of view of an older theory, the living things found in nature are themselves artificial—they are the results of artifice. The design argument for the existence of God asserts that organisms and the complex adaptations we observe them to have are brought into existence by intelligent design. True, the designer in question is supposed to be far more intelligent, powerful, and benevolent than mere human beings. Nonetheless, as William Paley argued in *Natural Theology*, the vertebrate eye and the watch found on the heath are both artifacts.[1]

The design argument for the existence of God took a probabilistic turn in the seventeenth and eighteenth centuries. Earlier versions, such as Thomas Aquinas's fifth way (set forth in his *Summa Theologica*, part 1, question 2, article 3), usually embraced the premise that goal-directed systems (things that "act for an end" or have a function) *must* have been created by an intelligent designer. This idea—which we might express by the slogan "no design without a designer"—survived into the seventeenth and eighteenth centuries,[2] and it is with us still in the writings of many creationists. The new version of the argument, inspired by the emerging mathematical theory of probability, removes the premise of necessity. It begins with the thought that goal-directed systems might have arisen either by intelligent design or by chance; the problem is to discern which hypothesis is more plausible. With the epistemic concept of plausibility characterized in terms of the mathematical concept of probability, the design argument was given a new direction.

The new probabilistic perspective did not extinguish the older idea of "no design without a designer." The two conflicting approaches coexisted, often with less than perfect clarity about their fundamental difference, and the details of how the probabilistic perspective ought to be articulated emerged only slowly. These characteristics of uneven development are exemplified in a debate that took place in the eighteenth century among three probabilists—John Arbuthnot, Nicholas Bernoulli, and Abraham DeMoivre—on the proper explanation of human sex ratio. Arbuthnot proposed a probabilistic version of the design argument, claiming that intelligent design, not chance, provides the better explanation of why slightly more boys than girls are born each year. Bernoulli rejected Arbuthnot's argument. DeMoivre defended Arbuthnot against Bernoulli's criticisms.

The problem of explaining sex ratio, both in human populations and in the rest of nature, experienced another transformation after 1859—it became a problem for the theory of evolution by natural selection. Darwin addressed the question in the first (1871) edition of The Descent of Man but withdrew his suggestion in the second (1874) edition. Carl Düsing (1884) provided a mathematical model that moved beyond Darwin's explanation. R. A. Fisher (1930) added the new idea of parental expenditure. George C. Williams (1966) argued that the sex ratios found in nature provide the opportunity to test hypotheses of group selection against hypotheses of individual selection. W. D. Hamilton (1967) constructed a model that generalizes Fisher's approach and represents the effects of both group and individual selection.[3] The theory of natural selection gradually developed the ability to predict sex ratios, not just explain them post hoc.

Sex ratio thus provides an interesting case study of the problem of whether we should regard living things as artifacts or as the result of mindless natural processes. If they are artifacts, then we must explain their features by describing the goals and abilities of the artificer. If they are the result of mindless natural processes, which processes are relevant, and how should they be conceptualized? The history of these ideas is not just history; these ideas are with us still in the form of the ongoing controversy about creationism and intelligent design. I briefly consider the contemporary controversy at the end of this essay.

AN EIGHTEENTH-CENTURY DEBATE

Arbuthnot on "the exact balance that is maintained between the numbers
of men and women . . . that the Species may never fail, nor perish"

John Arbuthnot, physician to Queen Anne, was the highly regarded satirist
who invented the character John Bull. He also translated Christian Huygen's
Ratiociniis in aleae ludo (Arbuthnot's title was *Of the Laws of Chance; or, A Method of
Calculation of the Hazards of Game*), to which he added a preface, whose examples, Ian Hacking says, are "characteristic of a bawdy age."[4] Jonathan Swift
said of Arbuthnot that if there had been a dozen men like him in England,
that he, Swift, could have burned *Gulliver's Travels*.[5]

Arbuthnot's "An Argument for Divine Providence, taken from the constant regularity observ'd in the births of both sexes" appeared in the *Philosophical Transactions of the Royal Society* for 1710.[6] He provides a tabulation of
eighty-two years of London christening records (see figure 8.1), noting
that more boys than girls are listed in each year. Arbuthnot takes this difference at face value; he must have realized that not every birth gets recorded,
but he nonetheless assumes that the records reflect a real difference in the

(190)

	Chriſtened.				Chriſtened.	
Anno.	Males.	Females.		Anno.	Males.	Females.
1667	5616	5322		1689	7604	7167
68	6073	5560		90	7909	7302
69	6506	5829		91	7662	7392
70	6278	5719		92	7602	7316
71	6449	6061		93	7676	7483
72	6443	6120		94	6985	6647
73	6073	5822		95	7263	6713
74	6113	5738		96	7632	7229
75	6058	5717		97	8062	7767
76	6552	5847		98	8426	7626
77	6423	6203		99	7911	7452
78	6568	6033		1700	7578	7061
79	6247	6041		1701	8102	7514
80	6548	6299		1702	8031	7656
81	6822	6533		1703	7765	7683
82	6909	6744		1704	6113	5738
83	7577	7158		1705	8366	7779
84	7575	7127		1706	7952	7417
85	7484	7246		1707	8379	7687
86	7575	7119		1708	8239	7623
87	7737	7214		1709	7840	7380
88	7487	7101		1710	7640	7288

FIGURE 8.1 | A segment of Arbuthnot's 1710 tabulation
of male and female births in London in each of eighty-
two years. Royal Society, *Philosophical Transactions*, 1710.

frequencies of male and female births. The main part of the paper is devoted to calculating the probability that this pattern would obtain if the sex ratio were due to chance. By "chance," Arbuthnot means that each birth has a probability of 1/2 of being a boy and 1/2 of being a girl. According to this hypothesis, the probability that more boys than girls will be born in a given year is the same as that for more girls than boys in that year; the Chance hypothesis also allows for a third possibility, namely, that exactly as many girls as boys will be born:

Pr(more boys than girls born in a given year | Chance) =
Pr(more girls than boys born in a given year | Chance) >>
Pr(exactly as many boys as girls born in a given year | Chance) = e.

Arbuthnot goes to the trouble of explaining how e might be calculated. The details of his calculation do not matter to the argument; the point is just that for each of the years surveyed, e is tiny.

Arbuthnot concludes that the probability of there being more boys than girls in a given year, according to the Chance hypothesis, is just under 1/2 and that the probability of there being more boys than girls in each of eighty-two years is therefore less than $(1/2)^{82}$. He further asserts that if we were to tabulate births in other years and other cities, we would find the same male bias. So the probability of all these data—both the data that Arbuthnot presents and the data that he does not have but speculates about—is "near an infinitely small quantity, at least less than any assignable fraction." The conclusion is obvious—"that it is Art, not Chance, that governs."

Arbuthnot's calculation of the probability of the observations under a concretely specified Chance hypothesis is a notable achievement.[7] Also notable is his claim that males have higher mortality than females, so that the male bias at birth gradually gives way to an even sex ratio at the age of marriage. "We must observe," he says, "that the external accidents to which males are subject (who must seek their food with danger) do make a great havock of them, and that this loss exceeds far that of the other sex, occasioned by diseases incident to it, as experience convinces us. To repair that loss, provident Nature, by the disposal of its wise creator, brings forth more males than females." At the end of the paper, Arbuthnot adds, as a scholium, that "polygamy is contrary to the law of nature and justice, and to the propagation of the human race. For where males and females are in equal number, if one man takes twenty wives, nineteen men must live in celibacy, which is repugnant to the design of nature, nor is it probable that twenty women will be so well impregnated by one man as by twenty." Here Arbuthnot shifts from explaining what is the case to urging what should be the case—a knowledge of God's intentions evidently underwrites both these claims.

What form of argument is Arbuthnot deploying? It has struck most commentators that Arbuthnot is constructing something like a Fisherian significance test, wherein a hypothesis is rejected on the ground that it says that what we observe is very improbable:

Data
Pr(Data | Chance) is tiny.

(Prob Modus Tolens)   ═══════════════════════════

We should reject the Chance hypothesis.

I draw a double line separating the conclusion from the premises of this argument to mark the fact that the argument is not deductively valid. I call this form of argument "probabilistic Modus Tolens" because it generalizes a perfectly valid principle of deductive logic:

D is true.
If C is true, then D is false.

(Modus Tolens)   ─────────────────────

C is false.

Whereas Modus Tolens tells you to reject C if something happens that it says will not, *probabilistic* Modus Tolens tells you to reject C if something happens that it says *probably* will not.

A second possible interpretation is that Arbuthnot is constructing a likelihood inference, in which two hypotheses are compared:

Data
Pr(Data | Chance) is tiny.
Pr(Data | Intelligent Design) is large.

(Likelihood Inference)   ──────────────────────────

The data favor Intelligent Design over Chance.

Here the conclusion deductively follows from the premises (and so I use a single line to separate them) once we add the following principle, which Hacking calls the Law of Likelihood:[8]

> The data favor hypothesis H_1 over hypothesis H_2 if and only if $Pr(\text{data} \mid H_1) > Pr(\text{data} \mid H_2)$.

"Favoring" means differential support; the idea is that the evidence points toward the hypothesis that renders that evidence more probable.

Should we understand Arbuthnot as employing probabilistic Modus Tolens or as constructing a likelihood inference? The question may be unanswerable, since the distinction was not one that Arbuthnot had at his fingertips. However, it may be relevant to bear in mind that probabilistic Modus Tolens requires that one attend only to a single hypothesis, whereas likelihood inference is essentially comparative, involving at least two. The fact that Arbuthnot discusses what an intelligent designer would do inclines me to view him as having likelihood instincts. Arbuthnot believes that a benevolent deity, if such a being existed, would seek to insure an even sex ratio at the age of marriage. Seeing that males die more frequently than females before the age of marriage, he would achieve his goal by ensuring a male-biased sex ratio at birth. If Arbuthnot's argument has as a premise not just the assertion that $Pr(\text{Data} \mid \text{Chance})$ is tiny but also the claim that $Pr(\text{Data} \mid \text{Intelligent Design})$ is large, then we should understand him as advancing a likelihood inference.

Separate from the question of what Arbuthnot intended, we can ask which form of inference makes more sense. Although Hacking follows standard practice in frequentist statistics in regarding *probabilistic Modus Tolens* as a valid form of inference, I am inclined to agree with Hacking's earlier point of view, and with that of Anthony Edwards and Richard Royall, that *probabilistic Modus Tolens* is invalid.[9] Perfectly plausible hypotheses sometimes confer extremely low probabilities on the observations, especially when the observations are numerous. Consider, for example, a coin that is tossed a million times. If the coin is fair, the exact sequence of heads and tails that this experiment produces has a probability of $(1/2)^{1,000,000}$, but that is no reason to reject the hypothesis that the coin is fair.[10] When he introduced the idea of significance testing, Fisher noted that the occurrence of an event that the hypothesis under test says is very improbable licenses the conclusion that a disjunction is true—either the hypothesis is false or something very improbable has occurred.[11] The disjunction is correct; what does not follow is that the hypothesis should be rejected.

A further issue, of which Arbuthnot also seems to have been unaware, became clearer some fifty years later, when Thomas Bayes's paper was published posthumously.[12] Construed as a likelihood inference, Arbuthnot's premises do not allow one to conclude that the observed sex ratio was *prob-ably* brought into being by an intelligent designer. What Arbuthnot discusses are probabilities of the form Pr(data | hypothesis), not probabilities of the form Pr(hypothesis | data). To reach a conclusion about what he calls the "probable cause" of the male bias at birth, Arbuthnot would need additional assumptions about the prior probabilities Pr(Chance) and Pr(Intelligent Design). This is because Bayes's theorem entails that

$$\text{(B)} \quad \frac{\text{Pr(Design | Data)}}{\text{Pr(Chance | Data)}} = \frac{\text{Pr(Data | Design)}}{\text{Pr(Data | Chance)}} \times \frac{\text{Pr(Design)}}{\text{r(Chance)}}$$

If the left-hand side of this equality is to be greater than unity, that is, if Pr(Design | Data) > Pr(Chance | Data), then the right-hand side must exceed unity as well. We can add a premise to the materials that Arbuthnot assembles to ensure that this is so. If Pr(Data | Intelligent Design) = i (where i is large), and Pr(Data | Chance) = j (where j is tiny), then one can conclude that intelligent design has the higher posterior probability, if one is prepared to endorse the further assumption that Pr(Design)/Pr(Chance) > j/i.[13]

Arbuthnot's argument has to be understood in a larger historical context. Like many other writers of his time and place, Arbuthnot sought to debunk Epicureanism.[14] The Epicurean hypothesis holds that particles whirling at random in the void sooner or later form stable combinations, some of which exhibit great order, complexity, and functional appropriateness. In 1735, Arbuthnot published "A Poem—Know Yourself," in which he expresses his rejection of Epicureanism by posing a series of rhetorical questions:

What am I? how produced? And for what end?
Whence drew I being? To what period tend?
Am I the abandoned orphan of blind chance,
Dropt by wild atoms in disordered dance?
Or from an endless chain of causes wrought?
And of unthinking substance, borne with thought?[15]

It takes probabilistic tools to discover what is wrong with Epicureanism. After all, it is possible, as we now would say, for monkeys pounding at random on typewriters to eventually produce the works of Shakespeare.[16] The problem is that this outcome, given some fixed number of monkeys and typewriters and a limited amount of time, is very improbable. Before typewriters were invented, other metaphors had to be found to convey this point.

Arbuthnot's friend Jonathan Swift provides a nice one in book 3 of *Gulliver's Travels*, where he describes a distinguished professor at the Grand Academy of Lagado (a stand-in for the Royal Society) who sought to "improve speculative knowledge by practical and mechanical operations." His innovation was to produce random arrangements of words by twiddling the handles of a device that resembles a foosball game (illustrated in plate 5 of *Gulliver*; reproduced here in figure 8.2). The probability of successfully generating a well-formed sentence of the language—and one that is a new and useful contribution to speculative knowledge as well—is not zero; rather, it is exceedingly tiny. It is not impossible that chance should produce this result, just very improbable.[17] When Hume has Philo invoke the Epicurean hypothesis as a possible alternative to the hypothesis of Intelligent Design in the *Dialogues Concerning Natural Religion*,[18] he is trotting out an old warhorse that almost everyone took to be risible. The probability of success via intelligent design was supposed to be much larger, which is why the evidence was taken to favor intelligent design over chance.

Bernoulli's Objection

In the second edition of his *Essai d'analyze sur les jeux de hazard*, Montmort reprints some correspondence between himself and Nicolas Bernoulli.[19] In one letter, Bernoulli says that he feels "obliged to refute" Arbuthnot's argument.[20] His central idea is that Arbuthnot did not cast his net widely enough. There are other Chance hypotheses besides the one that assigns to each male birth a probability of 1/2. Bernoulli considers the possibility that the probability is 18/35 and argues that Arbuthnot's argument falls to pieces once this new Chance hypothesis is explored. To understand Bernoulli's criticism, we must consider the two claims that Arbuthnot made about the data he cited. First, as already explained, he argued that the persistent sex ratio bias is evidence against the Chance hypothesis that the probability of a male birth is 1/2. In addition, he contended that the variation in sex ratio from year to year is too modest, if that hypothesis is correct. Although Arbuthnot did a calculation to help support the first of these claims, he did nothing to demonstrate the second. Bernoulli's crisp reply is that if the probability of a male birth is 18/35, and fourteen thousand babies are born in a given year, then the probability is 300/301 that the resulting sex ratio will fall between the upper and lower figures in Arbuthnot's table of data. Bernoulli concludes that Arbuthnot's data provide no argument at all for Divine Providence; the data are perfectly in accord with the Chance hypothesis that the probability of a male birth is 18/35.

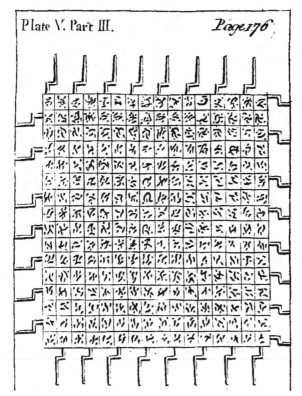

FIGURE 8.2 | Plate 5 of Swift's *Gulliver's Travels* depicts a device for randomly generating sentences, used in the Kingdom of Lagado to "improve speculative knowledge by practical and mechanical operations."

Bernoulli seems to have understood Arbuthnot as arguing via probabilistic *Modus Tolens*; Bernoulli's point was to show that the Chance hypothesis cannot be rejected on the basis of the data that Arbuthnot considers, at least not when the Chance hypothesis is formulated in the right way. However, both Arbuthnot's argument and Bernoulli's can be stated in terms of the law of likelihood. The method of *maximum likelihood estimation* instructs us to choose the estimate that maximizes the probability of the observations. To estimate the probability of a male birth at 18/35 strains our credulity less than an estimate of 1/2, because the former renders the data more probable. As it happens, 18/35 is not the best estimate either, but it is better than the one that Arbuthnot discusses.[21] Understood in this way, Bernoulli's criticism does not contradict Arbuthnot's initial point—that we should be very surprised at the eighty-two years of consistent male bias if the probability of a male birth were 1/2. However, instead of following Arbuthnot in preferring Intelligent Design to Chance = 1/2, the recasting of Bernoulli's argument as a likelihood inference instructs us to conclude only that Chance = 18/35 is preferable to Chance = 1/2. We see here a familiar property of the law of likelihood. The fact that the data favor H_1 over H_2 does not settle how each of these hypotheses compares to a third hypothesis, H_3.[22]

Arbuthnot thought that the year-to-year variation in the data is too narrow if the Chance = 1/2 hypothesis is correct. Bernoulli replied that the range of variation is perfectly in accord with the Chance = 18/35 hypothesis. According to Anscombe, both were wrong; there is too much variation, regardless of what value is assigned to the probability of a male birth.[23] Anscombe suggests that the christening records were a biased reflection of the real sex ratios; Hald bolsters this conjecture by noting that there are trends in Arbuthnot's data that point to a political explanation. Before 1642, the number of christenings is around 10,000. During the 1650s it declines to about 6,000 and then increases to about 15,000 in 1700. The years with the most extreme sex ratios were 1659–61; as Hald remarks, "the turning point is about the Restoration."[24] Apparently, the data do reflect the influence of intelligent design, though the designers were of human form.

DeMoivre's Defense of Arbuthnot—"if we blind not ourselves with metaphysical dust"

In *The Doctrine of Chances*, DeMoivre comes to Arbuthnot's defense.[25] He begins with a general affirmation of the soundness of the Design argument:

> As it is thus demonstrable that there are, in the constitution of things, certain Laws according to which Events happen, it is no less evident from Observation, that those Laws serve to wise, useful and beneficent purposes; to preserve the steadfast Order of the Universe, to propagate the several Species of Beings, and furnish to the sentient Kinds such degrees of happiness as are suited to their State.
>
> But such Laws, as well as the original Design and Purpose of their Establishment, must all be from without; the Inertia of matter, and the nature of all created Beings, rendering it impossible that any thing should modify its own essence, or give to itself, or to any thing else, an original determination or propensity. And hence, if we blind not ourselves with metaphysical dust, we shall be led, by a short and obvious way, to the acknowledgement of the great Maker and Governor of all; Himself all-wise, all-powerful and good.[26]

DeMoivre then says that Bernoulli, though "a very learned and good man . . . was led to discard and even to vilify this argument from final causes" when he rejected Arbuthnot's argument.

DeMoivre then describes Bernoulli's calculations concerning 18/35, after which he abruptly says what he thinks is wrong with Bernoulli's reasoning:

> To which the short answer is this: Dr. Arbuthnot never said "that supposing the facility of the production of a Male to that of the production of a Female already fixed to nearly the Ratio of equality, or to that of 18 to 17; he was amazed that the Ratio of the numbers of Males and Females born should, for many years, keep within such narrow bounds": the only Proposition against which Mr. Bernoulli's reasoning has any force.
>
> But he might have said, and we do still insist, that "as, from the observations, we can, with Mr. Bernoulli, infer the facilities of production of the two Sexes to be nearly in a Ratio of equality, so from this Ratio once discovered, and *manifestly serving to a wise purpose*, we conclude the Ratio itself, or if you will, the Form of the Die, to be an effect of Intelligence and Design."
>
> As if we were shewn a number of Dice, each with 18 white and 17 black faces, which is Mr. Bernoulli's supposition, we should not doubt but that those Dice had been made by some Artist; and that their form was not owing to Chance, but was adapted to the particular purpose he had in View.[27]

Just as Paley later pressed the analogy between a watch and the human eye,[28] so DeMoivre urged the similarity of a 35-sided die (with 18 white faces and 17 black) and the human reproductive machinery. Both exist only because an intelligent designer made them.

Has DeMoivre rescued Arbuthnot's line of reasoning? To answer this question, we should separate DeMoivre's argument from Bernoulli's by using the distinction that evolutionary biologists now draw between "proximate explanation" and "ultimate explanation."[29] The word *ultimate* is perhaps misleading, since the real point is to separate a relatively proximate from a more distal cause of some effect, as depicted in figure 8.3. If we begin with the sex ratio data that Arbuthnot cites, we can draw an inference about the proximate mechanism in the human reproductive system that produces the consistent pattern of male bias. This argument, based on the law of likelihood, is the one we associated with Bernoulli's reasoning: it is better to view the chance mechanism that decides a baby's sex as having a probability of 18/35 of making the baby a boy than to view that probability as having a value of 1/2. Having settled the matter of proximate mechanism, we can move to a second problem. It might

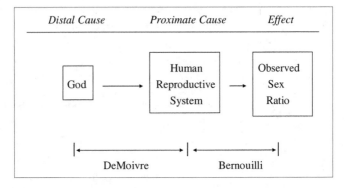

FIGURE 8.3 | Bernouilli offered a proximate explanation of Arbuthnot's sex ratio data, while DeMoivre offered a more distal explanation.

be suggested that the 18/35 chance setup is plausibly explained by the hypothesis of Intelligent Design. This is the ultimate explanation that De-Moivre proposes, but what is the nature of the argument he offers in its defense? No likelihood argument is given—DeMoivre does not compute the probability of the reproductive mechanism's having this 18/35 setting (or any other) if it arose by chance. Instead, DeMoivre invokes the older idea of "no design without a designer."

The problem with Arbuthnot's argument is that he does not keep the tasks of proximate and ultimate explanation separate. When he compares the hypotheses of Chance and Intelligent Design, the Chance hypothesis that he considers purports to describe the proximate mechanism at work, while his hypothesis of Intelligent Design provides a possible ultimate explanation.[30] This is an apples-and-oranges comparison. When the arguments are reconfigured to avoid this confusion, we have a likelihood argument concerning the proximate mechanism, which does not mention intelligent design, and DeMoivre's nonprobabilistic argument concerning the ultimate mechanism, which does. However, it is not difficult to supplement the probabilistic argument concerning proximate mechanism with a similar argument concerning ultimate explanation. What is the probability that the human reproductive system would confer on each birth a probability of 18/35 of being male if that system were the result of chance? Very low. What is the probability that this 18/35 arrangement would arise if it were produced by an intelligent designer? Very high. Likelihoods now play a role twice over.

DeMoivre was too generous to Arbuthnot. Arbuthnot *did* make claims that Bernoulli was able to confute. And the argument that DeMoivre claims is sound is not a restatement of Arbuthnot's argument, but a new argument altogether (albeit in an old style). Yet DeMoivre came to accept a compromise that Bernoulli and 'sGravesande (the person from whom Bernoulli first learned of Arbuthnot's argument) worked out together—that Bernoulli's Chance = 18/35 hypothesis explains the christening data, and Intelligent Design explains why that Chance hypothesis is true.[31]

EVOLUTIONARY SEX RATIO THEORY

These eighteenth-century discussions of sex ratio are interesting in part because of what happened to the problem after 1859. Symmetrically, the new evolutionary perspective is interesting in part because of the light it throws on the eighteenth-century debate. In what follows I do not attempt to give anything like a full account of the history of evolutionary thinking about sex ratio. Rather, I discuss some of the important episodes.

Darwin's Argument from Monogamy and His Retraction

In the first edition of The Descent of Man, Darwin begins his discussion in the section headed "Numerical Proportion of the Two Sexes" by recording what has actually been observed.[32] Although "the materials are scanty," he provides a "brief abstract" of what is known, giving more details later in a "supplementary discussion." Darwin says that his interest is to ascertain the "proportion of the sexes, not at birth, but at maturity" (264). His overall conclusion is that "as far as a judgment can be formed, we may conclude from the facts given in the supplement, that the males of some few mammals, of many birds, of some fish and insects, considerably exceed in number the females." He might have added that his supplement also describes eight groups of insect in which females outnumber males (314–15). The supplement also provides data on human populations (300–302). They are consistently male-biased at birth, although the degree of bias varies across populations. At the other end of the life span, "the females in all old-settled countries, where statistical records have been kept, are found to preponderate considerably over the males." He also notes that male mortality exceeds female mortality in utero and for the first four or five years of life. If male bias at birth gives way to female bias later in life, there must be a cross-over point at which the human sex ratio is even; perhaps this occurs "at maturity," though Darwin does not say this.

So much for the tangle of observations—imperfect as they are and failing to exemplify any single, simple, pattern. How should these observations be explained? Darwin begins his discussion of this question under the heading "On the Power of Natural Selection to Regulate the Proportional Numbers of the Sexes, and General Fertility" by noting that an uneven sex ratio might be "a great advantage to a species," and "might have been acquired through natural selection, but from their rarity they need not here be further considered." "In all ordinary cases," he adds, "an inequality would be no advantage or disadvantage to certain individuals more than to others; and therefore it could hardly have resulted from natural selection" (315–16). It is difficult to know how seriously to take Darwin's contrast between advantage to a species and lack of advantage to any individual. A modern biologist might be tempted to conclude that Darwin is saying that group selection can produce an uneven sex ratio but that purely individual selection cannot, and that Darwin is announcing his skepticism about group selection and advocating a purely individualistic approach. Is this interpretation Whiggish? As we will see, Darwin's explanation of even sex ratios invokes what is now usually classified as individual, not group, selection. But even so, we should remember that, in other contexts, Darwin slides easily between talking about benefit to the species and benefit to individuals, eliding a distinction that post-1960s biology has found to be exceedingly important. So perhaps we should not make too much of the present passage. What we can say is that Darwin chooses to focus on how natural selection might produce an even sex ratio, even though he grants that selection in some circumstances might yield an uneven sex ratio, and even though his data include many species in which the sex ratio is uneven.[33] One is reminded of the old saw about the man who looks for his keys under a lamppost. When asked why he is searching there, he answers that that is where the light is.

Having identified an even sex ratio at maturity as his proper *explanandum*, Darwin then states his argument for why "natural selection will always tend, though sometimes inefficiently, to equalise the relative numbers of the two sexes" (318):

> Let us now take the case of a species producing . . . an excess of one sex—we will say of males—these being superfluous and useless, or nearly useless. Could the sexes be equalized through natural selection? We may feel sure, from all characters being variable, that certain pairs would produce a somewhat less excess of males over females than other pairs. The former supposing the actual number of the offspring to remain constant, would necessarily produce more females, and would therefore be more productive. On the doctrine of chances a greater number of the offspring of the more productive pairs would survive; and these

would inherit a tendency to procreate fewer males and more females. Thus a tendency towards the equalization of the sexes would be brought about. (316)

What does Darwin mean by an "excess" of one sex, where this excess can be due to there being more males than females, or more females than males? He seems to be assuming that reproduction is purely monogamous. When individuals form up into mating pairs, each individual has exactly one partner if the sex ratio is even. However, if the sex ratio is uneven, the formation of mating pairs will prevent some members of the majority sex from finding a partner. These unpaired individuals are what Darwin means by "excess." The assumption of monogamy requires that some individuals must fail to reproduce if the sex ratio is uneven.[34] Darwin briefly remarks that his argument applies to polygamous species, "if we assume the excess of females to be inordinately great" (317).[35] His point seems to be that selection will reduce a very extreme degree of female bias to one that is more modest if a polygamous mating scheme is already in place. For example, if each male has three mates, then there will be "excess females" when there are more than 75 percent females at reproductive age, and so selection will reduce that greater figure to 75 percent, whereupon there is no longer an "excess." This interpretation of what Darwin meant by the argument's applying to polygamy clashes with his commitment to address only the evolution of an even sex ratio, but I cannot see any other way to interpret it.

Does Darwin's explanation invoke individual or group selection? As noted above, this distinction is more vivid now than it sometimes was in Darwin's own writings, though it certainly is very clear in some of what he wrote. The assumption of monogamy entails that a group with an even sex ratio will be more productive than a group of the same size that has an uneven sex ratio; given monogamy, an even sex ratio thus provides an advantage to the group. Though this is true, it is not Darwin's argument. Darwin is thinking about a single group and asks us to consider the different mating pairs within it. He seeks to identify the best reproductive strategy that a parental pair can use in determining its mix of sons and daughters. If the next generation contains an excess of males, the best strategy for a mating pair is to have more daughters, and if the next generation contains an excess of females, the best strategy is to have more sons. If the mix of sons and daughters produced by a parental pair were regulated just by the mother, or just by the father, it would be perfectly clear that this is an argument about individual advantage. The selection takes place among the individuals in a single population, not among populations, so no group selection is involved. But what if both parents influence their mix of daughters and sons? Is selection among the mating pairs in a population an instance of group selection, the groups each containing two members? Contemporary biologists who dislike the idea of group selection will be loath to see mating pairs as "groups" in the sense of group selection, but those more comfortable with the idea will see nothing amiss in classifying Darwin's model in this way.[36] This fine point aside, it is clear that Darwin's argument does not invoke "the good of the species" in any substantive sense.

In the second edition of *The Descent of Man*, Darwin retracts his analysis and substitutes a disclaimer: "I formerly thought that when a tendency to produce the two sexes in equal numbers was advantageous to the species, it would follow from natural selection, but I now see that the whole problem is so intricate that it is safer to leave its solution to the future."[37] He admits that although there are circumstances in which one sex ratio or another would be advantageous to the species, "in no case, as far as we can see, would an inherited tendency to produce both sexes in equal numbers or to produce one sex in excess, be a direct advantage or disadvantage to certain individuals more than to others; . . . and therefore a tendency of this kind could not be gained through natural selection." Here again one needs to consider whether Darwin meant to contrast what is good for the species with what is good for the individual or was using these phrases interchangeably. But even if Darwin is demanding that the evolution of sex ratio be explained without invoking the good of the species,[38] it is not clear why this led him to doubt his earlier argument.

Darwin does not state his reasons for retracting, but chapter 20 of *The Descent of Man* provides a good reason for him to do so. Darwin notes that although orangutans are monogamous, gorillas, chimps, and baboons are not. And the same holds of human beings, both present and past: "Judging from the social habits of man as he now exists, and from most savages being polygamists, the most probable view is that primeval man aboriginally lived in small communities, each with as many wives as he could support and obtain, whom he would have jealously guarded against all other men. Or he may have lived with several wives by himself, like the Gorilla" (pt. 2, 361–62). This supplements his earlier remark that "many mammals and some few birds are polygamous" (pt. 1, 266). A fundamental feature of Darwin's approach in the first edition is that mating scheme explains sex ratio: monogamy leads an even sex ratio to evolve, and polygamy leads an uneven sex ratio to evolve. If there are polygamous species with even sex ratios, this is a problem for his account.[39] Darwin did not have to look far from home to find such problem cases.

The Düsing Model: Monogamy Drops Out

Biologists nowadays often think of R. A. Fisher as having been the first biologist to grapple substantively with the problem of explaining sex ratio in terms of natural selection, but this is a mistaken impression. We have already seen that Darwin had a substantive and interesting take on the problem. Unfortunately, Fisher mentions Darwin's later bewilderment, but not his earlier clarity.[40] And Fisher does not mention the work of Carl Düsing at all; Düsing produced an algebraic argument that selection will lead to the evolution of an even sex ratio at reproductive age.[41] Seger and Stubblefield say that Düsing's account is "perhaps the first mathematical model in evolutionary biology."[42] How is Düsing's argument related to Darwin's formulation? Edwards and Seger and Stubblefield see Düsing as putting into mathematical language the ideas that Darwin had thought through only qualitatively in his first edition.[43] I see a difference in substance, not just in the introduction of algebraic symbols. As I have emphasized, Darwin's 1871 explanation of an even sex ratio rests on the assumption of monogamy. Düsing does without that assumption.

Düsing's initial idea makes explicit what was only implicit in Darwin—that we need to consider three generations: (1) the parental generation, (2) the generation of offspring, and (3) the generation of grandoffspring. Here is a simplified version Düsing's argument. If there are m males and f females in generation 2, and if they together produce the N individuals who exist in generation 3, then the average male in generation 2 has N/m offspring in generation 3, and the average female in generation 2 has N/f offspring in generation 3. This means that the minority sex in generation 2 will be more reproductively successful. A parent in generation 1 who wishes to maximize the number of grandoffspring she has in generation 3 should therefore produce offspring in generation 2 who are exclusively of the minority sex. An equilibration process is thus set in motion—if the population is male-biased, selection will favor the overproduction of females, and if it is female-biased, selection will favor the overproduction of males. The population reaches equilibrium when the sex ratio at reproductive age is even, at which time a parent who has one mix of daughters and sons does no better or worse than a parent who has any other. Only when the sex ratio is uneven will there be fitness differences among the different possible "sex ratio strategies" (a modern term) that a parent might follow.

Notice that there is no need to impose the requirement of monogamy to see what happens to the average male and the average female in generation 2. Even with polygamy, it remains true that the male average is N/m, and the female average is N/f. True, some males have more than N/m offspring, while others have none at all, and it also may be true that the female variance is less than the male variance. But Düsing saw that this does not matter: "Whilst the female sex shows a much greater constancy in the strength of reproduction, the widest variation may occur in the case of the male individuals. But in our calculation it is not a matter of how far any extremes deviate, but what the average number of offspring is, and this number is of the same magnitude for male and female individuals at normal [i.e., even] sex ratios." Here Düsing underscores his earlier remark that "it is true that in each individual case [the number of offspring produced] is subject to considerable variation, but if one wants to illustrate and calculate the total effect in an example one must naturally use the average number."[44] Whereas Darwin considers what happens to *each* male and *each* female in the second generation, Düsing considers just the *average* male and the *average* female.

When there is monogamy, Darwin and Düsing both think of natural selection as "aiming" at an even sex ratio at reproductive age. If males die more frequently than females before reproductive age, Darwin and Düsing both can make sense of the fact that there is a male-biased sex ratio at birth. But Darwin and Düsing part ways over polygamous species that have even sex ratios at reproductive age; these are a problem for Darwin, but not for Düsing. On the other hand, an uneven sex ratio at reproductive age is a problem for Düsing, whereas Darwin's argument makes room for uneven sex ratios when there is polygamy.

Fisher and Parental Expenditure

Fisher gives a characteristically compressed treatment of the evolution of sex ratio in his landmark book *The Genetical Theory of Natural Selection*.[45] Like Darwin's account, Fisher's is purely verbal, with no mathematical symbols in sight. How is Fisher's theory related to the theories of Darwin and Düsing? Edwards says that Düsing gave a mathematical account "based on the same argument that Darwin had advanced," and that Fisher "gave a verbal account of the argument in *The Genetical Theory of Natural Selection*."[46] I have already tried to separate Düsing's account from Darwin's; I now want to separate Fisher's from Düsing's.

The simple point is that Fisher's and Düsing's models predict different equilibrium values—different endpoints toward which the process of natural selection will tend to "push" a population. Whereas Düsing argues that natural selection will lead to an even sex ratio at "the time of reproduction,"[47] Fisher's fundamental idea is that natural selection leads to equal "parental expenditure," and equal expenditure need not manifest itself as equal numbers of the two sexes at reproductive age. As already noted, Düsing had room in his account for male-biased sex ratios at birth. What Düsing rules out, but Fisher does not, is a biased sex ratio at reproductive age.

Fisher's concept of parental expenditure encompasses both the creation of sons and daughters and the rearing of those offspring to independence. If each mother in a population has a package of resources to spend on creating and sustaining her sons and daughters, what percentages of that package should she devote to sons and to daughters? That is, what division will natural selection favor? The standard way to think about this question is to suppose that all the parents in a population use one division of resources and then to ask when a novel parent using a different division will do better, in the sense of having higher fitness, which in this context means having more grandoffspring.

Suppose that each mother in the population has a total package of resources T and that she devotes pT to build and maintain her sons and $(1-p)T$ to build and maintain her daughters. She does this during a time that begins with conception and ends when her children begin to live independently; perhaps a bit more time passes before her offspring reproduce. Suppose that each mother in the population spends c_m on the average son and c_f on the average daughter, and that the average son brings her b_m units of benefit (in terms of providing her with grandoffspring), and the average daughter brings her b_f. Then a mother's total fitness, taking account of both her costs and her benefits, is

$$b_m[pT/c_m] + b_f[(1-p)T/c_f].$$

The symbols inside the square brackets represent the number of sons and daughters she has, and the whole expression sums the benefit she receives from her sons and the benefit she receives from her daughters.

The evolutionary question can be posed as follows: When will a mutant mom have higher fitness than these resident females by producing some other mix (p^* and $[1-p^*]$) of sons and daughters? The mutant mom's total fitness is

$$b_m[p^*T/c_m] + b_f[(1-p^*)T/c_f],$$

so our question is: When it will be true that

$$b_m[p^*T/c_m] + b_f[(1-p^*)T/c_f] > b_m[pT/c_m] + b_f[(1-p)T/c_f].$$

This simplifies to

$$(b_m/c_m - b_f/c_f)(p^*-p) > 0.$$

For the product on the left-hand side to be positive, both product terms must be positive or both must be negative. This means that the mutant mom has higher fitness than a resident mom precisely when either

$$[b_m/c_m > b_f/c_f \text{ and } p^* > p] \text{ or } [b_m/c_m < b_f/c_f \text{ and } p^* < p].$$

In other words, if sons have a higher benefit-to-cost ratio than daughters, then for the mutant mom to be fitter than the others, she should produce a higher proportion of sons than the resident moms do; on the other hand, if sons have a lower ratio than daughters, then the mutant should produce a lower proportion of sons than the residents produce. But notice something simpler: the mutant cannot do better than a resident if $b_m/c_m = b_f/c_f$. When this equality holds, the selective evolution of sex ratio stops. But what mix of males and females does this equality represent?

To answer this question, we have to further clarify the meaning of the benefit terms b_m and b_f. Suppose that the offspring generation produced by the original generation of parents has m males and f females at the end of the period of parental care. Not all of these individuals need live to reproductive age, but the fact remains that if there are N offspring in the third generation, the average second-generation male at the end of the period of parental care has N/m offspring, and the average female has N/f. These are the benefits that sons and daughters provide. So the mutant mom will have the same fitness as the residents precisely when

$$N/mc_m = N/fc_f.$$

Fisher's principle of equal expenditure is simply the idea that $mc_m = fc_f$; at equilibrium, the parental population's total investment in sons equals its total investment in daughters. Notice that this expression also can be written as

$$m/f = c_f/c_m.$$

An even sex ratio (m = f) at the end of the period of parental care is an equilibrium if $c_f = c_m$, but if $c_f > c_m$, the equilibrium will require more males than females. The investment made in the average male will be less than the investment made in the average female if males have a higher mortality rate during the period of parental care. So greater male mortality during the period of dependence predicts a male-biased sex ratio at the end of that time.

It may seem that Düsing's model is a special case of Fisher's, with Düsing's derived from Fisher's by letting $c_f = c_m$. This is not correct. Düsing calculates the sex ratio "at the time of reproduction,"[48] whereas Fisher's model predicts the sex ratio at the time of independence, which may come earlier. Stipulating that $c_f = c_m$ in Fisher's model entails an even sex ratio at the end of the period of parental care, which is not what Düsing claims.[49] As Fisher notes, "The sex ratio at the end of the period of expenditure thus depends upon differential mortality during that period. ... [I]t will not be influenced by differential mortality during a self-supporting period; the relative numbers of the sexes attaining maturity may thus be influenced without compensation by differential mortality during the period intervening between the period of dependence and the attainment of maturity."[50]

Natural selection may of course reduce the rates of mortality that each sex experiences between independence and maturity, but this is irrelevant to how natural selection affects the mix of sons and daughters that parents produce. For Fisher, selection for sex ratio should be understood in terms of a prediction concerning the sex ratio that obtains at the age of independence, not the sex ratio that obtains at reproductive age. Darwin and Düsing had addressed the wrong question.

Williams and Hamilton: Group versus Individual Selection

The subject of sex ratio evolution took another step forward in 1966, when George C. Williams published *Adaptation and Natural Selection*. Williams's main subject was the difference between the processes of group selection and individual selection. As noted earlier, group selection occurs when there is competition among groups; it promotes the evolution of traits that reduce a group's risk of extinction and increase its productivity. Individual selection occurs when there is competition among the individuals in a group; it promotes the evolution of traits that enhance the viability and fertility of the individuals that have them. The concept of altruism shows that there can be a conflict of interest between what is good for the group and what is good for an individual. Consider a now-standard (and simplified) example: it may help a group of prairie dogs if an individual issues a warning cry when a predator approaches, but the warning cry may endanger the sentinel by attracting the attention of the predator; from that sentinel's selfish point of view, it does better by remaining silent. Group selection may lead to the evolution of sentinel behavior, but individual selection will drive that altruistic behavior to extinction.

Williams recognized that Fisher's treatment of sex ratio uses the idea of individual selection. As already noted, Fisher asked what *an individual parent* should do to maximize the number of grandoffspring she has. Williams then posed a different question: What sex ratio should we observe if the trait evolved by the process of group selection? His answer was that individuals should be able to adjust the mix of sons and daughters they produce. When groups have the opportunity to grow larger, there should be a female-biased sex ratio, since this maximizes group productivity. But when the population is in danger of overexploiting its environment and crashing to extinction, the population should shift to a male-biased sex ratio, since this will reduce its productivity. Williams's idea was that the hypothesis of group selection could be tested against the hypothesis of individual selection by seeing what sex ratios are present in nature. He reports that "in all well-studied animals of obligate sexuality, such as man, the fruitfly, and farm animals, a sex ratio close to one is apparent at most stages of development in most populations."[51] This statement is incorrect, not just as a claim about nature, but as a claim about what was known about nature in 1966. In any event, Williams concluded that sex ratio data provide strong empirical evidence against group selection. It is curious that Williams saw no difficulty in placing this empirical argument side by side with other arguments against group selection that are nearly *a priori* in character.[52] The purpose of *Adaptation and Natural Selection* was to decisively discredit the idea of group selection. For Williams, the hypothesis was not just factually mistaken—it was a product of fuzzy thinking and a deplorable encouragement for more of the same.[53]

A year later W. D. Hamilton published his landmark paper "Extraordinary Sex Ratios," in which he provided a mathematical model according to which evolutionary outcomes very different from the ones treated by Fisher can occur.[54] Unlike Williams, Hamilton was aware that the sex ratios in many insects are strongly female-biased. His approach to understanding how this arrangement might evolve was to consider a hypothetical species of parasitic wasp in which one or more fertilized females lay eggs in a host; when the eggs hatch, the offspring reproduce with each other, and the fertilized females then take flight to find new hosts to parasitize. Hamilton asked the same three-generation question that Darwin, Düsing, and Fisher posed, but he got a different answer. Suppose each host is parasitized by a single fertilized female. What mix of sons and daughters should she produce in order to maximize her number of grandoffspring? Clearly, she should produce the smallest number of sons needed to fertilize all her daughters. Because there is strong inbreeding, a female-biased sex ratio will evolve.

Hamilton's point was not to refute Fisher's model but to show that it rests on special assumptions. Fisher's model is correct when there is random mating in a large population. In fact, Hamilton's model has Fisher's as a special case. Hamilton's model describes what sex ratio will evolve if groups are founded by one, two, or n fertilized females. As the number of foundresses is increased, the predictions of Hamilton's model get closer to Fisher's.

Although Hamilton's model was framed in terms of inbreeding and local mate competition, it also can be framed in terms of within- and between-group selection, as Hamilton noted in a footnote and Williams later acknowledged.[55] In Hamilton's model, pure group selection favors a strongly female-biased sex ratio—this is what maximizes group productivity; pure individual selection favors the Fisherian solution of equal investment in the two sexes.[56] In the real world, organisms may simultaneously experience both group and individual selection; if there is both competition among groups and competition within groups, the resulting sex ratio will be a compromise between what happens in the two pure cases. An example in which groups are founded simultaneously by two fertilized females is analyzed in the appendix.

EVOLUTIONARY REFLECTIONS ON THE EIGHTEENTH-CENTURY DEBATE

According to the evolutionary ideas just summarized, what is good for the group (a female-biased sex ratio, if the group does best by maximizing its productivity) differs from what is good for the individual (equal investment in sons and daughters). With this thought in mind, we can return to Arbuthnot's argument and ask a new question. An even sex ratio at reproductive age, according to Arbuthnot, is good for the species ("that it may not perish") and good for individuals as well. Arbuthnot, I take it, did not know that there are many insect species in which the sex ratio is female-biased, but we nonetheless can ask what he and his latter-day epigones would say about this curious situation. Was God malevolent when he made these insects thus? Intelligent Design theorists, both ancient and modern, can be expected to reply in the negative—God was benevolently disposed to human beings when he gave us an (almost) even sex ratio and benevolently disposed to those insects when he made them female-biased. But what do Intelligent Design theorists mean by *benevolence* when the term is used so flexibly? To begin with, we cannot automatically assume that God would think of what is good for insects or human beings in terms of what maximizes the number of grandoffspring. And even if the concept of goodness is given a purely reproductive reading, it still is not clear what *good* means. Does it mean good-for-the-individual, or good-for-the-species, or good-for-the-ecosystem? The term *good*, as used by Intelligent Design theorists, apparently means good-for-something-they-know-not-what; the word is irremediably vague. The hypothesis that an omnipotent, omniscient, and benevolent intelligent designer produced the sex ratios of different species in fact makes no predictions at all about the sex ratios we should observe.[57] One half of Arbuthnot's likelihood argument—his claim that Pr(Data | Intelligent Design) is large—involves an undefended, and still indefensible, assumption. Arbuthnot *assumes* that God wants the sex ratio at the age of marriage to be even, but how does Arbuthnot know that? And even if God does want this outcome, why does he achieve it by making the sex ratio at birth uneven, rather than by reducing the male mortality rate, or increasing the female rate? An answer to this question can doubtless be *invented*; the point is that there is no *independent evidence* that the invented story is *true*.

Modern sex ratio theory, in contrast, makes testable predictions about the sex ratios we should observe. However, those predictions need to be understood probabilistically. It would be a mistake to think that Fisher's model predicts that random mating will *always* be associated with equal investment, or that Hamilton's model predicts that strong inbreeding will *always* be associated with female bias. Nor would it be correct to substitute "usually" for "always." Rather, we should regard the models as saying that different breeding structures are "positive causal factors" for different sex ratios.[58] Random mating raises the probability of equal expenditure, and inbreeding raises the probability of female bias. Breeding pattern is to sex ratio as smoking is to cancer.

The two causal hypotheses may be tested in the same way. The prediction for the smoking hypothesis is that smokers should get cancer more often than nonsmokers who are otherwise similar; the prediction for the sex ratio hypotheses is that this or that sex ratio should occur more frequently in populations with one breeding structure than it does in populations with another that are otherwise similar. Spelling out how this test should be structured would bring in a number of interesting complications, but I lack space here to provide details; in any event, they are irrelevant for present purposes, since the main point is this: The models that currently comprise evolutionary sex ratio theory can be tested by observing breeding structures and sex ratios (as well as other biological variables) in different populations. It might turn out that a model makes accurate predictions about sex ratios in one group of species but inaccurate predictions about sex ratios in another. Evolutionary theory involves no *a priori* commitment to the thesis that a trait found in different branches of the tree of life always evolves for the same reason. If contemporary sex ratio theory makes inaccurate predictions about a group of organisms, other evolutionary explanations will have to be found, and they will have to be testable if they are to pass scientific muster.

Historians will probably be less interested in criticizing eighteenth-century sex ratio theory than in understanding its conceptual structure. Even so, the evolutionary perspective developed in the twentieth century helps us recognize an important property of Arbuthnot's thinking. This eighteenth-century writer, like many of his contemporaries, was fundamentally committed to the idea of the harmony of nature. That commitment had the effect of rendering conflicts of interest invisible. Arbuthnot thought that what is good for each individual is also good for the species, just as Adam Smith thought that if each individual in a market economy pursues his own selfish advantage, a side effect, wrought "as if by an invisible hand," would be an improvement in collective well-being.[59] Although Darwin often fell into the rhetoric of assuming that traits that help an individual survive and reproduce are automatically good for the species, the logic of his theory undermines that automatic assumption.[60] Selection processes at different levels of organization promote different evolutionary outcomes.[61] Individual selection can cause a population to evolve a configuration that drives it straight to extinction, just as group selection can lead to a configuration in which some individuals drastically reduce their own prospects for surviving and reproducing. When both processes influence the evolution of a given trait, the character of the resulting compromise will depend on contingent details. Those who believe only in individual selection may think of natural selection as "aiming" exclusively at the evolution of traits that promote an individual's survival and reproduction. However, from the point of view of multilevel selection theory, there is no such thing as the one and only kind of trait that selection always promotes.[62] The idea of conflicts of interest makes it much harder to think of the living world as due to a designer's benevolence, not because there is so much moral evil in the world (though there is), but because we do not know what benevolence even means in this connection.

APPENDIX: EXAMPLE OF HAMILTONIAN SEX RATIO EVOLUTION IN GROUPS WITH TWO FOUNDRESSES

Suppose there are two sex ratio strategies a female wasp can follow: Even (producing five sons and five daughters) and Biased (producing one son and nine daughters). If each group is founded by two fertilized females who simultaneously lay their eggs, there are three types of group—those founded by two Even females, those founded by an Even and a Biased female, and those founded by two Biased females. After the eggs in a nest hatch, there is mating exclusively among nest-mates; the fertilized females then disperse, and pairs of fertilized females form at random and establish a new generation of nests, as before. Notice that the three types of group produce different numbers of grandoffspring: 100 (two Evens), 140 (one Even and one Biased), and 180 (two Biased). The Biased trait is advantageous to the group; the more Biased foundresses the better, as far as group productivity is concerned. Group selection therefore favors the evolution of this trait.

What is the nature of the selection process that occurs at the individual level—at the level of individuals who live in the same group? Homogeneous groups (Even-Even and Biased-Biased) contain no variation, so no individual selection occurs there. However, individual selection does occur in mixed groups. When an Even and a Biased individual together found a group, which will have the larger number of grandoffspring? The Even female has 5 of the 6 sons in the group and 5 of the 14 daughters. The Biased female has 1 of the 6 sons and 9 of the 14 daughters. If the 6 males and 14 females in the second generation mate at random, what will be the expected pedigrees of the individuals in the third generation? That is, to which of the two foundresses can they be expected to trace back? These expected pedigrees are depicted in figure 8.4.

The grandoffspring in the upper-right and lower-left cells of figure 8.4 trace back to both foundresses (to one through their father and to the other through their mother), while the grandoffspring in the upper-left cell come exclusively from the Even foundress, and those in the lower-right cell come exclusively from the Biased foundress. It should be clear from this that the Even foundress has been more successful in producing grandoffspring.

		Pedigree frequencies of females in the second generation	
		5/14 Even	9/14 Biased
Pedigree frequencies of males in the second generation	5/6 Even	25/64	45/64
	1/6 Biased	5/64	9/64

FIGURE 8.4 | A group is founded by two fertilized females; the Even female has five sons and five daughters, while the Biased female has one son and nine daughters. If there is random mating within this offspring generation, what pedigrees should we expect the individuals in the grandoffspring generation to have? The four cells in the illustration indicate the percentage of individuals in the third generation that can be expected to trace back to just one foundress, or just to the other, or to both.

Selection at the individual level—within mixed groups—therefore favors the Even trait.[63]

It follows that Even is a selfish trait, and Biased is altruistic. The Biased trait enhances the fitness of the group, but is disadvantageous to individuals in mixed groups.[64]

NOTES

I am grateful to David Bordwell, James Cortada, Anthony Edwards, Daniel Hausman, Jonathan Hodge, John Lazarus, Ann Pollock, Armin Schulz, Howard Weinbrot, David S. Wilson, James L. Wood, Ed Zalta, and especially Steve Orzack for useful suggestions.

1. W. Paley, Natural Theology, or, Evidences of the Existence and Attributes of the Deity, Collected from the Appearances of Nature (London: Rivington, 1802).

2. For example, Newton says, in a letter to Bentley, that the fact that the planets all orbit in the same direction and in the same plane "could not spring from any natural cause alone, but were impressed by an intelligent Agent" (H. David and A. Edwards, eds., Annotated Readings from the History of Statistics [Berlin: Springer, 2001], 8). He repeats this argument in the 2nd edition of the Opticks, as does Bentley in his inaugural Boyle lectures.

3. C. Darwin, The Descent of Man and Selection in Relation to Sex (London: Murray, 1871; 2nd ed., 1874); C. Düsing, Die Regulierung des Geschlechtsverhältnisses (Jena: Fischer, 1884). Selections from this work are translated in A. Edwards, "Carl Düsing on the Regulation of the Sex-Ratio," Theoretical Population Biology 58 (2000): 255–57. See also R. Fisher, The Genetical Theory of Natural

Selection (1930; repr., New York: Dover, 1957); G. C. Williams, *Adaptation and Natural Selection* (Princeton, NJ: Princeton University Press, 1966); and W. Hamilton, "Extraordinary Sex Ratios," *Science* 156 (1967): 477–88.

4. I. Hacking, *The Emergence of Probability* (Cambridge: Cambridge University Press, 1975), 166.

5. H. H. Williams, *The Correspondence of Jonathan Swift* (Oxford: Clarendon Press, 1963–65), 3:104; M. Kendall, "Measurement in the Study of Society," in *Studies in the History of Statistics and Probability*, ed. M. Kendall and R. Planckett (London: Charles Griffin, 1972), 35.

6. J. Arbuthnot, "An Argument for Divine Providence, taken from the constant regularity observ'd in the births of both sexes," *Philosophical Transactions of the Royal Society* 27 (1710): 186–90. Reprinted in Kendall and Planckett, *History of Statistics and Probability*, 30–34.

7. L. Beattie, in *John Arbuthnot: Mathematician and Satirist*, Harvard Studies in English, vol.16 (Cambridge, MA: Harvard University Press, 1967), 341–42, shows that Arbuthnot got his data and conclusion from John Graunt's *Natural and Political Observations, Mentioned in a following Index and made upon the Bills of Mortality* (1662), though Beattie concedes that Arbuthnot's method of reasoning was original.

8. Ian Hacking, *The Logic of Statistical Inference* (Cambridge: Cambridge University Press, 1965), 59.

9. Hacking, *Emergence of Probability*, 168; Hacking, *Statistical Inference*; A. Edwards, *Likelihood* (Cambridge: Cambridge University Press, 1972); R. Royall, *Statistical Evidence: A Likelihood Paradigm* (Boca Raton, FL: Chapman and Hall, 1997); E. Sober, "Intelligent Design and Probability Reasoning," *International Journal of Philosophy of Religion* 52 (2002): 65–80, and "The Design Argument," in *The Blackwell Guide to the Philosophy of Religion*, ed. W. Mann (Oxford: Blackwell, 2004), 117–47.

10. Defenders of significance testing will want to say that what should be considered as a description of the experimental outcome is not the exact sequence of heads and tails but just the proportion of heads. Why this weakened description of the evidence captures everything that is relevant is a mystery from the point of view of that testing philosophy, though attending just to the proportions can be justified from the likelihood point of view (see Hacking, *Statistical Inference*, 80–81).

11. R. Fisher, *Statistical Methods and Scientific Inference* (London: Oliver and Boyd, 1956), 39.

12. Thomas Bayes, "An Essay Toward Solving a Problem in the Doctrine of Chances," *Philosophical Transactions of the Royal Society of London* 53 (1763): 370–418.

13. Proposition (B) provides an additional reason to regard probabilistic Modus Tolens as invalid. The fact that $Pr(Data \mid H_1)$ is tiny does not settle whether $Pr(Data \mid H_2)$ is even smaller, nor does it settle how the prior probabilities of the two hypotheses are related. This means that probabilistic Modus Tolens can lead one to reject H_1 and to not reject H_2, even though H_1 has the higher posterior probability. It also means that this form of inference can lead one to reject each of an exhaustive set of hypotheses if each one says that the observations are very improbable. Frequentists should not dismiss these objections on the grounds that they are "Bayesian." I say this because frequentists ought to grant that prior probabilities have an objective basis in *some* situations (even if not in all), and that is enough to lend weight to these criticisms.

14. T. Mayo, *Epicureanism in England* (College Station, TX: Southwest Press, 1934).

15. H. David and A. Edwards, "The First Formal Significance Test: Comments on Arbuthnott 1710," in David and Edwards, *Annotated Readings from the History of Statistics*, 11.

16. The earliest source I have been able to find for the monkeys-and-typewriters analogy is E. Borel, "Mécanique Statistique et Irréversibilité," *Journal de Physique*, 5e série, 3 (1913): 189–96: "Concevons qu'on ait dressé un million de singes à frapper au hasard sur les touches d'une machine à écrire et que, sous la surveillance de contremaîtres illettrés, ces singes dactylographes travaillent avec ardeur dix heures par jour avec un million de machines à écrire de types variés. Les contremaîtres illettrés rassembleraient les feuilles noircies et les relieraient en volumes. Et au bout d'un an, ces volumes se trouveraient renfermer la copie exacte des livres de toute nature et de toutes langues conservés dans les plus riches bibliothèques du monde. Telle est la probabilité pour qu'il se produise pendant un instant très court, dans un espace de quelque étendue, un écart notable de ce que la mécanique statistique considère comme la phénomène le plus probable" (192). A. Eddington had the same thought when he said that "if an army of monkeys were strumming on typewriters they might write all the books in the British Museum" (*The Nature of the Physical World* [Cambridge: Cambridge University Press, 1928], 72). The analogy is sometimes attributed to Huxley, in his 1860 debate with Wilberforce, but there is no transcript of that debate to consult.

17. Richard Bentley also constructed an argument from "linguistic combinatorics" to deride the atheistic Epicurean hypothesis. In his inaugural Boyle lectures of 1692, he asks what the probability would be that a male and a female of the same species should each arise by chance. He answers by proposing an analogy, derived from Cicero's *Natura Deorum*, between the gigantic number of sequences that can be constructed from the Latin alphabet of twenty-four letters and the still greater number of arrangements there can be of the one thousand or more parts that comprise the human body. See E. Shoesmith, "The Continental Controversy over Arbuthnot's Argument for Divine Providence," *Historia Mathematica* 14 (1987): 133–46.

18. D. Hume, *Dialogues Concerning Natural Religion*, ed. N. Kemp Smith (1779; repr., Oxford: Oxford University Press, 1935).

19. P. Montmort, *Sur les jeux de hazard*, 2nd ed. (1713; repr., New York: Chelsea, 1980), 371–75, 388–94.

20. Montmort, *Essai d'analyze*, quoted in I. Todhunter, *A History of the Mathematical Theory of Probability* (New York: Chelsea Publishing, 1865), 130–31.

21. See E. Shoesmith, "Nicholas Bernoulli and the Argument for Divine Providence," *International Statistical Review* 53 (1985): 255–59.

22. Although he does not fault Arbuthnot for using probabilistic Modus Tolens, Hacking does think that Arbuthnot reasons invalidly when he infers that the hypothesis of intelligent design is true (see Hacking, *Emergence of Probability*, 167). Note that Arbuthnot is rescued from this criticism if he is interpreted as making a likelihood argument in which Intelligent Design and Chance = 1/2 are compared; his mistake would then be that he failed to consider a third hypothesis, not that he reasoned invalidly from his premises.

23. F. Anscombe, *Computing in Statistical Science through APL* (Berlin: Springer-Verlag, 1981), 301.

24. A. Hald, *A History of Probability and Statistics and Their Applications Before 1750* (New York: John Wiley, 1990), 284.

25. A. DeMoivre, *The Doctrine of Chances*, 3rd ed. (London: Millar; New York: Chelsea, 1967), 252–54.

26. Ibid., 252.

27. Ibid., 253.

28. Paley, *Natural Theology*.

29. E. Mayr, "Cause and Effect in Biology," *Science* 134 (1961): 1501–6.

30. Arbuthnot's favored hypothesis is not that an intelligent designer intervenes directly in each birth; rather, he says that "there seems no more probable cause to be assigned in physicks for this equality of births, than that in our first Parents seed there were at first formed an equal number of both sexes" ("An Argument for Divine Providence"). Arbuthnot's preformationism is interesting here; his reference to "equality" seems to be a slip.

31. Shoesmith, "Continental Controversy," 144.

32. Darwin, *The Descent of Man*, 1st ed., 263ff. Subsequent citations of this edition are given by page number in the text.

33. S. Orzack, "Sex Ratio and Sex Allocation," in *Evolutionary Ecology: Concepts and Case Studies*, ed. C. Fox, D. Roff, and D. Fairbairn (New York: Oxford University Press, 2001), 169–70.

34. See A. Edwards, "Natural Selection and the Sex Ratio: Fisher's Sources," *American Naturalist* 151 (1998): 564–69. When Darwin considers the evolution of an even sex ratio when there is an excess of females, he says that excess females "from not uniting with males would be superfluous and useless" (317), which clearly indicates that he is assuming monogamy.

35. Recall Arbuthnot's scholium on polygamy, in which the existence of an even sex ratio at the age of marriage is used to argue for the normative appropriateness of monogamy. Darwin reasons somewhat in reverse—from the assumed fact of monogamy to the existence of an even sex ratio—except that his argument involves no ethical judgment.

36. See, for example, E. Sober and D. S. Wilson, *Unto Others: The Evolution and Psychology of Unselfish Behavior* (Cambridge, MA: Harvard University Press, 1998).

37. Darwin, *The Descent of Man*, 2nd ed., 267–68.

38. This interpretation appears in J. Seger and J. Stubblefield, "Models of Sex Ratio Evolution," in *Sex Ratios: Concepts of Research Methods*, ed. I. Hardy (Cambridge: Cambridge University Press, 2002), 6.

39. I clarify the nature of this problem in the last section of this paper.

40. Fisher, *Genetical Theory of Natural Selection*, 158.

41. Düsing, *Die Regulierung des Geschlechtsverhältnisses*. (See note 3 above.)

42. Seger and Stubblefield, "Sex Ratio Evolution," 8.

43. Edwards, "Natural Selection and the Sex Ratio," and "Carl Düsing on the Regulation of the Sex-Ratio"; Seger and Stubblefield, "Sex Ratio Evolution."

44. Translated in Edwards, "Carl Düsing on the Regulation of the Sex-Ratio," 256–57.

45. Fisher, *Genetical Theory of Natural Selection*, 158–60.

46. Edwards, "Carl Düsing on the Regulation of the Sex-Ratio."

47. Ibid., 256.

48. Ibid.

49. Of course, Düsing's claim that selection leads to the evolution of an even sex ratio at reproductive age will be a consequence of Fisher's model if we assume that the costs of sons and daughters while they are being reared is the same and that there is zero time between independence and reproductive age. However, the time between independence and maturity is not a variable in Fisher's model.

50. Fisher, *Genetical Theory of Natural Selection*, 159–60.

51. Williams, *Adaptation and Natural Selection*, 151.

52. The incoherence of claiming both that group selection is *a priori* impossible and that it is empirically disconfirmed is discussed in E. Sober, *The Nature of Selection*, 2nd ed. (Chicago: University of Chicago Press, 2000).

53. For further discussion of sex ratio theory and its relation to the units of selection problem, see Sober and Wilson, *Unto Others*.

54. Hamilton, "Extraordinary Sex Ratios."

55. Ibid., 487, n. 43; G. C. Williams, *Natural Selection: Domains, Levels, and Challenges* (New York: Oxford University Press, 1992).

56. In his "Extraordinary Sex Ratios," Hamilton does not explore Williams's suggestion that group selection will favor organisms that facultatively adjust their sex ratios. However, his model does describe the situation in which there is variation in the number of females who found a nest. If females can detect how many other foundresses there are or will be, they should adjust the mix of sons and daughters they produce. The parasitic wasp *Nasonia vitripennis* is a case in point; for empirical details on how closely this organism conforms to the predictions of Hamilton's model, see S. Orzack, E. Parker, and J. Gladstone, "The Comparative Biology of Genetic Variation for Conditional Sex Ratio in a Parasitic Wasp, *Nasonia vitripennis*," *Genetics* 127 (1991): 583–99.

57. See Sober, "Intelligent Design and Probability Reasoning," and "The Design Argument."

58. Sober, *The Nature of Selection*; S. Orzack and E. Sober, "Adaptation, Phylogenetic Inertia, and the Method of Controlled Comparisons," in *Adaptationism and Optimality* (Cambridge: Cambridge University Press, 2001), 45–63.

59. For a discussion of the historical context in which Smith developed his views concerning the relationship of individual self-interest and the wealth of nations, see A. Hirschman, *The Passions and the Interests: Political Arguments for Capitalism Before Its Triumph* (Princeton, NJ: Princeton University Press, 1977).

60. E. Sober, "Progress and Directionality in Evolutionary Theory," in *Creative Evolution?!*, ed. J. Campbell (Boston: Jones and Bartlett, 1994), 19–33.

61. Fisher notes that his fundamental theorem of natural selection applies to traits that provide an "advantage to the individual," but that "it affords no corresponding explanation for any properties of animals or plants which, without being individually advantageous, are supposed to be of service to the species to which they belong." He then adds that "this

distinction was unknown to the earlier speculations to which the perfection of adaptive contrivances naturally gave rise. For the interpretation that these were due to the particular intention of the Creator would be equally appropriate whether the profit of the individual or of the species were the objective in view" (*Genetical Theory of Natural Selection*, 49).

62. Sober and Wilson, *Unto Others*.

63. The same conclusion would follow if the Biased trait produced any other unequal mix of sons and daughters. Furthermore, in a competition between two strategies that show different degrees of bias, the one that is closer to producing an even mix of sons and daughters will be favored at the within-group (individual) level.

64. R. Colwell, "Group Selection Is Implicated in the Evolution of Female-Biased Sex Ratios," *Nature* 290 (1981): 401–4.

It has occasionally been observed that android automata of the eighteenth century were about equally male and female, while those of the nineteenth century were mostly female. When they were not female, they were typically other sorts of uncanny or exotic creatures: talented children, blacks, acrobats, monkeys, magicians, and others (see figure 9.1). The poster of figure 9.2, announcing an 1827 exhibition in Hull by the impresario Schmidt, collects several of these mysterious performers, including versions of two of the famous automata of the Jacquet-Droz family in Switzerland: the Juvenile Artist (*dessinateur*) and the Musical Lady (*musicienne*). Their mechanisms were so exquisite as to seem magical, supposedly defying "all the first-rate Mechanics of Europe" in the case of the magician, who provided sagacious answers to every question.

I am interested here in the automaton as a reservoir of embodied meanings. What sort of thing was it? What was the significance of its female or otherwise exotic nature; and how was this nature related to the new sciences of the nineteenth century?[1] A literary example from the end of my story will set the stage.

Recall the Eloi and the Morlocks in H. G. Wells's 1895 novel *The Time Machine*. The Eloi and the Morlocks are the remains of human civilization that the time traveler finds living in the valley of the Thames, formerly London, in the year 802,701. The Eloi are a species living aboveground in the light of day, while the Morlocks live in darkness underground and emerge only at

FIGURE 9.1 | (a) Musical lady of Jacquet-Droz (Neuchâtel, Musée Historique), from Alfred Chapuis and Edmond Droz, *Les automates: Figures artificielles d'hommes et d'animaux, histoire et technique* (Neuchâtel: Griffon, 1949), frontispiece. (b) Fruitseller, 1880; Nationaal Museeum van Speelklok tot Pierement, Utrecht, Netherlands. (c) Japanese lady and Chinese man, 1885; Nationaal Museeum van Speelklok tot Pierement, Utrecht, Netherlands. (d) Juggler with fan, from Chapuis and Droz, *Les automates*, facing p. 304.

FIGURE 9.2 | Exhibition poster, from Alfred Chapuis and Edmond Droz, *Les automates: Figures artificielles d'hommes et d'animaux, histoire et technique* (Neuchâtel: Griffon, 1949), 291.

night. Both are forms of automata. They are the end result of the class struggle between capitalists and laborers of the nineteenth century, which was already consigning workers to sunless lives in factories, mines, and subway tunnels and had ultimately produced a state of perfect symbiosis between a refined intellectual aristocracy aboveground and the subterranean laborers who ran the machinery required to maintain an advanced civilization. "The rich had been assured of his wealth and comfort, the toiler assured of his life and work."[2] This had been the pinnacle of human evolution.

It had been a state of perfect adaptation between the two species and of complete subjugation of nature: no biting insects, no weeds, no disease, only butterflies, abundant fruits, and flowers. But that meant no struggle for existence. Progressive evolution ceased, and degeneration set in. "For countless years," the time traveler judged of life aboveground, "there had been no danger of war or solitary violence, no danger from wild beasts . . . no need of toil. For such a life what we should call the weak are as well equipped as the strong, are indeed no longer weak." The once magnificent Eloi had declined in size, strength, and intelligence. They had become effeminate, androgynous, and uniformly childlike, "for the strength of a man and the softness of a woman, the institution of the family, and the differentiation of occupations are mere militant necessities of an age of physical force." The life of the Eloi was now one of sameness and repetition, without purpose and without intelligence—in Wells's terms, *automatic.* "An animal perfectly in harmony with its environment is *a perfect mechanism*," observed the time traveler. "There is no intelligence where there is no change and no need of change."[3] To obtain an initial sense of why Wells should have used the phrase "perfect mechanism" for the Eloi, consider how the term *mechanism* alone, in figure 9.3, had served already in 1811 to advertise the Draftsperson and Musical Lady, as exhibited in London by the Jacquet-Droz associate Henri Maillardet. Apparently the prospective audience would immediately understand "Mechanism" to connote the more esoteric "automaton." The same usage appears in figure 9.2 for the Juvenile Artist of the "Mechanical and Magical Theatre."

FIGURE 9.3 | Exhibition poster, from Alfred Chapuis and Edmond Droz, *Les automates: Figures artificielles d'hommes et d'animaux, histoire et technique* (Neuchâtel: Griffon, 1949), 311.

For Wells, both the Eloi and the Morlocks seem to have evolved into androgynous automata, perfectly adapted to a life of "mere mechanical industry." Of course the Morlocks had already been feminized in the nineteenth century in their role as the masses. But now they had evolved into quite disgusting subterranean creatures, with this all-important advantage: they had retained the bare minimum of intelligence necessary to service their machinery. This capacity was enough that, when a disruption in their food supply occurred, the "bleached, obscene, nocturnal Things" could reverse the former dominance of their pretty but feeble overlords and begin to prey on them as a ready food supply, ultimately farming them, as ants herd aphids on roses.[4]

If Wells's characterizations of automaticity vividly portray the automaton as a feminized being and a perfect mechanism within Darwinian evolutionary theory, they also place the automaton firmly within British political economy. Those two primary sources for Wells were deeply connected historically. To develop this theme of the nature of automata in British science and culture, I begin with machinery in the factory economy and proceed through evolutionary interpretation to the physiology of the brain.

MACHINES AND POLITICAL ECONOMY

When Wells used the term *perfect mechanism*, he was using a standard term of art in the classification of machinery. It will be useful to develop its meaning from Charles Babbage's *Economy of Machinery and Manufactures* (1832) and Andrew Ure's *Philosophy of Manufactures* (1835). The large cotton factory of figure 9.4a appeared as Ure's frontispiece and can serve here to make a basic point.[5] The smokestack for the boilers of the great steam engines driving the whole system is distinctly separated from the textile factory full of other sorts of machinery, a distinction that we may take as emblematic of the distinction between engines and other machines. In fact, the engine house, with its two ninety-horsepower engines, was attached to the main building (figure 9.4b), but separated from the galleries of looms, spinning mules, and other working machines by a heavy, solid wall (no doubt for fire protection). Thus the structure itself sharply demarcates the two types of machines.

Writing on the "division of the objects of machinery" Babbage remarked, "There exists a natural, although, in point of number, a very unequal division amongst machines: they may be classed as; 1st, *Those which are employed to produce power*; and as, 2dly. *Those which are intended merely to transmit force and execute work*." Thus the division is between *engines*, which produce power, and trains

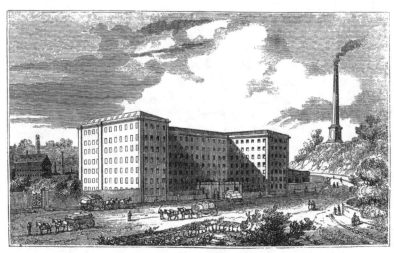

ORRELL'S *Cotton Factory, Stockport.*

A

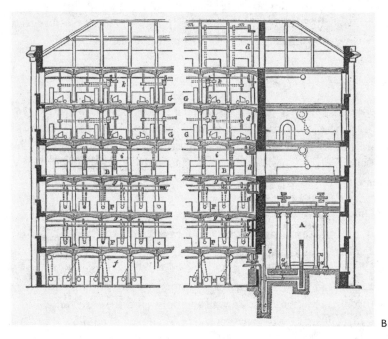

B

FIGURE 9.4 | (a) Orrell's Mill, Stockport, from Andrew Ure, *The Philosophy of Manufactures* (1835; repr., New York: Kelley, 1967), frontispiece. (b) Section, Orrell's Mill, Stockport, from Andrew Ure, *A Dictionary of Arts, Manufactures, and Mines, Containing a Clear Exposition of Their Principles and Practice*, enlarged ed., 2 vols. (New York: Appleton, 1863), 1:502.

of *mechanism*, which transmit the power from its production by the engine to its expenditure as work done by a working machine (occasionally also called an "engine" and treated by Ure and others as a separate class of machines).[6] Although the distinction was sometimes subtle and never entirely consistent, *engine* always implied productive power, as in referring to the differential calculus as "an engine of analysis," while "mechanism," in its more specific sense, referred to a device for executing a particular form of motion, typically repetitive.

This distinction was developed originally by French engineers writing in the 1820s and 1830s, who aimed to unite the action of machines with the production of value in the political economy. As a crucial conceptual achievement, they fully articulated the concept of work and identified work done with the value of commodities produced, that is, their labor value in classical political economy. Secondarily, but importantly here, they separated the question of how work is produced and expended from that of the transmission of motion per se.[7] In Britain, these notions appeared first in textbooks on engineering mechanics and then were incorporated into the formal structure of theoretical mechanics. Thereby, the distinction of a pure mechanism from an engine would become that between the geometry of motion, independent of productive capacity—labeled *kinematics*—and the causes of motion, or forces doing work—labeled *dynamics*.

An important exemplar for this history is a closely linked pair of Cambridge engineering textbooks, both derived from French sources, by Robert Willis and William Whewell. Willis wrote the purely kinematical part in the *Principles of Mechanism*, considering all of the geometrically possible ways in which motion could be converted from one speed and direction to another in a train of mechanism by connections of various kinds, using bevel gears, pulleys, shafts, rack-and-pinion gears, Hooke's joints, and many other contrivances. Meanwhile, Whewell wrote the causal part (dynamics) in *The Mechanics of Engineering*, in which he adopted the concept of work (called "labouring force") as the basic analytic concept for the effect of forces acting in machines, or better, circulating through a factory as what Poncelet called "mechanical money." These engineering concepts grounded the transformation of force mechanics into energy mechanics after mid-century. The classic text is the *Treatise on Natural Philosophy* (1867) of William Thomson and Peter Guthrie Tait, which would establish the canonical distinction between kinematics and dynamics.[8]

The basic division of machines into engines and mechanisms found ready extension to the organization of labor inside the factory itself, providing images for the differentiation of status and gender. An interior drawing of

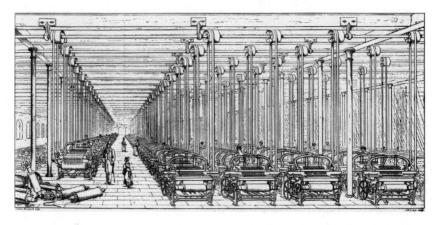

FIGURE 9.5 | Loom gallery in Stevenson's Mill, Stockport, from Andrew Ure, *The Philosophy of Manufactures* (1835; repr., New York: Kelley, 1967), facing p. 1.

a power loom gallery makes this apparent (see figure 9.5). Women are the "minders" who tend the self-acting machines, while the prominent man in a top hat is the overseer (or overlooker). These differential roles of males and females are summarized for a particularly striking but non-British case in figure 9.6, from Thomas Dublin's analysis in *Women at Work*. The table presents the organization of labor in a cotton factory in Lowell, Massachusetts. It makes the ratio of female to male workers 6 to 1, gives the average daily pay ratio as $0.60 to $1.05, and shows that jobs were segregated by sex. Men held all positions of oversight and machine repair. They also operated machines for the initial processes of picking and carding cotton. Women held all other machine-tending jobs. In Dublin's words, "The work of all women was repetitious, and almost always machines, not women, performed the basic operations of production."[9] Generalizing, while women's work was associated with mechanism, with the repetitive motion of shafts, pulleys, belts, spinning machines, and looms moving forever in a cycle, men were the metaphorical engines driving and controlling the system of production.

In Britain the separation of women's from men's work was much less strict, but similar conclusions apply. Ure gives the breakdown for more than forty-eight thousand workers in 151 mills in Lancashire in 1833 (figure 9.7), showing nearly equal total numbers of male and female workers (60 percent adult, 41 percent children under eighteen) and overlapping employment of adult males and females in some positions, such as weavers. But all overseers in every department were male (including 82 men to oversee

Men		Women	
Job	Mean daily pay	Job	Mean daily pay
Overseer	$2.09 (16)	Speeder	$0.66 (44)
Second hand	1.20 (12)	Drawer	0.52 (60)
Operative[a]	0.85 (38)	Spinner	0.58 (134)
Machinist	1.27 (25)	Weaver	0.66 (316)
Watchman	1.10 (6)	Dresser	0.78 (38)
		Warper	0.73 (23)
		Drawing in	0.66 (26)
		Sparehand	0.44 (185)
Overall[b]	$1.05 (149)	Overall[b]	$0.60 (881)

[a] Includes hands in the picking and carding departments.
[b] Overall totals are greater than the sums for specific jobs because only major jobs are listed.

FIGURE 9.6 | Distribution of work and pay in Lowell, Massachusetts, from Thomas Dublin, *Women at Work: The Transformation of Work and Community in Lowell, Massachusetts, 1826–1860* (New York: Columbia University Press, 1979), 66.

1,123 female throstle spinners), as were the "overseers" of virtually all construction and machinery (engineers, mechanics, firemen, etc.), all operators of carding machines ("engines" in Ure's usage), and most mule spinning machines, which were still powered manually and required coordinated muscular exertion to turn a drive wheel with the right hand while moving a carriage with the left. All of these jobs paid from two to two and a half times as much as the best jobs in which women also were employed, as warpers and weavers. Thus women's work in Britain, too, was associated with repetitive mechanism and men's with productive engines and oversight.[10]

The point can be made in a somewhat different and more general form by considering the anthropomorphic language that Ure used to celebrate the term *automatic*, which he employed for the self-acting machines in a factory, itself labeled an *automaton*, driven by a great steam engine.

A well-arranged power mill combines the operation of many work-people, adult and young, in tending with assiduous skill a system of productive machines continuously impelled by a central force. . . . In our spacious factory apartments the benignant power of steam summons around him his myriads of willing menials, and assigns to each the regulated task, substituting for painful muscular effort upon their part, the energies of his own gigantic arm, and demanding in return

Denomination of Process in which employed.	Class of Operatives.	Classification as respects Sex and Age.	Total Number of Persons employed.	Total Number of Hours worked by them, during the Month ending 4th May, 1833.	Average Number of Hours worked by each.	Aggregate Net Earnings for the Month ending 4th May, 1833.	Average Weekly Net Earnings of each Individual in each Process, calculated for 69 Hours.
						£. s. d.	Pence.
Carding . . .	Carders or overlookers.	Male adults	376	103,495·9	275·2	1,762 17 3½	282·06
	Jack-frame tenters . .	Principally female adults .	696	190,385·0	273·5	1,103 10 3½	95·98
	Bobbin-frame tenters . .	Ditto ditto . .	945	261,650·1	276·8	1,414 3 5	89·50
	Drawing tenters . .	Ditto ditto . .	1,931	532,287·3	275·6	2,885 3 8½	89·76
Mule-spinning .	Overlookers	Male adults	145	40,018·6	275·9	843 2 2	350·95
	Spinners	Male and female adults, but principally the former .	3,797	1,046,252·0	275·5	19,454 3 5	307·91
	Piecers	Male and female adults and non-adults, but principally the latter . . .	7,157	1,966,804·8	274·3	7,688 14 8½	64·73
	Scavengers	Male and female non-adults	1,247	340,019·1	272·6	712 2 11½	34·68
Throstle-spinning	Overlookers	Male adults	82	22,371·9	272·8	362 14 11½	268·51
	Spinners	Female adults and non-adults	1,123	305,712·4	272·2	1,716 17 6½	93·00
Weaving . . .	Overlookers	Male adults	400	109,577·0	273·9	2,088 1 4	315·56
	Warpers	Male and female adults . .	332	90,660·2	273·0	805 5 0	147·08
	Weavers	Male and female adults, male and female non-adults, but chiefly females . .	10,171	2,784,258·7	273·7	21,835 9 6½	129·87
	Dressers	Male adults	836	230,771·0	276·0	4,650 4 11	333·69

FIGURE 9.7 | Average net earnings among 48,645 hands in the Lancashire mills, from Andrew Ure, *The Cotton Manufacture of Great Britain*, 2 vols. (1836; repr., New York: Johnson Reprint, 1970), 1:347.

only attention and dexterity to correct such little aberrations as casually occur in his workmanship.

The menials governed by this very masculine engine have been relieved not only of manual labor but also of the productive power of their craft, reduced to tending the automatic machines with the "attention and dexterity" traditionally ascribed to women. In this sense, they have all been feminized, whether male or female. (The passage referred originally to the loom gallery of figure 9.5.)[11]

That male and female roles had differential status in the hierarchy of labor can come as no surprise. It is rather the gender differences inscribed in the classification of machinery that merit attention here. This inscription reached far and wide. Even the later Prime Minister Benjamin Disraeli, in his novel *Coningsby* of 1844, has his hero, on a visit to the Manchester cotton mills, marveling at the vast halls full of machines, where "he beheld in long continued ranks, those mysterious forms full of existence without life, that perform with facility, and in an instant what man can fulfil only with difficulty and in days." His gendered perception of these "supernatural slaves" mirrors the categories of mechanism and engine: "Why should one say that the machine does not live . . . has it not a voice: Does not the *spindle* sing like a merry girl at her work, and the *steam-engine* roar in jolly chorus, like a strong artisan handling his lusty tools, and gaining a fair day's wages for a fair day's

FIGURE 9.8 | Weavers in Lowell Massachusetts, from
Thomas Dublin, *Women at Work: The Transformation of Work and
Community in Lowell, Massachusetts, 1826–1860* (New York:
Columbia University Press, 1979), fig. 8.

toil?"[12] Figure 9.8 shows two such "merry girls," but holding shuttles rather
than spindles as the emblem of the mechanisms they tended.

This hierarchy extends much deeper into Babbage's account of the division of labor in manufactories. Babbage introduced into political economy a
principle of division of labor considerably more nuanced than that commonly
discussed since Adam Smith. Smith had attended only to the division of a
process such as pin-making into a serial set of tasks carried out by different
people—I will call this "horizontal" division. Even more important, Babbage
emphasized, was what I will call "vertical" division, the separation of tasks

demanding more skill and a higher wage from those demanding only strength or time and a lower wage. Babbage generalized this hierarchical organization as the "division of mental labor," with mere skill at the bottom and general knowledge and intelligence at the top (figure 9.9a). In his optimistic liberalism, he envisaged machines taking over at the bottom as workers climbed the ladder of expertise, responsibility, and civilization. Nevertheless, his concept of division of mental labor effectively established as a principle of "the interior economy of a manufactory" the stereotypically gendered hierarchy of bodily and intellectual labor, as exemplified in his own description of the work in a pin factory.[13] The capacities of typical working women, according to the measure of customary jobs and wages, were, next to those of children, closest to the capacities of repetitive mechanism. With increasing division of mental labor, in fact, they would be taken over by mechanism.

As has often been noted, Babbage understood the division of mental labor in the factory in the same terms as he understood his famous calculating "engines" (now called computers), having developed them in parallel from the early 1820s.[14] He modeled both on the hierarchical organization employed by Gaspard de Prony in Paris in the 1790s to calculate tables of logarithms and trigonometric functions, which Prony in turn claimed had been inspired by Adam Smith's division of labor. Prony's organization consisted of three levels of work (figure 9.9b): five or six eminent mathematicians who designed the basic formulae to be used, reducible to sums and differences of constants; seven or eight competent mathematicians who calculated the constants and oversaw the remainder of the work; and seventy or eighty computers, knowing only the rudiments of arithmetic, who actually carried out the enormous numbers of additions and subtractions required to construct the tables.[15]

In Babbage's earliest calculating engine, the Difference Engine, whose construction began in 1822 and ended incomplete in 1834, he expected

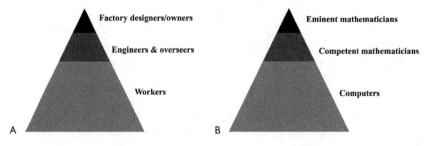

FIGURE 9.9 | (a) Representation of Babbage's principle of division of mental labor. (b) Corresponding division of calculating labor.

only to replace Prony's lowest-level workers with the machine. But he then began work on the Analytical Engine, which would possess higher intellectual capacities like foresight. Babbage and his interpreter, Ada Lovelace, saw such mechanization idealistically as demonstrating the mechanical basis of all intelligence and as raising the status and intelligence of all workers in the long run. The reality, of course, was a bit different and reinforced the gender hierarchy of mechanism and engine.

That reality can be seen in Babbage's own account of having set up two machines for the admiration of guests attending the Saturday evening soirées at his London house. In his drawing room Babbage presented on a pedestal a cherished female automaton with a one-foot-high body of silver who danced with an animated bird on her forefinger. (She has apparently been lost.) He had originally met this Silver Lady as a boy, when he found her eyes and movements irresistible to his "boyish passions," and had later purchased and lovingly restored her. To clothe her graceful form, he supplied a wardrobe of elegant dresses, or rather, since this "would have been rather difficult for a philosopher," he enlisted two of his fair friends who "generously assisted with their own peculiar skill and taste at the *toilette* of their rival Syren." As he remarked, "This piece of *Mechanism* formed a striking contrast with the . . . Difference Engine, No. 1, which was placed in the adjacent room" (figure 9.10).

Babbage found himself contemplating this contrast one memorable evening when the Silver Lady was "in brilliant attire." All of his friends, "relaxing from their graver pursuits," were captivated by her dance, while the calculating engine engaged the "deep thought" of only an American visitor and a Dutch professor.[16] The distinction between pure mechanism and an engine has here become the distinction in high society between superficial and profound intelligence: a mere mechanism of feminized entertainment versus an engine of masculinized rational action.

It will be observed, however, that following the distinction of engine and mechanism, Babbage's "engine" is actually pure mechanism, for it does not produce its own power but has to be cranked. The conundrum finds its resolution in the analogy between the calculating engine and a factory. Indeed the factory itself consists largely of pure mechanism. But considered as a whole, the factory is an engine of production, producing goods that carry the value of the power transmitted through its mechanism. Similarly, the calculating engine is a manufactory of numbers. The same resolution applies in reverse to automata as mechanism, for they contain coiled springs or weights to drive themselves but produce nothing in their repetitive action.

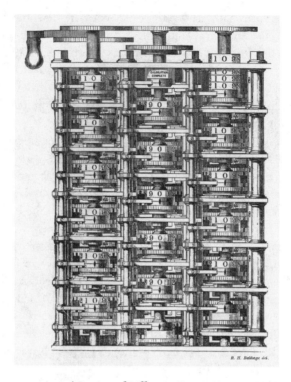

FIGURE 9.10 | Portion of Difference Engine No. 1, 1833, from Charles Babbage, *Babbage's Calculating Engines*, ed. H. P. Babbage (1889; repr., Los Angeles: Tomash, 1982), first plate.

In summary, Babbage's division of machines into engines and mechanisms found parallel expression in his vertical division of labor in a factory and in his distinction between his calculating engines and a typical automaton, always carrying with it the masculine and feminine qualities of progressive action versus repetition. These connotations became perfectly standard in British culture. The example of Disraeli's *Coningsby* may make that assertion sufficiently plausible initially.

EVOLUTION

Turn now to the evolutionary context that structures Wells's *Time Machine*, for it follows directly on Babbage's scheme and was fundamental to contemporary physiology. Recall the first immensely popular exposition of an evolutionary theory in Britain, the *Vestiges of the Natural History of Creation*, published anonymously by Robert Chambers in 1844. Chambers was able to use Babbage's exposition of the properties of his engines as a crucial resource in arguing that a basic change in behavior—here a jump from the production of one series of numbers to a different series—could follow from a mechanical law. If that were true for numbers, he asked, taking up a suggestion that Babbage himself had made in his *Ninth Bridgewater Treatise*, why not for organisms? Did the progression in complexity from lower to higher animals not also derive from a perfectly determinate law? Drawing especially on the work of the medical lecturer John Fletcher and of the well-known physiologist William Carpenter (including his views on the automatism of animals, below) Chambers created his grand scheme for a generative law that would control the transmutation of species, beginning from the earliest and simplest forms of life and progressing to the most complex.[17]

Chambers's scheme (figure 9.11) instantiated the recapitulation thesis (ontogeny recapitulates phylogeny). On the far left is a progressive ordering of the animal kingdom. In the middle is the corresponding order of their

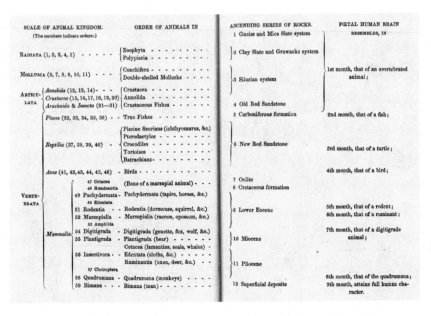

FIGURE 9.11 | Table illustrating how ontogeny recapitulates phylogeny, from [Robert Chambers], *Vestiges* (1844), reproduced in James A. Secord, ed., *Vestiges of the Natural History of Creation* (Chicago: University of Chicago Press, 1994), 226–27.

appearance as fossils in geological strata, and on the right is the development of the human brain during the gestation period of a fetus. The fetal brain develops through all of the basic stages of evolution of the order of animals. Figure 9.12 shows Carpenter's representation of the rudimentary "vesicles" for this development at six weeks.

Chambers's text prominently displays the most bizarre accompaniment of the recapitulation thesis, that a shortened gestation period, halting the fetal development of an animal prior to full development, or subjecting the mother to adverse conditions during the gestation period, will produce animals with decreased capacity. Paradigmatically, it will produce females. Chambers, following Carpenter, cites the differences among the bees in a hive. The queen's gestation period is only 16 days, while workers continue to 20 days, and the fully developed males to 24 days. The unity of nature suggested that a similar process occurs in the reproduction of the higher animals. If so, though Chambers did not explicitly say so, even women would

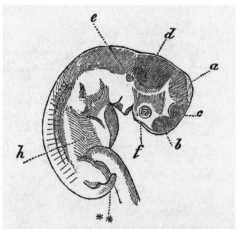

Human embryo of sixth week, enlarged about three times:—*a*, vesicle of corpora quadrigemina; *b*, vesicle of cerebral hemispheres; *c*, vesicle of thalami optici and third ventricle; *d*, vesicle for cerebellum and medulla oblongata; *e*, auditory vesicle; *f*, olfactory fossa; *h*, liver; ** caudal extremity.

FIGURE 9.12 | Human fetus at six weeks, from William B. Carpenter, *Elements of Physiology, Including Physiological Anatomy, for the Use of the Medical Students* (Philadelphia: Lea & Blanchard, 1846), 496.

represent the arrested development of men, in an evolutionary sense, just as African Negroes did in Chambers's view.[18]

Much of this continues in Darwin's evolutionary theory, although in a much more sophisticated form, for Chambers had no actual mechanism to explain the transition from one species to another through the sort of fixed law of generation that he took from Babbage's calculating engines. Darwin, in replacing all such bare analogies with the two principles of variation and natural selection, dropped Babbage's computational model and employed instead something like Baggage's principle of vertical division of labor in a factory, but in the form in which Milne-Edwards had described the physiological division of labor in an organism. From this concept, Darwin obtained his "principle of diversification," accompanied by the only illustration in the *Origin of Species*. This is the principle that explains why, in the course of time, any limited range of species will increasingly diversify under the pressure of natural selection, evolving a full spectrum from simple to complex, though not necessarily in that order. Using Darwin's example, if a few species of grass are sown on a plot of ground, they will originally yield a certain limited quantity of produce, but as they diversify under variation and natural selection, they will occupy more and more available niches in the polity of nature, thereby maximizing the total produce. It is this principle of diversification that gives a direction to time in Darwin's theory. Starting with homogeneity, it will yield heterogeneity; starting with lower organisms, it will yield higher ones, much like Babbage's vertical division of labor, but only so long as the competitive pressures of natural selection continue to operate.[19]

So far there is nothing in this to explain the differences between the sexes. Famously, Darwin coupled natural selection with sexual selection to account for sexual dimorphism. Of this story we need only recall that it incorporated many of the gender stereotypes of the age, including those that associated femininity with sentiment, imitation, habit, and repetition, and masculinity with strength, intelligence, and progress. In *The Descent of Man and Selection in Relation to Sex* (1871), Darwin produced his most infamous statements on the subject.[20] We tend today to identify sexual selection with females choosing males, like peahens choosing peacocks. But Darwin cautioned against extending the sense of beauty in the lower animals to the complex ideas of civilized man. Although some vestiges of the animal sense of beauty remain in savages, he thought, the primary means of sexual selection in all mammals is combat between males: "With mammals the male appears to win the female much more through the law of battle than through the display of his charms."[21]

Sexual selection provided an asymmetric mechanism of inheritance (based on Darwin's notion of pangenesis), such that continually striving and competing males evolved more rapidly than more passive females, at least in those characteristics favorable for battle. And this process of sexual selection joined with natural selection to produce higher reasoning powers and a higher degree of individuality, which mature males passed on to their male offspring. Meanwhile, females remained at the lower evolutionary level of intuition, imitation, habit, and repetition, somewhat closer to savages.

> It is generally admitted that with woman the powers of intuition, of rapid perception, and perhaps of imitation, are more strongly marked than in man; but some, at least, of these faculties are characteristic of the lower races, and therefore of a past and lower state of civilization. . . . We may also infer . . . that if men are capable of a decided pre-eminence over women in many subjects, the average of mental power in man must be above that of woman.

Darwin's subsequent summary, considered in reverse, will recall Wells's account of the degeneration of the Eloi after natural and sexual selection ceased to preserve the traditional gender roles in the family.

> We may conclude that the greater size, strength, courage, pugnacity, and energy of man, in comparison with woman, were acquired during primeval times, and have subsequently been augmented, chiefly through the contests of rival males for the possession of the females. The greater intellectual vigour and power of invention in man is probably due to natural selection, combined with the inherited effects of habit, for the most able men will have succeeded best in defending and providing for themselves and for their wives and offspring.[22]

It was precisely this need for defense and provisioning that had been eradicated at the pinnacle of civilization and that had led to the degenerate evolution of the Eloi.

One final feature of Darwin's theory of the lower evolutionary status of the sentiments and imitative traits of women is directly relevant to the gender of automata. The subject is rhythm. Think of musical automata, repetitive mechanism, and Disraeli's line "Does not the spindle sing like a merry girl at her work?" Rhythm, Darwin believed, was the evolutionary basis for imitative behavior. He observed that the capacity for rhythmic motion and even for sympathetic response to a musical note extends "low down in the animal scale: thus Crustaceans are provided with auditory hairs of different lengths, which have been seen to vibrate when the proper musical notes are struck." Through sexual selection, these capacities for rhythm had developed into the musical tones and melodies observed in the courtship of insects, fishes, amphibians, and birds. Ascending the evolutionary scale through simians (figure 9.13) and primitive humans to civilized man, such musical expression aroused the most profound feelings of tenderness, love, devotion, triumph, and ardour for war. Quoting Spencer and others, Darwin evoked the mysterious and primitive power of the emotions still aroused by music and rhythmic oratory (figure 9.14). "The negro in Africa when excited often bursts forth in song; 'another will reply in song, whilst the company, as if touched by a musical wave, murmur a chorus in perfect unison.'" The repetitive cadence of music and oratory thus pass by sympathetic imitation to animate a crowd, "like mental reversions to the emotions and thoughts of a long-past age."[23]

Darwin's argument may sound like mere popular rhetoric, but he was relying here not only on Carpenter's earlier *Physiology* but on the research program of Michael Foster, founder and director of the famous Cambridge

FIGURE 9.13 | Simian musician, from Jean Prasteau, *Les Automates* (Paris: Grund, 1968), 96.

FIGURE 9.14 | Negro playing the flute, from Jean
Prasteau, *Les Automates* (Paris: Grund, 1968), 89.

school of experimental physiology from 1870. Foster's research centered
on the heart, that venerable metaphor for the feminine virtues, as opposed
to masculine intelligence and will. He sought the source of its rhythmic
beat and found it in the protoplasm of each cell, independent of any gan-
glial connections, and in the harmony among cells established in a mimetic
manner, "as if each cell just felt by its primeval protoplasmic sensibility the
throb of its neighbour cells, and as if that throb were the keynote by which
all its own molecular processes were pitched."[24]

Fond of the musical metaphor, Foster extended it to hearts of all kinds, including those of fully developed birds and mammals with their much more
highly organized muscles and nerves. In fact, he had earlier suggested that
the simple throb of the heart revealed the underlying character of all our
repetitive vital actions, "such as our inbreathing and outbreathing, our sleep-
ing and waking, our working and resting." For Foster, "our whole life [is]

one throb, the sequence of our fathers', the forerunners of our children's."[25] Rhythmicity derives originally from the nutritive processes in an undifferentiated mass of protoplasm. Evolution occurs by differentiation through "physiological division of labour," beginning with the rudimentary functions already present in protoplasm.

The details of Foster's research, which I omit here, show how deeply rooted in evolutionary physiology was his separation of the repetitive functions, coordinated mimetically, from the continuous actions of centralized control mechanisms. This perspective put evolutionary distance between involuntary and voluntary functions and between sensibility and volition. By implication—or better, by association with the explicit arguments of Darwin, T. H. Huxley, and Herbert Spencer—it also grounded in experimental physiology the separation of primitive female sentiments and imitation from advanced and individualized masculine reason, just as it distinguished,

FIGURE 9.15 | Harp player, from Jean Prasteau, *Les Automates* (Paris: Grund, 1968), 187.

via physiological division of labor, repetitive manual or machine labor from mental labor, as Babbage had advocated in 1832 and had embodied in his calculating engines. At the level of rhythm and music, automata met the autonomic nervous system (figure 9.15).

AUTOMATON PHYSIOLOGY

Huxley the physiologist had originally wanted to be a mechanical engineer and remained faithful to that desire, calling physiology "the mechanical engineering of living machines."[26] It is not surprising, therefore, to find him using the distinction between mechanism and engine to characterize different functions of the human "machine." In his 1868 essay, "On a Liberal Education and Where to Find It," he remarked:

> That man, I think, has had a liberal education who has been so trained in youth that his body is the ready servant of his will, and does with ease and pleasure all the work that, as a *mechanism*, it is capable [of]; whose intellect is a clear, cold, logic *engine*, with all its parts of equal strength, and in smooth working order; ready, like a steam engine, to be turned to any kind of work, and spin the gossamers as well as forge the anchors of the mind.[27]

The question arises, therefore, whether women are also capable of such a liberal education of body and mind, mechanism and engine. We can find Huxley's answer in his notorious Belfast address to the British Association for the Advancement of Science in 1874, "On the Hypothesis That Animals Are Automata, and Its History."

The essay co-opts Descartes as the hero of the automaton theory, citing passages from his treatise *On Man*, but it actually depends much more heavily on the modern research of Foster and especially Carpenter, from whom most of Huxley's argument—right up to a crucial juncture for humans—was apparently taken. Already in his *Elements of Physiology* (1839), Carpenter had characterized most of the actions of the nervous system as being like those of an automaton, from the production of simple reflex movements, independent of sensation, through many kinds of instinctive and "consensual" action, stimulated either directly or indirectly by sensations, and including all those that had depended originally on conscious thought but had become automatic by habituation, such as walking a familiar route unconsciously while meditating on a problem of philosophy. This acquired, or "secondary," automatism, would become increasingly important in Carpenter's later work and would mark both a displacement of the engine-mechanism distinction and a sharp break from Huxley's views.[28]

For Carpenter, the three basic functions of reflex action, consensual action, and reasoning and volition involved three different "instruments" of the nervous system, which had been added successively in the progression from simpler to more complex animals: the spinal column and medulla oblongata, the sensory ganglia, and the cerebrum (or cerebral hemispheres). The cerebrum, superadded in the vertebrates (figure 9.16), gradually replaced many reflex and consensual functions with intelligence in the progression to humans. The action of the emotions, however, so closely analogous to the instinctual movements of the lower animals and to consensual action, remained rooted, even in the higher animals, in the sensory ganglia as their instrument rather than in the thinking cerebrum.[29]

The location of the emotions had consequences for the place of women in relation to men in this progression of the nervous system. Throughout his career Carpenter believed that some of the characteristic features of the female nervous system were just those of the automaton, especially imitation and emotion, with attendant disorders of an overexcited emotional system, like hysteria. "The tendency to *imitation* is a most powerful cause in Hysterical subjects; the mere sight of a paroxysm in one young *female*, being

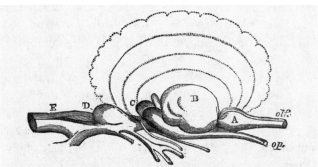

Fig. 6.—BRAIN OF TURTLE, with diagrammatic representation of the increased development of the Cerebrum in higher Vertebrata ;—A, Olfactory ganglia; B, Cerebral hemispheres; C, Optic ganglia; D, Cerebellum; E, Spinal cord; *olf*, Olfactory nerve; *op*, Optic nerve.

FIGURE 9.16 | Structure of turtle brain, with indication of development in higher animals, from William B. Carpenter, *The Doctrine of Human Automatism* (London: Sunday Lecture Society, 1875), 17.

often sufficient to produce a similar attack in a whole room-full of her companions." Males, too, might become subject to imitative hysteria, although the intellect governed their behavior to a higher degree.[30]

In his early writings Carpenter had insisted that automaticity, as an attribute of mechanism, could not extend to the mind, for "the changes which mental operations produce in the cerebral fibres, cannot be imitated, as changes in other motor fibres may be, by physical impressions." But by the early 1870s he had changed his view dramatically, extending the theory of secondary automatism to include ideas and relations of ideas acquired by learning, beginning with the ideas of animals and extending to humans. Only the will now escaped the capacity of the cerebral automaton, for Carpenter believed that "animals simply governed by ideas and feelings are to be regarded as 'thinking automata,' and that it is only when the action of these is controlled and regulated by the Will, as in the fully developed Human being, that the highest form of Intelligence can be properly said to exist."[31]

Carpenter's new belief in the "thinking automaton" appeared along with his increasing commitment to evolution, especially to a Lamarckian form of inheritance of acquired characteristics in human evolution. He had also come firmly to believe (perhaps having reflected on Francis Galton's arguments for hereditary genius) that "the aptitude of mind for the acquirement of knowledge . . . [and] the disposition to act upon right principles . . . is, to a very great degree, hereditary." But these capacities for mental and moral action, if hereditary, had to be embodied in material states of the brain, in its automatic mechanisms. That recognition seems to have expanded Carpenter's view of "secondary" automatism to include even the capacity for the most complicated reasoning of such people as mathematicians, lawyers, and bankers—capacities acquired by rigorous training and most efficient when operating unconsciously—for only then could training and discipline have their ultimate effect "upon the race, as well as upon the individual."[32]

Although training and discipline could embed right thinking in the thinking automaton, the same could not be said for the controlling action of the will, Carpenter's final barrier against automatism. "It is by this exercise of the will, therefore, in training and disciplining the mind, that it acquires that method by which it will work of itself." Such self-control required a "determinate effort," and it was this effort of conscious choice, of selective attention, that gave "the mark and measure of the independent exertion of the will," for otherwise the effort would not be exhausting.[33]

Interestingly, in developing the idea of the thinking automaton, Carpenter found the metaphor of the brain as engine inappropriate, and changed both engine and brain into "mere mechanism." The real source of the work done by an engine, he observed, is not the engine itself but the force (energy) of the heat supplied. "The steam engine furnishes the mechanism; the work is done by the force [energy]. Now in the same manner, the brain serves as the mechanism of our thought; and it is only in that sense that I speak of the work of the brain. But there can be no question at all that it works of itself, as it were—that it has an automatic power."[34] Carpenter apparently had something like the following scenario in mind. When thinking on its own, unconsciously, without any exertion of the will, the working brain would automatically draw the energy it needed from chemical reactions in its tissues, like an engine controlled by a governor, automatically regulating its own supply of steam. If instead, the will were exercising its conscious effort of attention and choice, like an engineer exercising direct control of the engine in producing more work, it would be directing the mechanism of the brain to draw more energy from its tissues; the brain would tire sooner. Although the will could not create the extra energy expended in conscious thought, it could direct its automaton to expend that energy.[35] Thus the ultimate "engine" of thought, in the sense of the agency that either trained the thinking automaton to act on its own or controlled it directly, was the will.

It was precisely at this point of distinguishing a thinking automaton—a mere mechanism—from a thinking automaton controlled by the will—a proper engine—that Huxley intervened in Belfast in 1874. His argument rests on a series of experiments on frogs and reports on humans (largely drawn from Carpenter) in which the spinal cord is damaged at levels successively closer to the brain and then up to the frontal lobes (cerebrum). Modern research, Huxley claimed, gives every reason to suppose that the vital functions are effects of the mechanism of the nerves, the molecular motion of nerve-substance. The spinal cord itself, and right on up to the frontal lobes, although "incapable of giving rise to consciousness" or "of carrying out volitions," has remarkable powers for coordinating movement and responses, such as swimming, maintaining balance with the delicacy of a rope-dancer (a standard automaton; see figure 9.17), and of responding to visible objects. All of these actions in frogs are performed mechanically, without consciousness, by molecular changes in the nervous system.[36]

A dramatic case from a brain-injured patient suggested that a similar hypothesis would be justifiable for humans, since the patient was able in his abnormal states to carry out elaborate pantomimes of skirmishing with a sword, singing songs, and writing letters, all without apparent consciousness,

FIGURE 9.17 | Tightrope equilibrist, from Jean
Prasteau, *Les Automates* (Paris: Grund, 1968), 119.

like one of Vaucanson's automata. Yet rather than conclude with Descartes
that brutes have no consciousness, no rational soul, Huxley argued from the
"doctrine of continuity"—here Darwinian evolutionary continuity but also
the continuity of physical causation—that they have a state of consciousness
that is evolutionarily prior to but continuous with that of humans; and we,
like them, are "conscious automata."[37]

"Conscious automaton" sounds very like Carpenter's thinking automaton, but in making the automaton conscious—a contradiction in terms to
Carpenter—Huxley imposed a new metaphysics.[38] He subjected all mental
states, including volition, to the physical causation of the body, arguing that
"all states of consciousness in us, as in them [animals], are immediately
caused by molecular changes of the brain-substance." He insisted, furthermore, that no evidence existed for the reverse action, of mental states acting

on brain states, which would necessarily require, he thought—in contrast to Carpenter—that the mind be material in order for it to cause a change of material motion in the brain.[39] No doubt Huxley had in mind here the mechanical, reductionist electrophysiology of Hermann Helmholtz and Emil du Bois-Reymond, along with Helmholtz's formulation of energy conservation in terms of atoms and central forces, which John Tyndall, Huxley's fellow popularizer of "scientific naturalism" and then president of the British Association for the Advancement of Science, highlighted in an attack on supernatural religion in his presidential address at the Belfast meeting.[40] Their collaborator, the mathematician W. K. Clifford, basing his argument explicitly on energy conservation, quickly extended Huxley's own manifesto in a polemic against the "nonsense" that "the will influences matter," treating that belief as "the crude materialism of the savage."[41]

Knowingly or not, Clifford thereby set himself up as an authority who trumped not only Carpenter but leading British physicists like William Thomson, James Clerk Maxwell, and Peter Guthrie Tait, who believed that mind might exert its directing power without any energy exchange, somewhat like a pointsman switching railroad tracks, in Maxwell's analogy, or more profoundly by acting at what he called "singular points" in continuum mechanics, where a literally infinitesimal difference could have large consequences.[42] In contrast, the scientific naturalists held that the physics of energy conservation precluded the possibility of mind acting on matter, on the ground that such action would require a transfer of energy, implying either that the mind could create energy out of nothing or that it was itself material. In writing that brain states *cause* mental states, Huxley had actually compromised this logic, even though he surely did not intend to imply any transfer of physical energy to the mind. Clifford quickly corrected the error: since mind was not material, no causal relation at all could exist between mind and matter. Mental states appeared only as the inevitable accompaniment of brain states, as parallel phenomena.[43]

Huxley, too, unwilling to adopt the alternative materialist position that mental states are nothing other than brain states, had already concluded that "our mental conditions are simply the symbols in consciousness of the changes which take place automatically in the organism." We have consciousness, but it is only a "collateral product," and it cannot act back on the brain, which goes on automatically. "We are conscious automata." The difference is only a matter of degree, of relative location on the evolutionary scale of brain function, between the automatic mechanism in the brains of lower animals and the equally automatic engine of conscious intellect and will in humans. Indeed, as Huxley would put it later, the evolution of sentient nature, represented in terms of human thought, is a "materialized logical process, accompanied by pleasures and pains, the incidence of which, in the majority of cases, has not the slightest reference to moral desert."[44]

This view has ethical implications. Since animals are conscious, they suffer as we do in their struggle for existence. We ought, therefore, to treat our domestic animals, at least, as we would any conscious beings: we should "deal with them as weaker brethren."[45] The same reasoning informed Huxley's progressive politics with respect to "savages" and women. Although he did not discuss them here, he had presented his views already in 1865 in a little paper on "Emancipation—Black and White." "Black" refers to the emancipation of American slaves and "White" to "the 'irrepressible' woman question." His main argument is that, although neither blacks nor women are the equal of white men, they nevertheless deserve to be emancipated and educated, in accordance with "the doctrine of equal natural rights." But one should not expect too much. "Women will find their place, and it will neither be that in which they have been held, nor that to which some of them aspire. . . . The big chests, the massive brains, the vigorous muscles and stout frames of the best men will carry the day, whenever it is worth their while to contest the prizes of life with the best women."[46]

Arguing for equal educational opportunity, Huxley damns the current system as enhancing all the natural defects of women—lack of physical prowess, emotional excitability, timidity, and dependence—teaching them that in contrast to our brother, "our sister is to be left to the tyranny of authority and tradition." Thus "girls have been educated either to be drudges or toys beneath man; or a sort of angels above him." In Huxley's view, "the female type of character is neither better nor worse than the male, but only weaker."[47] In sum, men should treat women like weaker brethren, for they are conscious automata on the evolutionary scale between animals and men.

CONCLUSION

I have attempted to show that the gender of automata, at least in Britain, was rooted in the natural sciences of the nineteenth century and that the relation deepened considerably as those sciences matured. I began with a poster from early in the century, advertising Maillardet's show in 1811 of two remarkable automata (figure 9.3) under the banner of "Mechanism." One of the two was a very talented draftsperson, exhibited either as a child

A

B

FIGURE 9.18 | (a) Maillardet's draftswoman/writer. (b) Mechanism of the draftswoman/writer, both from Alfred Chapuis and Edmond Droz, *Les automates: Figures artificielles d'hommes et d'animaux, histoire et technique* (Neuchâtel: Griffon, 1949), 312, 314.

(the Juvenile Artist of figure 9.2) or a young woman (figure 9.18a). She could produce such elaborate drawings as a Middle Eastern temple with palm trees. Her mechanism (figure 9.18b) was indeed impressive.

Between this point in 1811—when "mechanism" already carried something of the sense of repetitive, mimetic action—and Wells's *Time Machine* of 1895, those connotations had successively become grounded in the classification and analysis of machinery, the political economy of industry, Darwinian evolution, and brain physiology. With such deep roots in some of the most successful sciences of the century, the gender stereotype acquired something like the status of an unquestionable fact of nature, a solid rock of truth rather than a malleable ideology. If this interpretation is correct, it helps to explain why so many android automata of the nineteenth century were either gendered female or represented blacks, acrobats, magicians, and other uncanny characters.

NOTES

1. This article derives in part from M. Norton Wise, "Time Discovered and Time Gendered in Victorian Science and Culture," in *Energy to Information: Representation in Science, Art, and*

Literature, ed. Bruce Clarke and Linda Henderson (Stanford, CA: Stanford University Press, 2002), 39–58.

2. H. G. Wells, *The Time Machine: An Invention*, ed. Nicholas Ruddick (Peterborough, Ontario: Broadview, 2001), 141.

3. Ibid., 92, 88–89, 101–2, 141.

4. Ibid., 142, 107.

5. Charles Babbage, *On the Economy of Machinery and Manufactures* (Philadelphia: Carey & Lea, 1832); Andrew Ure, *The Philosophy of Manufactures; or, An Exposition of the Scientific, Moral, and Commercial Economy of the Factory System of Great Britain* (London: Knight, 1835; repr., New York: Kelley, 1967), 109–12. Ure gives an extended discussion, with vertical and horizontal sections, of Mr. Orrell's advanced Lancashire mill, built by the famed Manchester engineer William Fairbairn, in Andrew Ure, *The Cotton Manufacture of Great Britain*, 2 vols. (1836; repr., New York: Johnson, 1970), 1:295–304, and plates 1 and 2.

6. Babbage, *Economy of Machinery*, 26. Babbage himself is not entirely consistent in using "mechanism" to refer only to those machines that do not produce power. Indeed, all engines consist of mechanism arranged to take up the power of wind, water, steam, and so forth, so that the distinction is a matter of the object of interest. Ure, *Philosophy of Manufactures*, 27.

7. A standard text was Jean Victor Poncelet, *Traité de mécanique industrielle, physique ou expérimentale*, 2nd ed., 3 vols. in 2 (Bruges: Schiwel, 1844), vol. 3, *Des Machines et des moteurs*. See also M. Norton Wise, "Mediating Machines," *Science in Context* 2 (1988): 81–117.

8. Robert Willis, *Principles of Mechanism* (London: Parker, 1841). This is the same Willis who partially unmasked the automaton chess player called the Turk, taking measurements with an umbrella, and insisted that no mechanism could usurp the faculties of the human mind, in *An Attempt to Analyse the Automaton Chess Player of Mr. von Kempelen* (London: Booth, 1821); William Whewell, *Mechanics of Engineering* (Cambridge: Parker and Deighton, 1841). A brief account of the kinematics-dynamics distinction appears in M. Norton Wise, "William Thomson and Peter Guthrie Tait, *Treatise on Natural Philosophy* (1867)," in *Landmark Writings in Western Mathematics*, ed. Ivor Grattan-Guinness (Amsterdam: Elsevier, 2005), 521–33.

9. Thomas Dublin, *Women at Work: The Transformation of Work and Community in Lowell, Massachusetts, 1826–1860* (New York: Columbia University Press, 1979), 66.

10. Ure, *Cotton Manufacture*, supplementary appendices B, D, H. Ure details the operation of the various machines under "Cotton Manufacture" in his *A Dictionary of Arts, Manufactures, and Mines; Containing a Clear Exposition of Their Principles and Practice*, 2 vols. (New York: Appleton, 1863), 1:505–34, carding engine, 509–13. The self-acting spinning mule (532–35), which came into use in the 1830s, was called an Iron Man and could be tended by juveniles. On age, wages, and sex, see also Ure, *Philosophy of Manufactures*, 473–76.

11. Ure, *Dictionary of Arts*, 1:108–9, adapted from *Philosophy of Manufactures*, 13, 18, with reference to figure 9.5 above, and where "automaton" describes the factory. Ure's entry "automatic" in the *Dictionary* immediately precedes "automaton" (109–13) for that special class of automatic machines made to imitate living beings with a concealed source of power, typically a spring. Complex automata were no longer made, he observed, because

they were expensive and soon satisfied curiosity, presumably because of their repetitive action. Ure uses "mechanism" throughout to refer to the works of automata, including their driving springs.

12. Benjamin Disraeli, Coningsby; or, The New Generation, ed. Thom Braun (1844; repr., Harmondsworth, UK: Penguin Books, 1983), chap. 2, 179–80.

13. Babbage, Economy of Machinery, 141.

14. For a penetrating analysis in relation to automatism, see Simon Schaffer, "Babbage's Intelligence: Calculating Engines and the Factory System," Critical Inquiry 21 (1994): 203–27.

15. Babbage, Economy of Machinery, 140–45.

16. Charles Babbage, Passages from the Life of a Philosopher (London: Longman, 1864), 273–74. On the "Silver Lady" and this event, see Simon Schaffer, "Babbage's Dancer and the Impresarios of Mechanism," in Cultural Babbage: Technology, Time, and Invention, ed. Francis Spufford and Jenny Uglow (London: Faber & Faber, 1996), 52–80, esp. 53–58.

17. Charles Babbage, The Ninth Bridgewater Treatise: A Fragment, 2nd ed. (1838), in The Works of Charles Babbage, ed. Martin Campbell-Kelly (London: Pickering, 1989), 4–11; [Robert Chambers], Vestiges of the Natural History of Creation, ed. James A. Secord (Chicago: University of Chicago Press, 1994), xvii, 206–11; James Secord, Victorian Sensation: The Extraordinary Publication, Reception, and Secret Authorship of "Vestiges of the Natural History of Creation" (Chicago: University of Chicago Press, 2000), 63–64. The recapitulation theory structures William Carpenter's Elements of Physiology, including Physiological Anatomy, for the Use of the Medical Student (1839; repr., with minor revisions, Philadelphia: Lea & Blanchard, 1846), esp. chap. 12 on the nervous system, including a diagram of a human fetus at six months, par. 870.

18. [Chambers], Vestiges, 213–18.

19. Charles Darwin, On the Origin of Species, in Darwin, ed. Philip Appleman, 3rd ed. (New York: W. W. Norton, 2001), 124–29. (Orig. pub. 1859.)

20. Charles Darwin, The Descent of Man and Selection in Relation to Sex (New York, 1913), 632–33. (Orig. pub. 1871; 2nd rev. ed., 1874.)

21. Ibid., 215, 511.

22. Ibid., 576, 618–19.

23. Ibid., 581, 584, quoting Winwood Reade, The Martyrdom of Man (London: Trübner, 1872), 441, and African Sketch Book, 2 vols. (London: Smith, Elder, 1873), 2:313.

24. Quoted in Gerald Geison, Michael Foster (Princeton, NJ: Princeton University Press, 1978), 203.

25. Ibid., 193.

26. Thomas Henry Huxley, Autobiography and Essays, ed. Brander Matthews (1919; repr., New York: Kraus, 1969), 30.

27. Thomas Henry Huxley, "On a Liberal Education and Where to Find It," in Collected Essays of T. H. Huxley, vol. 3, Science and Education (1893; repr., Bristol: Thoemmes Press, 2001), 76–110, on 86; my emphasis.

28. Carpenter, Elements of Physiology, pars. 842, 858, 876, 921.

29. Ibid., pars. 847, 867–68, 906–7.

30. Ibid., par. 908, and more extensively in Carpenter's lectures in Manchester, December 1 and 8, 1871, published in *Unconscious Action of the Brain and Epidemic Delusions*, Half-Hour Recreations in Popular Science, ed. Dana Estes (Boston: Estes & Lauriat, 1873), 224–28.

31. William B. Carpenter, *On Recent Investigations into the Different Parts of the Brain, Being the Substance of Two Lectures Delivered Before the Sunday Lecture Society, 2nd and 9th November, 1873* (London: Sunday Lecture Society, 1873), 8.

32. Carpenter, *Elements of Physiology*, par. 920, and *Unconscious Action*, 208–12, 236–38, 250–51; William B. Carpenter, *The Doctrine of Human Automatism. A Lecture (with additions) Delivered Before the Sunday Lecture Society, 7 March 1875* (London: Sunday Lecture Society, 1875), 26–27.

33. Carpenter, *Unconscious Action*, 212–15, and *Doctrine of Human Automatism*, 28–32.

34. Carpenter, *Unconscious Action*, 207–8.

35. William B. Carpenter, "The Limits of Human Automatism," *Fortnightly Review*, November 1874, reprinted as the preface to his *Principles of Mental Physiology*, 4th ed. (London: King, 1876), and in his *Nature and Man: Essays Scientific and Philosophical* (London: Kegan, Paul, Trench, 1888), 284–315, on 286.

36. Thomas Henry Huxley, "On the Hypothesis That Animals Are Automata, and Its History," in *Science and Culture and Other Essays* (London: Macmillan, 1881), 199–245, on 218. Adrian Desmond gives a political-social account of Huxley's automaton tactics in *Huxley*, vol. 2, *Evolution's High Priest* (London: Michael Joseph, 1997), chap. 3, "Automatons," 51–80.

37. Huxley, "Animals Are Automata," 232–34, 237, 239.

38. Carpenter rejected Huxley's (and W. K. Clifford's) extension of the thinking automaton to include volition in a variety of venues in 1874–76, including *The Doctrine of Human Automatism*, "The Limits of Human Automatism," and "On the Doctrine of Automatism," in Carpenter, *Nature and Man*, 261–83.

39. Huxley, "Animals Are Automata," 237–39.

40. In an extensive analysis, Daan Wegener, "Conscious Automata and Energy Conservation in Victorian England (1874–1898)," argues that energy conservation was the main argument of the scientific naturalists for the conscious automaton theory (in preparation; contact at Dept. of Physics, Utrecht University, The Netherlands, e-mail: F.D.A.Wegener@phys.uu.nl). For a characterization of scientific naturalism, see Frank Turner, *Between Science and Religion: The Reaction to Scientific Naturalism in Late Victorian England*, Yale Historical Publications, Miscellany, 100 (New Haven, CT: Yale University Press, 1974), 8–37; see also John Tyndall, "Presidential Address," in *Victorian Science: A Self-Portrait from the Presidential Addresses of the British Association for the Advancement of Science*, ed. G. Basalla, W. Coleman, and R. H. Kargon (Garden City, NY: Anchor Books, 1970), 436–78, esp. 462.

41. William K. Clifford, "Body and Mind," delivered to the Sunday Lecture Society, November 1, 1874, in *Lectures and Essays*, by William K. Clifford, 3rd ed., 2 vols. (London: Macmillan, 1901), 2:1–51, esp. 30–33.

42. Carpenter employs the metaphor of "the self-acting points of a railway" in "Limits of Human Automatism," 305. For Thomson, Maxwell, and Tait, see Crosbie W. Smith and M. Norton Wise, *Energy and Empire: A Biographical Study of Lord Kelvin* (Cambridge: Cambridge University Press, 1989), 612–33.

43. Huxley and Clifford may have derived their metaphysics in part from the philosopher Shadworth H. Hodgson, who argued in The Theory of Practice that nerve movements and states of consciousness form parallel series (2 vols. [London: Longman, Green, Reader, and Dyer, 1870], 1:417).

44. Huxley, "Animals Are Automata," 236–39. It is curious, given the significance of evolution to Huxley's and Clifford's argument for the radical separation of brain states and mental states, with no actual function for the mind, that no one seems to have attacked their doctrine from an evolutionary perspective until William James pointed out, in a devastating critique, that if consciousness had no utility, it could hardly have evolved on the basis of natural selection. Other aspects of James's account, such as the importance of attention and choice and the likely role of consciousness as a factor in evolution, owed much to Carpenter. See William James, "Are We Automata?" in Essays in Psychology (1879; Cambridge, MA: Harvard University Press, 1983), 38–61, and "The Automaton Theory," in The Principles of Psychology, 2 vols. (1890; Cambridge, MA: Harvard University Press, 1981), 1:132–47. See also Thomas Henry Huxley, "The Struggle for Existence in Human Society" (1888), in Collected Essays, 9:195–236, on 202.

45. Huxley, "Animals Are Automata," 233.

46. Thomas Henry Huxley, "Emancipation—Black and White," in Collected Essays, 3:66–75, on 73–74.

47. Ibid., 71–72.

10 Techno-Humanism: Requiem for the Cyborg

TIMOTHY LENOIR

Since the mid-1980s, when Donna Haraway published her "manifesto,"[1] the cyborg has been a featured actor in the field of cultural studies of contemporary technoscience. I come to sing the cyborg's praises. I also come to bury it and declare that it's time to move on.

First defined in 1960 by scientists interested in adapting humans to space flight through control and redesign of physiological and psychological regulatory systems, the term *cyborg* has been applied with a great deal of flexibility since then. Nearly any form of human-machine coupling and the engineered union of separate organic systems has counted. One author or another has classified persons with implants, such as pacemakers, prostheses, and even immunization by vaccination, along with bioengineered transgenic organisms as cyborgs. From this perspective some would argue that we (humans) have always been cyborgs.[2]

Such usages can be so broad as to be meaningless. I prefer the suggestion by Chris Hables-Gray in his *Cyborg Handbook*: within the broad field of human/machine coevolution we should confine the use of *cyborg* to the relatively recent phenomenon, constituting a new—possibly the last—stage in human evolution, in which human/machine coevolution has been managed by cybernetics, the science outlined by Norbert Weiner that interprets both machinic and organic processes as parts of informational systems.

In this (last) stage of evolution the transition is made to what many have called the "posthuman," bridging the discontinuity between humans and machines.[3] N. Katherine Hayles has provided the most useful diagnosis of the posthuman condition in her pathbreaking volume, *How We Became Posthuman:Virtual Bodies in Cybernetics, Literature and Informatics*. According to Hayles, the posthuman view holds that consciousness is not the effect of a gathered, unified entity but is rather a distributed phenomenon. It "configures human being so that it can be seamlessly articulated with intelligent machines." Most important, the posthuman condition understands the body as simply a prosthesis we have learned to use, capable of being extended or replaced by other technological prostheses as they develop. "In the posthuman, there are no essential differences or absolute demarcations between bodily existence and computer simulation, cybernetic mechanism and biological organism, robot teleology and human goals."[4]

The posthuman is manifest in the cyborg. For both Hayles and Haraway, the critical figure of the cyborg is not just a referent to the inventions of technoscience. We are all cyborgs, with no choice but to live inside the material and imaginative apparatuses that make us cyborgs. These critics use the cyborg as a figure for exploring the material, technological, and imaginative apparatuses that construct and ground our situations, shaping us as subjects. For Haraway the cyborg captures the ways in which "fact and fiction, rhetoric and technology, analysis and story-telling are all held together by a stronger weld than those who eschew taking narrative practice seriously in science—and in all other sorts of 'hard' explanations—will allow."[5] Like Latour's Janus-face of science, which contrasts "science in the making" with "ready-made science,"[6] the cyborg has been a useful device for preventing us from falling prey to nature/technological determinism on the one hand and naive social construction on the other. Haraway, who has a much deeper agenda, tells us she is engaged in an exercise of civics, aiming to unravel these layers of construction, mutate the narratives, and reshape the apparatuses for producing what counts as "us." Haraway's cyborg is a figure for exploring those inventions, whom they serve, and how they can be reconfigured. The aim of her project is to be in power-sensitive engagement with other materializations of the world.

For Hayles, the conjunction of technology and discourse in the cyborg is crucial. Were the cyborg only a technological object, it could be confined to such technical fields as bionics, medical prostheses, and virtual reality. Were the cyborg only a product of discourse, it would be of interest to sci-fi aficionados, without carrying wider cultural and political importance. Manifesting itself as both technological object and discursive formation, the cyborg partakes of the power of the imagination as well as the actuality of technology; it is both a product of technoscientific processes of production and a signifier for those processes themselves.

For Hayles, as for Haraway, cyborgs are simultaneously entities and metaphors, living beings and narrative constructions. Hayles describes this figure as a marginal one, recalling Latour's figure of Janus: "Standing at the threshold separating human from the posthuman the cyborg looks to the past as well as the future. It is precisely this double nature that allows cyborg stories to be imbricated within cultural narratives while still wrenching them in a new direction."[7]

While the possibility of wrenching cultural narratives in new directions is exciting, I am wary of claiming too deep an impact for this kind of study. In practice Hayles's work takes the form of fairly traditional literary analysis, privileging narratives over technological objects. What can this kind of criticism do to shape our possible posthuman future in desirable ways? Can we really, as Hayles and Haraway propose, engage this subject appropriately and effectively by exposing the narratives and discursive frameworks that sustain it?

Because Hayles would like to have the cyborg located in the world as well as in words, foremost on her agenda is the necessity of critiquing the dominant narrative handed down by the cybernetics tradition: the view that information is disembodied. One of the central threads of discourse constructing the posthuman, according to Hayles, is the notion that information consists of a pattern independent of a specific material medium, capable of being rewritten into different substrates. Hayles shows how this notion grew in tandem with deep concerns about preserving the liberal humanist subject dear to the creators of the first wave of cybernetics. Cybernetics was envisioned by scientists and engineers such as Weiner, McCulloch, and their colleagues at the Macy Conferences as a way to maximize human potential in a chaotic and unpredictable postwar world. They wanted to insure human beings a position of mastery and control removed from noise and chaos. The vision of the posthuman emerging from the work of Moravec, Minsky, and others, Hayles has argued, simply reinscribes the liberal humanist subject—conceived as a rational, self-regulating, free, and autonomous individual with clearly demarcated boundaries and a sense of agency linked with a belief in enlightened self-interest. For those convinced that this view of the subject underpins a sense of manifest destiny to dominate and control nature—a view detrimental to women; to other, particularly non-Western cultures; and to other life-forms—engaging the techniques through which the posthuman is emerging is a way to contend for a different vision. Rather than simply acquiescing in a view of the posthuman as an apocalyptic erasure of human subjectivity, the posthuman can be made to stand for a positive partnership among nature, humans, and intelligent machines.[8]

Hayles does not delve into the details of how that partnership between intelligent machines and humans might evolve. She has instead pointed to resources other than cybernetics for generating narrative alternatives to treating information as disembodied. Her main resources have been those of the so-called second-generation cognitive sciences, the sciences of the *embodied* mind.[9] Rejecting Cartesian assumptions of a disembodied mind, Antonio Damasio,[10] Francisco Varela,[11] and other neuroscientists and cognitive scientists have shown that human consciousness is not localized in a set of neural connections in the brain alone but is highly dependent on the material substrate of the biological body, with emotion and other dimensions as supportive structure. Similarly, philosophers George Lakoff[12] and Mark Johnson[13] have argued that metaphors for embodied interactions with the world are the sources of higher-level representations of language and thought. Carrying this idea further, Hayles applauds Andy Clark's notion of extended mind, which treats body boundaries as fluidly intermingled with technological affordances, and she adopts Clark's notion that we are cyborgs "not in the merely superficial sense of combining flesh and wires, but in the more profound sense of being human-technology symbionts."[14] Like Clark, Hayles argues that in the world of smart appliances it becomes harder and harder to say where the world stops and the person begins.

Hayles and other critics seek to preserve the "human" in the "posthuman" future by arguing that information—to be information at all—must be embodied and, specifically, that it must be embodied in a human agent/interpreter. But here we encounter what seems to me another dilemma. For just as second-generation cognitive scientists have emphasized the embodied mind, so second-generation builders of artificial intelligence, robots, and autonomous agents for the last decade have recognized embodiment as crucial to the design of intelligent agents. In fact, they have already moved on from cybernetics to deploy the very cognitive science models that Hayles is using in her critique of their project.

In agreement with the views of Damasio, Lakoff, and Johnson, Rodney Brooks has made situatedness and embodiment the two fundamental principles of his construction of humanoid robots.[15] But while Brooks agrees that humanoid intelligence must evolve through embodied interaction with the environment and other creatures, he sees nothing mystical about carbon-based matter. Brooks writes, "My own beliefs say that we are machines, and from that I conclude that there is no reason, in principle, that it is not possible to build a machine from silicon and steel that has both genuine emotions and consciousness."[16] This statement hints at the powerful technical advances in computing over the past decade, along with rich new ideas about the nature of computation and amazing progress in both biotechnology and nanotechnology. Brooks's view, which represents not only the view of Artificial Intelligence (AI) scientists and engineers but what is also rapidly becoming the view we all silently share, fuses perspectives from computing, communications technology, biotechnology, and nanotech into a powerful new technoscience. In this view, there are only assemblages of machines, whether in the domain of consciousness, intelligence, or other biological and material systems; and these constructions are all to be understood as different forms of computation. According to this view, the world is a collection of machines—indeed, a computer.[17]

Second-generation AI researchers such as Brooks all agree on the question of our posthuman future: it is happening. Yet Hayles's cyborg critique, waged through discourse analysis rather than technological practice and based on cybernetic models a decade out of date, is, to paraphrase a recent reflection by Bruno Latour, "running out of steam."[18] At the same time, huge and pressing questions and concerns are arising out of contemporary AI work that call out for investigation and perhaps intervention. For instance, opinions differ about the technical goals to be achieved, the means for reaching them, and the place of humans in a future heavily populated with intelligent agents. Not everyone on the frontlines of robotics research shares Moravec's optimism about mapping neural structures with sufficient completeness to enable a migration of consciousness to other media, such as silicon. Danny Hillis, designer of the world's fastest computer, the Connection Machine, believes that we may never be able to understand and map natural intelligence into a wiring diagram. Nevertheless, for Hillis, this ultimate limitation does not imply that we cannot engineer an artificial intelligence eventually superior to human intelligence. Hillis argues that intelligence is really an emergent phenomenon, a complex behavior that self-organizes as a consequence of billions of tiny local interactions. This conception of intelligence leads Hillis to predict, "We will not engineer an artificial intelligence; rather, we will set up the right conditions under which an intelligence can emerge. The greatest achievement of our technology may well be the creation of tools that allow us to go beyond engineering—that allow us to create more than we can understand."[19] From this perspective, the future direction in question is not elimination of the human but its coevolution with artificial agents.

During the 1980s and 1990s, arguments rejecting a possible merger of human and machine intelligence were sustained by what seemed insurmountable differences between the reasoning capabilities of computers and humans as well as by the limitations of silicon-based computer architectures. But major breakthroughs in several areas—including single-electron transistors, quantum dots, quantum computing, and the beginnings of biological computing—have renewed the optimism of those who greet a posthuman future. Leonard Adelman launched the field of biomolecular computing in 1994 with a demonstration that it was possible to build a DNA computer that solved intractable mathematical problems.[20] The first decade's work in this area offered no serious competition with silicon computing, but by 2003 a number of milestones marked the way for serious in vivo computing with biological cells. For example, as part of an effort to learn how to control the chemical mechanisms of the cell for purposes of protein engineering, Tom Knight of the MIT Artificial Intelligence Lab and Ron Weiss of Princeton have worked on implementing digital circuits in DNA and inserting them into biological cells. These molecular processes are quite slow and will not compete with silicon for speed, but the key point is that computation controls some internal processes of the cell. The resulting logic technology allows Knight and Weiss to engineer the chemical behavior of cells for use as sensors and effectors. Connecting these with cellular sensors and actuators gives rise to a cellular robot. Among the envisioned uses of such cellular robots is the construction of neural network modular implants for enhancing human capabilities in numerous areas such as vision and even reasoning.[21] Similar efforts are well under way to engineer nanoscale logic circuits using DNA lattices as a fabric on which to synthesize a template upon which specific metal deposition can produce a nanometer-scale electronic circuit.[22] Coupled with advancing experimental and theoretical studies on self-assembling DNA nanostructures for molecular computation and nanofabrication that have received major funding as part of the National Nanotechnology Initiative (NNI) since 2004, the era of silicon-based computing (with many of its limitations) seems on the path toward realizing the merger of biotech, computing, and information technology, including "new media" that Eugene Thacker has characterized by the term biomedia as a new technoscientific configuration that displaces the trope of the cyborg.[23]

Current work in cellular robotics illustrates the notion at the heart of Hayles's thesis: informatics requires embodiment. But recent directions in behaviorally based, evolutionary robotics regard the need to employ a human designer in the circuit as a temporary and annoying stopgap. Current experiments in the field are integrating evolutionary processes that depend on genetic algorithms and evolving neural networks, on self-organization and emergent complex adaptive systems rather than on the layering of behaviors carefully hand-crafted by the designer, as in the technique of "bottom-up" subsumption architecture pioneered by Rodney Brooks.

These researchers adopt the "embodied mind" as their operating principle by emphasizing that an organism is an embodied system living and acting in an environment. This position leads them to be critical of even some biologically inspired approaches, such as connectionism, which treats cognition as a structure that develops in an abstract environment. Evolutionary robotics insists that an organism is not only a collection of neurons; it is also immersed in a physical environment and uses its sensory-motor apparatus to actively extract information from the environment. Since behavior cannot be reduced to an internal mechanism alone, emergence is crucial to the evolutionary robotics approach. For example, the behavior of following a wall does not require the design of a module for wall-following within the agent; rather, basic processes are defined that together, in interaction with the environment, lead to wall-following: this is designing for emergence. This same strategy has been applied across the board to robots from insects all the way to humanoids such as Cog and Kismet (see the essay by Evelyn Keller in this volume), in which a minimally basic substrate of systems—a perceptual system, motivational system, behavioral system, and motor system—are built in, and the robot's overall behavior emerges from the interactions of these simpler behaviors, mediated through the world. Thus, seeking means to evolve higher forms of intelligence "on their own," without the supervision of human oversight, second-generation cyberneticists have embraced the strategy that the best way to develop intelligent artificial agents is simply to turn them loose and let them evolve.

In combining the input of code with that of the world, evolutionary robotics scientists are fusing the digital and the real, just as contemporary AI researchers fuse the human and the machine. For the last five years, my work has focused on tracking some of the many venues in which we are pursuing this kind of fusion of the digital and the real, a remaking of the world to enable a seamless interface between digital and physical reality. As examples of our increasing comfort with this merger, perhaps symptomatic of a posthuman future to come, we need look no further than the shift occurring in the habits of professional communities, such as surgeons. In the field of telesurgery, for instance—a "field" that a decade ago seemed like utter science fiction—surgeons work collaboratively with intelligent-agent technology and surgical robots to perform complex procedures beyond the capabilities of earlier advanced surgical technique. A prominent recent example is the completely closed-chest endoscopic bypass surgery performed with a tele-surgical system manufactured by Intuitive Surgical of Palo Alto, California. While not currently widely available, such systems are working their way into broad areas of medical practice. Such fusions of computer technology and surgical technique transform the skills, perceptions, and material practices of surgeons, eventually becoming standard procedure in their everyday work.[24]

Such arenas in medicine are developing today, but even these are too exotic to capture the pervasive nature of computer inscriptions and incorporations. We might consider the ordinary bar code, the universal identifier for nearly every manufactured item on the planet. Today's familiar bar codes are passive, registering information only when scanned by a laser that transmits the bar code's digital information. Large-scale initiatives are under way to introduce enhanced versions of bar codes that will actively transmit all sorts of data, a system called radio-frequency identification (RFID) tagging.[25] An RFID "tag" is a wireless semiconductor integrated circuit that stores an ID number in its memory and transmits that ID as well as potential access to other information through networked databases when accessed by, for instance, a Web browser.

A number of major corporations are experimenting with these technologies as a means for generating ubiquitous computing environments. One of the most interesting experiments is Hewlett-Packard's project Cooltown, pursued at multiple lab sites around the world. Cooltown's creators greet visitors to their Web site in this manner: "Welcome to Cooltown, our vision of a technology future where people, places, and things are first-class citizens of the connected world, wired and wireless—a place where e-services meet the physical world, where humans are mobile, devices and services are federated and context-aware, and everything has a web presence."[26] Don't mistake this for the scene from the 2002 film Minority Report, where characters in a futuristic world are personally greeted by the smart walls of a department store. That scene could happen today in Cooltown, or even in one of the new Rem Koolhaas–designed Prada stores. Through the use of RFID tagging, as well as other various types of "smart" materials, persons with network access in places like Cooltown will be able to "surf" reality. The virtual world of cyberspace and the physical environment are seamlessly connected. Soon the increasing numbers of artificial intelligent agents that assist us in a variety of Web-related tasks will be physically present to us as companions as we start the day at our local wireless Starbucks.

These projects are for the most part prototypes, not yet implemented on a wide scale but waiting in the wings if markets for them can be generated. It is to popular culture and the market itself that we must turn to see other examples of the fusion of the digital and the real. A stunning example is the development of immersive gaming, where the increasing convergence and mobility of digital network technologies have given rise to new, massively scaled modes of social interaction in which the physical and virtual worlds meet.

While it is not the first entry in this genre, perhaps the most spectacular example is a game called *The Beast*, which was not actually advertised as a game at all.[27] Most people discovered *The Beast* when they spotted a provocative credit ("Jeanine Salla, Sentient Machine Therapist") in a trailer for Steven Spielberg's 2001 film, *A.I.: Artificial Intelligence*. A Google search on Salla's name revealed a complex network of Web sites, many dealing with the technical, social, and philosophical problems of artificial intelligence and sentient machines, and all of which were set in the year 2142 CE.

The Beast was originally conceived by Microsoft, DreamWorks, and Warner Brothers as an elaborate interactive backstory to Spielberg's film *A.I.: Artificial Intelligence*. Although it soon outgrew this supportive marketing function to become a creation in its own right, *The Beast* drew heavily on the cognitive-technological themes of its associated film. These themes were manifest not only in the subject matter of game play, which included tracking down rogue sentient robots and participating in a human rights campaign for AI machines, but also in the structure of game play. *The Beast* was created not just as a commentary on, but also as an experiment in hybrid (organic-digital) artificial intelligence. According to its producers, it was specifically designed to spawn an online problem-solving community that modeled the emergent networked functioning of the human brain. Game play consisted of tracking and interpreting plot developments and evidence circulating mostly through Web sites and e-mails, but also through phone calls, faxes, television and newspaper ads, as well as occasional real-time and offline events. Players were also charged with cracking complicated and time-consuming puzzles that variously required programming, translating, and hacking skills; obscure knowledge of literature, history, and the arts; and brute computing force. The diverse skill and knowledge base required to solve the game's problems, as well as the magnitude of its unwieldy plot, made the formation of cooperative groups absolutely necessary.

The Beast's producers (Microsoft and DreamWorks) estimate that more than one million people from around the world played the game, many of whom worked in large online groups. One group, the Cloudmakers, was the most organized and high-profile collective, working literally around the clock; some players complained of losing not just sleep, but also jobs and friendships. The Cloudmakers group was founded in April 2001 by a twenty-four-year-old computer programmer named Cabel Sasser. Forty-eight hours after Sasser launched the group, there were 153 new members in the group investigating these mysterious sites. When the game ended three months later, the Cloudmakers group had grown to 7,480 members. The Cloudmakers provided new players and other online collectives with important tools for grappling with the game's complex narrative—comprising three core mysteries and a dozen rich subplots with nearly 150 characters—and for navigating the game's vast Web presence of nearly four thousand digital texts, images, flash files, and QuickTime videos. These tools included a 130-page walk-through guide to The Beast and a nearly perfect online archive of ephemeral and offline game content, such as audio recordings of voice-mail messages and digital photographs of clues left in public bathrooms in Chicago, New York, and Los Angeles.

The Beast's designers had originally developed a three-month game schedule with varying levels of difficulty: puzzles were planned to take a day or a week to solve, while others were designed to be all but insoluble. Demonstrating the power of collective intelligence, the Cloudmakers solved all of the puzzles on the first day. In response to this startlingly efficient collective play, the producers raised the bar of the game by making it even more challenging and sprawling, and by requiring even more cooperation. For example, clues required to access important game files were distributed separately at live events in multiple cities, and groups were required to assign players in each region to attend the events, where they communicated in real-time with players at home to piece together the necessary data.

Although the pervasive elements of The Beast (phone calls, PDA [personal digital assistant] downloads, e-mails, faxes, etc.) were the most hyped immersive component of the game, the proliferation of game Web sites was actually the largest and most powerful component of the immersive experience. The vast majority of game content was distributed via the Internet, on the Web sites of fictional characters, corporations, news services, and political action groups, as well as a fictional psychiatric clinic, weather bureau, coroner's office, and so on. These sites featured every functional hallmark of nonfictional sites, including pop-up warnings for software upgrades, banner ads for fictional companies, and incredibly deep links (many sites featured dozens of internal pages). Nowhere did these pages admit to being simply part of a game. The game took place in real time, with e-mails and datelines properly registered to match the clock in the player's world; even the source code and identifying personal information were rigorously monitored to eliminate any information that might link game content to its producers. Aesthetically, technologically, and phenomenologically speaking, there was no difference at all between the look, function, or accessibility of the in-game sites and nongame sites.

In this sense, nothing about this virtual play was simulated. The computer-driven alternate reality The Beast created was make-believe, but every aspect of the player's experience was, phenomenologically speaking, real. Hacking into the in-game coroner's office's fictional Web report, for example, was identical in practice to the process of hacking into a nongame coroner's office's Web site. This stands in stark contrast to other kinds of massively multiuser role-playing games, such as The Sims Online and EverQuest, in which the digital display of virtual worlds is clearly simulated and, although absorbing, provides a totally different mental and physical experience of being and acting than everyday life.

Online interactive gaming is fusing the virtual and the real in other ways—too many to explore here—as well. One particularly striking example is the online game America's Army. In the mid-1990s the military began to take advantage of the high level of interactivity, game play, and graphical realism in commercial games. Games like Hasbro's Falcon 4.0 were in fact so good that the military adapted them as flight training programs for F-16 fighter pilots. Subsequently, a large number of combat games have been adapted with little modification to military training purposes. While these games were useful, they did not include all the skills the military wanted to train, and so an effort was launched by Col. Casey Wardynski, director of the Office of Economic Manpower Assessment, to create a game company devoted to building a training scenario and recruiting device for the U.S. Army that would both provide an enjoyable gaming experience and at the same time teach players the values at the core of the U.S. Army experience.

In 2002, on Independence Day (the traditional summer blockbuster date in the entertainment industry), the U.S. military released its new video game, America's Army: Operations. Designed by the Modeling, Virtual Environments, and Simulation Institute (MOVES) of the Naval Postgraduate School in Monterey, California, the game is distributed free on the Internet.[28] Produced with brilliant graphics and the most advanced commercial game engine available (the Unreal game engine) at a cost of around eight million dollars, the game is a first-person, multiplayer combat simulation that requires players to complete several preliminary stages of combat training in an environment mirroring one of the military's own main training grounds—cyber boot camp. After training, recruits embark on missions of various sorts. In the new multiplayer online version released in January 2004, squads of players can coordinate their missions and play along with or against military personnel who are also playing the game, occasionally including military personnel taking R&R at centers outside Iraq. Just as simulation equipment is now folded seamlessly into new weapons systems purchased by the military, enabling training and mission coordination right up to the start of a mission, some people close to the military training and simulation effort predict that interactive online game/training scenarios will soon be integrated into command-and-control operations in the field.[29] Gaming environments may soon provide the command-and-control interfaces to real-world missions.

Let me return to the most salient feature of Hayles's characterization of the posthuman: "The posthuman view configures human being so that it can be seamlessly articulated with intelligent machines. In the posthuman, there are no essential differences or absolute demarcations between bodily existence and computer simulation, cybernetic mechanism and biological organism, robot teleology and human goals."[30] The unnamed concern aroused by this specter is, of course, the standard narrative of technology out of control, especially of computer technology as a dark force destined to enslave us, to rob humanity of its personal freedom, creativity, and individuality. Such rhetoric calls up scenes of 2001's HAL, the futile resistance to the communal intelligence of the Borg in Star Trek, and the pods of human "copper tops" in suspended animation powering the dream world of The Matrix.

But whether or not the future holds the grim scenario of a replacement of the human by an organic machine, in the meantime there are signs of hope. Some of these examples of our mergers with computer technologies actually seem to hold out the prospect of realizing a different future—one in which collective agency will play a much larger role than it has in our past. In the late 1960s and early 1970s, Doug Engelbart and his coworkers at Stanford Research Institute created the computer mouse, hypertext, online editing tools, and interactive video conferencing linked into a collaborative work environment. One of their goals was to create tools to enable groups to form working arrangements that would increase their collective IQ. As Engelbart explained it, these technologies would permit the collective leveraging of the bits and pieces of knowledge and expertise of individuals, in order to generate innovations and solutions to major problems.

The Wikipedia, the online free encyclopedia, is an example of the sort of leveraging of the collective IQ that Engelbart had in mind. Founded in

January 2001, more than 20,000 entries were created in the first year of its existence, at a rate of more than 1,500 articles per month. On August 30, 2002, the articles numbered 40,000. The rate of growth has more or less steadily increased since the inception of the project. The astonishing thing about *Wikipedia* is that its contents are created completely by individuals interested in freely contributing their knowledge on a topic. Moreover, the built-in editing features of Wiki software technology allow other people to modify and update the content of any article. The users are completely self-monitoring, so that if an error is published, it is usually ferreted out and corrected by the user community. Many people are beginning to use *Wikipedia*, which had more than 4.6 million articles in mid-2006, as their first source of information because of the extremely high quality of the work it includes and the up-to-date accuracy of the information.

Since the launch of the *Wikipedia*, other forms of participatory, peer-based production have been created. *OhmyNews* based in Seoul, Korea, is an extremely successful adventure in participatory journalism, to which, by June 2006, it is estimated that more than forty-one thousand citizen reporters will have contributed news stories and commentary for review, editing, feedback, and publication by *OhmyNews* professional journalists.[31] With a 30 percent rejection rate coupled with editorial workshops intended to improve the reporting and writing skills of its contributors, and summer conventions where selected groups of contributors from twenty-one countries meet face-to-face to learn more about each other and their historic contribution to new media, *OhmyNews* is a remarkable instantiation of Engelbart's original vision to use networked communication to leverage collective intelligence. Another contemporary example that illustrates how the Web and new participatory media are being used to empower collective intelligence, or "smart mobs," as Howard Rheingold has suggested, is the Web-based research and development initiative of Eli Lilly, called InnoCentive®.[32] InnoCentive is an online, incentive-based initiative created to tap a networked global R&D community to solve problems in the fields of pharmaceuticals, biotech, agribusiness, and basic and specialty chemicals. InnoCentive offers companies an opportunity to increase their R&D potential by enrolling as "Seekers," posting challenges to a confidential online forum, and gaining access to leading scientific minds. More than seventy thousand scientists and engineers from more than 170 countries around the world are registered with the InnoCentive network. Individually, and potentially in virtual, distributed teams, they tackle problems posed to the network with the possibility of winning financial rewards from ten thousand to more than one hundred thousand dollars for solving these challenges.

While Lilly's spin-off InnoCentive reminds us that organizations of corporate capital are quick to seize upon smart mobs to increase their efficiency in a global, networked economy, the examples of *The Beast*, the *Wikipedia*, and *OhmyNews* discussed above provide hopeful signs for the development of what I would call "techno-humanism": a new kind of critical approach that focuses humanistic aims through technology. If the humanities have a future in the current scene, which is increasingly shaped by a powerful technoscientific fusion of information technology, biotechnology, and nanotech, we will have to reorient our compass and rethink our methods. We can develop approaches that directly engage technology and treat it as more than simply a metaphor for examining the narratives of science fiction a generation out of date. I agree with and applaud Hayles, Hansen, Haraway, and others seeking a critical practice that will lead to a positive partnership among nature, humans, and intelligent machines rather than simply acquiescing in a view of the posthuman as an apocalyptic erasure of human subjectivity. But the cyborg or posthuman critique is already out of steam before it gets going, because it is second-order analysis dependent on the absorption of technological metaphors and their crafting as narrative by writers who are themselves reacting to and engaging the cultural imaginary of technology already in the making.[33] We are all witnesses to the fact that we are living in a period of unprecedented technological and scientific change. Such change is so rapid, in fact, that the cybernetic tradition providing the metaphoric stuff out of which Gibson's *Neuromancer*, Stephenson's *Snow Crash*, and Ridley Scott's *Blade Runner* were created was already criticized and rejected by second-generation AI workers, such as Brooks, who were rapidly constructing a new technical landscape based on some of the premises—such as the importance of embodiment—that Hayles, Haraway, and others were crafting as "critique" at the same time.

Moreover, while I deeply appreciate attention paid to literature, film, and images—to "old media"—I frankly think the critical spirit needs to be oriented toward new media such as video games, simulations, and database-driven streaming media: in short, to the media that actually use the technologies we want both to deploy and critique. If as humanists we cannot be active participants in the AI labs and the sites of nanotechnology and stem cell research, as "techno-humanists" we can be flies on the walls where these projects are created, so that we can render them public as fully as possible. Rather than examining the metaphoric use of technology in science fiction, film, print media, and ad campaigns, the path to that critical and productive engagement is to get *inside* the technology and use it for our own purposes.

This is where the current communications revolution can assist our projects, precisely in the ways suggested by collective enterprises such as the Cloudmaker detective group and the *Wikipedia* community. The success of decentralized modes of organization in the networked economy suggests that we can move from a command-and-control model of hierarchical, centralized organization and planning to new distributed modes of work that foreground cooperation and cultivation. These collaborative ways of working allow us to leverage the skills and expertise of individuals who can contribute to the solution of problems in which they have some valued stake. And we can deploy new media-communications technology to leverage the problem-solving capabilities of individuals across disciplinary boundaries for collective self-organizing objectives. This is the essence of techno-humanism.

What might such an effort in techno-humanism look like? I want to conclude this paper by offering a provisional example of how we might use new media technologies in the spirit of the *Wikipedia* and *OhmyNews* to publicize emerging developments in contemporary technoscience, and how in the process we might assist in developing more socially and ethically responsible forms of knowledge production. With more than three billion dollars in initial funding by the National Science Foundation (NSF) and other federal agencies, the National Nanotechnology Initiative (NNI) is one of the largest funding initiatives of new science and technology in history. These projects offer the context and opportunity for greater informed public participation in the start-up phase of emerging science and technology. Faced with potential erosion of public confidence in science and in governmental institutions charged with protecting the public's interests, science-studies researchers have been challenged to participate in this large-scale initiative to develop nanoscience and nanotechnology applications. Our role as science-studies researchers in these projects is to address issues related to the societal, economic, and ethical implications of the nanoscience and technologies resulting from government-funded efforts.

The reason for including science-studies researchers in these efforts is that the government agencies involved in promoting nanoscience and nanotechnology in the United States and in Europe have determined that attempts to communicate their scientific and technological visions to the public will be unsuccessful unless the cultural attitudes and preconceptions of the public are fully engaged at the outset. Independent studies done with experimental issue groups in the United States by researchers at the Woodrow Wilson Foundation have found that a staggering 95 percent of the participants in their studies expressed little or no trust in government or in industry to manage risks associated with nanotechnology. Moreover, higher education (college degree or higher) was related to low trust in government to manage any risks. No other demographic variable showed any significant link.[34] In order to prevent a negative climate from enveloping nanotechnology, the National Nanotechnology Initiative (NNI) has asserted that ethical, societal, and legal considerations are more likely to be relevant if they are not undertaken in reaction to research or its applications after the fact. It is important for researchers and the public to consider ethical and societal questions proactively, before the research is a *fait accompli*. Accordingly, the NNI has mandated that projects it funds should integrate concerns about societal, ethical, and legal issues (SEI) into, rather than allowing them to emerge in conflict with, nanotechnological research and applications in their earliest design phases, and that strategies be developed for involving the public more in the decision-making process by addressing concerns about the potential hazards and disadvantages of emerging technologies upstream, while the science is in the making.

I am currently part of a team of researchers working to develop the NSF-funded Center for Nanotechnology in Society sited at the NanoSystems Institute at the University of California, Santa Barbara. The aim of our center is to address the societal, ethical, economic, and legal implications of emerging nanotechnology. Part of my role in the project is to design software and Web-based tools for enabling informed public engagement with nanoscientists and engineers about their work and its potential ethical, legal, and economic implications. One of these tools is what we call "benchside consultation" for addressing areas of ethical and legal concern.

The ability to address the larger societal and ethical implications of nanotechnology and manufacturing will require building resources to enable scholars from the social sciences and humanities as well as the broader public to understand the nature and origins of nanotechnology—developments in science and engineering that have taken place at a breathtaking rate in less than two decades. Current work in nanotechnology is based on a convergent, synergistic combination of four major domains of science and technology, each of which is currently progressing rapidly: (1) nanoscience and nanotechnology; (2) biotechnology and biomedicine, including genetic engineering; (3) information technology, including advanced computing and communications; and (4) cognitive science, including cognitive neuro-science. This convergence of diverse technologies, known as NBIC, is based on material unity at the nanoscale and on technology integration from "the bottom up." To provide background and context for addressing the societal and ethical implications of current efforts in nanoscience and bionano-technology, the historians of science forming part of the Center for Nano-technology in Society[35] (namely, Patrick McCray, Cyrus Mody, and myself, along with others as our project grows) are researching the very recent his-tory of the key technologies and scientific developments that are enabling NBIC convergence. Building on our combined previous work in the histo-ries of MEMS (micro-electromechanical systems), instrumentation (such as the atomic force microscope and scanning tunneling microscope), bio-informatics, polymerase chain reaction (PCR), sequencing technology for genomics, fluorescent-activated cell sorting, and combinatorial chemistry, we are researching topics such as the development of MEMS technology from the late 1980s up through the development of the first "laboratory on a chip system" in 1997, just part of the dazzling explosion in develop-ment of all sorts of microanalyzers, from microchips, bioelectric chips, and gene chips to microarray technologies, technologies of microfluidics, and related systems that have set the stage for bionanofabrication. Our proj-ect will also concentrate on emergent technologies related to spintronics research, one of the major research areas of the NanoSystems Institute at UC Santa Barbara.[36] Relevant publications, technical documents, lectures and conference reports, videos, interview transcripts, and so on related to the recent history of bionanotechnology and bionanofabrication will form one key component of our digital archive, accessible via our Web site. Ap-propriate links will also be established between the electronic archive and the NanoBank being constructed at UCLA and the California NanoSystems Institute with facilities at UC Santa Barbara, UCLA, and the National Bureau of Economic Research in Cambridge, Massachusetts. (NanoBank will be an integrated database and Web-deployed digital library containing currently disparate data sets, such as articles, patents, firm financial reports and di-rectory listings, and university data.) In addition to the recent historical origins of nanoscience and nanofabrication, we will also explore and docu-ment the recent development of work on DNA engineering, such as the use of self-assembly of DNA nanostructures for biomolecular computing and nanofabrication.

While scientists frequently lead the way in calling attention to the po-tential benefits and risks of their work, it is difficult to impose on them the requirement that they imagine the policy context that might affect how their research is done or applied in addition to foreseeing their work's full range of societal consequences, intended or otherwise. In areas of convergent science and technology such as bionanomanufacturing, where the work is highly interdisciplinary, attending to all the possibly relevant perspectives is difficult if not impossible. Our goal is to assist nanoscientists and engineers in identifying and addressing potential societal and ethical implications of their work by bringing them together with groups of knowledgeable dialogue partners from the fields of ethics, research policy, law, intellectual property, and environmental studies who might help formulate and manage issues with potential ethical and social impact arising from their work in nanotech. We envision several interrelated project components to facilitate our goals of (1) assisting the nanoscientists we work with in proactively identifying and addressing potential ethical issues in their work and its broader societal impact, and (2) making the public aware of these issues and involving them as deeply as possible in the dialogue.

One approach to such dialogues is the NanoJury being piloted in Great Britain. The NanoJury UK brings together twenty randomly chosen people from different backgrounds who hear evidence about a wide range of possible futures and the role that nanotechnologies might play in them. The experiment began this past year. Over a period of five weeks, the jurors heard from a variety of witnesses with widely varying perspectives, upon which they will draw in formulating a set of recommendations to be made public. These recommendations will shape the debates on how this emerging and potentially revolutionary technology should develop. The NanoJury is co-sponsored by the Interdisciplinary Research Collaboration in Nanotechnology at Cambridge, Greenpeace UK, the *Guardian*, and the Policy, Ethics and Life Sciences Research Centre at the University of Newcastle.[37]

The aims of the NanoJury are similar to those of our own projects:

- to provide a potential vehicle for people's informed views on nanotechnology to have an impact on policy;
- to facilitate a mutually educational dialogue among people with diverse perspectives and interests, including critical and constructive scrutiny of the hopes and aspirations of those working in the nanotech-related sectors by a wider group of citizens; and
- to explore the potential for deliberative processes to broaden discussions about nanotechnology research policy—both in terms of the range of issues and the diversity of people who are given a say.

Our approach to achieving these goals is to couple the face-to-face meetings of the sort used by the NanoJury with new Web-based media technology aimed at facilitating wider collaboration, and in some cases virtual "town meetings" around specific topics. New media technologies allow us to couple the face-to-face meetings with a growing body of literature, legal opinions, blogs, and other related Web-based content through the combined use of video annotation, "podcasts" with audio-blogging, and semantic Web technology. Our approach to opening up the black box of the lab has two related components that are discussed below.

Our first concern is to get "inside the lab." A critical component of the Nanotechnology Initiative is to enable more direct public participation in the internal workings of the new science and technology while it is still in the making: to make publicly funded science visible and accessible. Among the distinctive ways our project attempts to address these imperatives is through a series entitled *Inside the Lab*. *Inside the Lab* will feature monthly interviews with the nanoscientists and engineers we are affiliated with in our project. We use this venue to learn about their work, their motivations for doing it, their approaches to their projects, and the impact they envision for their work. The interviews will be conducted by SEI staff and students working in the field of ethics in engineering and science, and will be made available on our Web site through a video annotation application (see figure 10.1) that allows viewers to comment upon the interviews and pose questions to the scientists and interviewers about their work. Through an audio-blogging application we have developed, interviews and town meetings can be posted on our site. People who want to comment on a specific point can annotate the podcast exactly as in our video annotation module. We are also developing the capability for listeners to call in their comments by phone or through a voiceover IP application, such as Skype. The voice annotations will be integrated into an "enhanced podcast."

We call a second exciting direction we are pursuing "benchside consultation." Analogous to clinical consultations where bioethicists and ethics committees meet at the bedside for real-time discussion, analysis, and resolution of ethical issues in the clinical setting, we are attempting to implement a series of face-to-face dialogues between SEI researchers and nanoresearchers about their work. These sessions involve one or more researchers in the projects we are attached to with SEI staff and outside area experts, including members of the wider community as appropriate. The consultation session will provide a forum in which to identify, discuss, and decide how to act upon potential ethical and policy issues that may arise

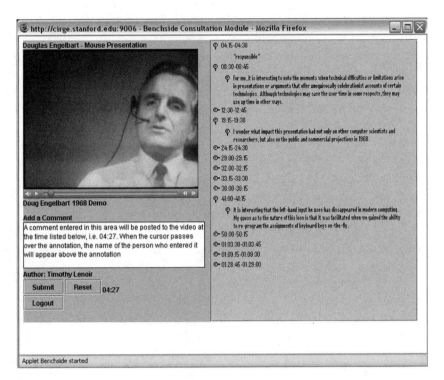

FIGURE 10.1 | Video annotation and benchside consultation tool. An mpeg4 streaming video of an event is hosted in a Wiki. Time-stamped comments and extended threaded commentary can be attached to specific time-locations in the video. Each commentator is allowed to edit his/her own annotations. As the video scrolls forward, existing annotations open automatically and close automatically after an extended period. The event commented upon here is Douglas Engelbart's 1968 demo of his "online system," which featured the world's debut of the computer mouse, hypertext, online editing, and collaboration with interactive video support. Copyright Tim Lenoir.

within ongoing or planned research projects but are not addressed by other institutional mechanisms. The entire session is recorded either as video or audio and made accessible initially to the nano and SEI researchers for annotation, comment, and reaction through the Web-based video/audio annotation module developed by our SEI team. A digital library of relevant supporting background documents will be assembled from ethics, law, and policy case studies in biomedical and engineering fields as guidance for establishing precedents. The outcome of a consultation would be a summary brief with a policy recommendation. These modules can be drawn upon for later use and, if they pass screening for issues of confidentiality and nondisclosure, made available to the wider public for viewing and participatory commentary. The most interesting among them will be expanded and shaped into publishable case studies and educational modules suitable for science and engineering ethics courses.

The cyborg has performed yeoman's service for science studies. Examining narrative constructions, metaphors, and discursive networks under the cyborg banner has powerfully exposed assumptions as well as political, military, and economic interests embedded within technoscience since the 1980s. I am not suggesting that we discontinue such valuable work. However, with the predicted convergence of nanoscience and nanotechnology, biotechnology and biomedicine, advanced computing and communications, cognitive science and neuroscience, fueled by the massive funding of the nanotechnology initiatives of several nations, we move toward a qualitatively different configuration of biological bodies and the human. When the biological and the digital are no longer ontologically distinct but inhere in one another, we will most likely confront a more elusive trope than the cyborg, conceived narrowly as a human-machine hybrid, where the machine is a (possibly dangerous) supplement, tool, and extension of the human. As Haraway has suggested, these new, "posthuman" machines will undoubtedly be much livelier, and they may just be "us." Future science-studies scholars will need to unpack those new metaphors, discursive networks, and tropes—whatever they turn out to be. My proposal has been that in the interest of serving (and preserving) humanistic interests, we should not wait until the new discursive frameworks have already formed beyond what Jessica Riskin metaphorizes in this volume as the "Sistine Gap." Rather, we should use new media technology to get inside the labs, make the science public, and engage this new technoscience in the making. The tools I have used to illustrate how this might look may not escape criticism and may ultimately not even be the right tools for the job, but my intent is to encourage the development of means for facilitating attention to societal, ethical, and environmental concerns at the sites of production of scientific knowledge.

NOTES

1. Donna Haraway, "Manifesto for Cyborgs: Science, Technology, and Social Feminism in the 1980s," *Socialist Review* 80 (1985): 65–108.

2. Andy Clark, "Natural-Born Cyborgs?" in *The New Humanists: Science at the Edge*, ed. John Brockman (New York: Barnes & Noble Books, 2003), 67–78. Anthropologists such as André

Leroi-Gourhan, among others, have argued that tool use, particularly tools of communication such as speech and other cognitive technologies, has been the defining force of human evolution. Expanding on this view, cognitive scientist Andy Clark defends the "extended mind" thesis that humans, more than any other creature on the planet, deploy nonbiological elements (instruments, media, notations) to complement our basic biological modes of processing; that our brains actively dovetail their problem-solving activities with a variety of nonbiological resources; and that these larger systems thus created evolve. Viewed in this light as thinking and reasoning beings whose minds and selves are spread across biological and nonbiological circuitry, we have always been human/technological symbionts, or, in Clark's phrase, "natural-born cyborgs."

3. This is frequently referred to as the Fourth Discontinuity, the other three being (1) the discontinuity between humans and the cosmos (overcome by Copernicus), (2) the discontinuity between humans and animals (overcome by Darwin), (3) the discontinuity between human consciousness and the human unconscious (overcome by Freud).

4. N. Katherine Hayles, *How We Became Posthuman: Virtual Bodies in Cybernetics, Literature, and Informatics* (Chicago: University of Chicago Press, 1999), 2–3.

5. Donna Haraway, "Cyborgs and Symbionts: Living Together in the New World Order," in *The Cyborg Handbook*, ed. Chris Hables-Gray (New York: Routledge, 1995), xi–xx.

6. Bruno Latour, *Science in Action: How to Follow Scientists and Engineers Through Society* (Cambridge, MA: Harvard University Press, 1988), 4.

7. N. Katherine Hayles, "The Life Cycle of Cyborgs: Writing the Posthuman," in Hables-Gray, *Cyborg Handbook*, 322.

8. For a similar discussion of the body as the primary sociocultural product and engagement with virtual reality as an opportunity for feminist rethinking of the production of alternative models, registers, alignments, perspectives, and corporealities other than the functioning male body under the name of the "neutral human," see Elizabeth Grosz, *Architecture from the Outside: Essays on Virtual and Real Space* (Cambridge, MA: MIT Press, 2001), especially chap. 2, "Lived Spatiality (The Spaces of Corporeal Desire)," 31–47.

9. Edwin Hutchins, *Cognition in the Wild* (Cambridge, MA: MIT Press, 1995).

10. Antonio Damasio, *Descartes' Error: Emotion, Reason, and the Human Brain* (New York: Putnam, 1994), and *The Feeling of What Happens: Body and Emotion in the Making of Consciousness* (New York: Harcourt Brace, 1999).

11. See especially Francisco J. Varela, Evan Thompson, and Eleanor Rosch, *The Embodied Mind: Cognitive Science and Human Experience* (Cambridge, MA: MIT Press, 1991).

12. George Lakoff, *Women, Fire and Dangerous Things: What Categories Reveal about the Mind* (Chicago: University of Chicago Press, 1989).

13. Mark Johnson, *The Body in the Mind: The Bodily Basis of Imagination, Reason, and Meaning* (Chicago: University of Chicago Press, 1987).

14. See Clark, "Natural-Born Cyborgs?" 4.

15. Brooks defines the terms *situatedness* and *embodied* as follows: "A situated creature or robot is one that is embedded in the world, and which does not deal with abstract descriptions but, through its sensors, with the here and now of the world, which directly influences the behavior of the creature. An embodied creature or robot is one that has a physical

body and experiences the world, at least in part, directly through the influence of the world on that body. A more specialized type of embodiment occurs when the full extent of the creature is contained within that body." See Rodney Brooks, *Flesh and Machines: How Robots Will Change Us* (New York: Pantheon, 2002), 51–52.

16. Brooks, *Flesh and Machines*, 180.

17. For a discussion of the computational view of nature see, Tommaso Toffoli, "Physics and Computation," *International Journal of Theoretical Physics* 21 (1982): 165–75; Norman Margolus, "Looking at Nature as a Computer," *International Journal of Theoretical Physics* 42 (2003): 309–27, also available at MIT's Web site: http://www.ai.mit.edu/people/nhm/looking-at-nature.pdf (accessed 18 January 2006); Neil Gershenfeld, *The Physics of Information Technology* (Cambridge: Cambridge University Press, 2000), chap. 15, and 252, 282–84. See also Neil Gershenfeld, *When Things Start to Think* (New York: Holt, 1999); and, of course, Stephen Wolfram, *A New Kind of Science* (Champaign, IL: Wolfram Media, 2002).

18. Bruno Latour, "Why Has Critique Run Out of Steam?: From Matters of Fact to Matters of Concern," *Critical Inquiry* 40 (Winter 2004): 225–48.

19. Daniel Hillis, *Pattern on the Stones* (New York: Perseus Books, 1999), 138.

20. Leonard M. Adleman, "Molecular Computation of Solutions to Combinatorial Problems," *Science*, 11 November 1994, 1021–24.

21. For examples of these directions of research, see Jessica O. Winter, Timothy Y. Liu, Brian A. Korgel, and Christine E. Schmidt, "Recognition Molecule Directed Interfacing Between Semiconductor Quantum Dots and Nerve Cells," *Advanced Materials*, 16 November 2001, 1673–77; Nadrian C. Seeman and Angela M. Belcher, "Emulating Biology: Building Nanostructures from the Bottom Up," *Proceedings of the National Academy of Sciences* 99, supp. 2 (30 April 2002): 6451–55; Seung-Wuk Lee, Chuanbin Mao, Christine E. Flynn, and Angela M. Belcher, "Ordering of Quantum Dots Using Genetically Engineered Viruses," *Science*, 3 May 2002, 892–95.

22. See, for instance, the work by John Reif and Thomas LaBean, with Hao Yan, Sung Ha Park, Gleb Finkelstein, John H. Reif, and Thomas H. LaBean, "DNA-Templated Self-Assembly of Protein Arrays and Highly Conductive Nanowires," *Science*, 26 September 2003, 1882–84; Jaidev P. Patwardhan, Chris Dwyer, Alvin R. Lebeck, and Daniel J. Sorin, "Circuit and System Architecture for DNA-Guided Self-Assembly of Nanoelectronics," *Proceedings of Foundations of Nanoscience: Self-Assembled Architectures and Devices* (April 2004): 344–58, also available at the Duke University Web site: http://www.cs.duke.edu/~alvy/papers/fnano.pdf (accessed 18 January 2006).

23. See Eugene Thacker, "What Is Biomedia?" *Configurations* 11 (2003): 47–79, esp. 52, 58–59, 75. Thacker has elaborated on these ideas further in his important book *Biomedia* (Minneapolis: University of Minnesota Press, 2004).

24. See Timothy Lenoir and Sha Xin Wei, "Authorship and Surgery: The Shifting Ontology of the Virtual Surgeon," in *From Energy to Information: Representation in Science, Art, and Literature*, ed. Linda Henderson and Bruce Clarke (Stanford, CA: Stanford University Press, 2002), 283–308.

25. For detailed information on this ongoing initiative, see the Web site of the collaborative MIT/Cambridge University project, AutoID (http://www.autoidcenter.org/), the

goals of which are described on the project's home page: "We are on the brink of a revolution of 'smart products' that will interconnect everyday objects, consumers, and manufacturers in a dynamic cycle of world commerce. In October 1999, the Uniform Code Council (creators of the UPC) joined with MIT, Procter & Gamble, and The Gillette Company to make their vision of forming one seamless global commerce network a reality. Corporate and industrial sponsors have invested their participation and funding on a variety of levels in recognition of their integral role in helping develop a technology that will shape a new world of optimally efficient and beneficial global commerce."

26. See the Web site at http://cooltown.hp.com/ (accessed 15 September 2005).

27. For an outstanding study of The Beast and pervasive gaming, see Jane McGonigal, "'This Is Not a Game': Collective Subjectivity and Immersive Entertainment," Digital Arts and Culture Proceedings, 2003, available online at the Web site of Sean Stewart, one of four puppetmasters for The Beast: http://www.seanstewart.org/beast/mcgonigal/notagame/paper.pdf (accessed 17 January 2006). I am indebted to Jane McGonigal's work for many insights in this part of my paper.

28. On the first day of its release, the military added additional servers to handle the traffic, a reported five hundred thousand downloads of the game. The site continued to average 1.2 million hits per second through late August 2002. Gamespot, a leading review, not only gave the game a 9.8 rating out of a possible 10, but also regarded the business model behind the new game as itself deserving of an award.

29. Jack Thorpe, "Perspectives on Distributed Simulation, Persistent Worlds, Command, and Control," paper presented at a conference at the MOVES Institute, Monterey, CA, 5–7 August 2003; available online at http://movesinstitute.org/openhouse2003-wslides .html (accessed 18 January 2006). Ret. Col. Thorpe is one of the original architects of the military SIMNET.

30. Hayles, How We Became Posthuman, 2–3.

31. See Jean Kim, "Journalism as a Conversation," Harvard Nieman Reports 59, no. 4 (Winter 2005): 17–19; available online at http://www.nieman.harvard.edu/reports/05-4NRwinter /05-4NFwinter.pdf (accessed 17 January 2006) and at the Web site of OhmyNews: http:// english.ohmynews.com/articleview/article_view.asp?menu=c10400&no=268058&rel_ no=1 (accessed 17 January 2006).

32. Howard Rheingold, Smart Mobs: The Next Social Revolution (New York: Basic Books, 2003).

33. By "second-order," I do not mean that the works of these authors are derivative from "secondary" or popular sources, as some readers of this essay have presumed. Far from it. Their works are pathbreaking in their use of so-called primary sources. Rather I am referring to the problem that in dealing with metaphor, narrative, and description as the focus of critique, this approach deals with the effects of a dynamic technoscientific infrastructure already in place and difficult to dislodge or divert.

34. See Jane Macoubrie, "Informed Public Perceptions of Nanotechnology and Trust in Government," in Report of the Project on Emerging Nanotechnologies (Woodrow Wilson International Center, September 2005), available online at http://www.nanotechproject.org/; Michael D. Cobb and Jane Macoubrie, "Public Perceptions about Nanotechnology: Risks, Benefits and

Trust," *Journal of Nanoparticle Research* 6 (2004): 395–405. For the launching of the national discussion on this theme, see M. C. Roco, "Broader Societal Issues of Nanotechnology," *Journal of Nanoparticle Research* 5 (2003): 181–89.

35. See the home page for the Center for Nanotechnology in Society: http://www.cns .ucsb.edu/ (accessed 18 January 2006).

36. For more on the goals of spintronics research, see J. A. Gupta, R. Knobel, N. Samarth, and D. D. Awschalom, "Ultrafast Manipulation of Electron Spin Coherence," *Science*, 29 June 2001, 2458–61; S. A. Wolf, D. D. Awschalom, R. A. Buhrman, J. M. Daughton, S. von Molnár, M. L. Roukes, A. Y. Chtchelkanova, and D. M. Treger, "Spintronics: A Spin-Based Electronics Vision for the Future," *Science*, 16 November 2001, 1488–95.

37. For information on the NanoJury UK, see http://www.nanojury.org/ and the discussion on the Web site of the Policy, Ethics, and Life Sciences Research Centre at the University of Newcastle: http://www.ncl.ac.uk/peals/dialogues/juries.htm (accessed 18 January 2006). The NanoJury project is discussed in "Citizens' Jury to Tackle Nanotech," *BBC News*, "Science and Nature," 23 May 2005, online at http://news.bbc.co.uk/1/hi/sci/ tech/4567241.stm (accessed 18 January 2006). For the Greenpeace Report on the Nano-Jury UK, see http://www.greenpeace.org.uk/MultimediaFiles/Live/FullReport/7333.pdf (accessed 18 January 2006).

11   Nanobots and Nanotubes: Two Alternative Biomimetic Paradigms of Nanotechnology

BERNADETTE BENSAUDE-VINCENT

Nanotechnology, the hottest engineering trend of the day, is being publicized with revolutionary claims of "shaping the world atom by atom." Science writers and Nobel Prize winners seem engaged in a rhetorical competition to describe a new era of complete "understanding and control" over matter, of "limitless possibilities to create new things," or of "the builder's final frontier."[1] The interdisciplinary character of nanoscience is an important facet of these claims to innovation without boundaries. The people who converged to work on the nanoscale came not only from traditional disciplines such as biology, physics, chemistry, but also from more recent fields such as computer science, materials science, and engineering. Indeed, it is a distinctive feature of emerging disciplines such as nanoresearch that they merge science and technology under the same umbrella. The overall impression of convergence is reinforced by the fact that nanoscientists and engineers derive a common inspiration from living organisms. Biomimetics, above all else, is the cement that holds together various groups exploring the potentials of the nanoscale.

In this small theatre of nanoscientific research, nanobots and nanotubes are the two principal actors. *Nanobot* is a term coined by Ray Kurzweil, a computer engineer who designed the first reading machine for the blind and who founded Kurzweil Computer Products, Inc., a firm renowned for its optical character-recognition systems. He also authored two books—*The Age of Intelligent Machines* and *The Age of Spiritual Machines*—that describe futuristic beings acting as extensions of the human brain. Kurzweil's nanobots still inhabit the domain of science fiction, but this has not prevented them from becoming a driving force in the field. With its goal of building smaller and smaller machines, nanotechnology appears to be the natural continuation of the miniaturization process initiated in the 1960s. In this capacity, nanotechnology claims to confirm and illustrate Gordon Moore's famous law, formulated in 1965, according to which the power of integrated circuits doubles every eighteen months. In contrast to nanobots, nanotubes live in the present. They are observable entities made of carbon (C_{60}) synthesized in laboratories and manufactured by a number of spin-off companies. The ball-shaped C_{60}—known as fullerene—was discovered by chance in the 1980s, but the technological potential of hollow nanostructures did not emerge until after the synthesis of carbon nanotubes in the 1990s.

It seems to me that nanobots and nanotubes emerged from two different approaches to the nanoscale. From an epistemological perspective, there is a striking heterogeneity beneath the alleged disciplinary convergence in the race toward control of the nanoscale. The contrast that I would like to emphasize in this paper is not one between fictitious and real entities, for both nanobots and nanotubes remain laboratory curiosities with a possible future as commercial products. Nor is it a contrast between popular representations and the so-called "serious" scientific understanding of nanoresearch. Both nanobots and nanotubes have been widely touted in the popular media, and each promises to bring about nothing less than a revolution in our daily life. Instead, I argue that this new research field, to the degree that it is defined in terms of the imitation of life, seems to be evolving according to two distinct scientific paradigms.[2]

The distinction between nanobots and nanotubes epitomizes the divergence of these schemes. Nanobots involve a mechanistic view of nature in which living organisms are understood as automata made of passive and uniform matter and informed by a program. For Ray Kurzweil, the objective of nanotechnology is to insert as much information as possible into a tiny volume of matter, thereby enabling the conversion of information into physical products. The resulting molecular constructs will be intelligent or even spiritual machines.[3] In contrast to this mechanistic approach, nanotubes make use of strategies, such as self-assembly, that are exemplified in life itself. Carbon atoms have supplied the essential material of this line of research due to their inherent disposition toward the constitution of balls and tubes. Richard Smalley, who was awarded the Nobel Prize for the self-assembly of nanotubes and who later started a nanotube production facility at Rice

University, emphasized the importance of the inner dynamics embedded in carbon atoms in his Nobel Lecture entitled "Discovering the Fullerenes": "The discovery that garnered the Nobel Prize was the realization that carbon makes the truncated icosahedral molecule, and larger geodesic cages, all by itself. Carbon has wired within it, as part of its birthright ever since the beginning of this universe, the genius for spontaneously assembling into fullerenes."[4]

Even though in most research practices the mechanistic and the organicist models often merge, the difference of philosophical assumptions is more striking in semipopular articles on nanotechnology. I use the popular works of Eric Drexler and Ray Kurzweil to illustrate the mechanistic approach, and the writings of Richard Smalley and a few other chemists to represent the organicist model.

A STANDARD HISTORY

In order to better understand the origins of bio-inspiration, let me first briefly sketch the standard historical narrative that appears in most papers and books about nanotechnology. Like most disciplines, nanoscience has shaped its history around heroic figures. Richard Feynman's 1959 speech "There Is Plenty of Room at the Bottom" is inevitably quoted as a prophetic inspiration of nanotechnology.[5] Already within this common starting point, however, there is the potential for two distinct historical narratives. A continuous narrative puts the emphasis on Feynman's striking claim that it should be possible to store the Library of Congress (twenty-four billion volumes) on the head of a pin. From this perspective, nanotechnology appears to continue the miniaturization process that drives information technologies. Hence the equally common reference to Gordon Moore's famous law on the ever-increasing power of nanocircuits, an allusion that views nanoscience in terms of an already existing technological momentum. More discontinuous narratives, however, place the emphasis on Feynman's suggestion of storing information by rearranging atoms as we please. From this angle, the shift from top-down to bottom-up strategies appears as a radical break or revolution in technology. Thus did Feynman turn the physicists' attention toward biological material, where enormous amounts of information could be stored in exquisitely small spaces.

While Feynman holds the conventional role of the prescient precursor of nanotechnology, Eric Drexler presented himself as the founder. His 1986 *Engines of Creation* was an extremely successful book promising that human culture would be drastically transformed by a new kind of engineering at the nanoscale.[6] Drexler clearly cast himself as the prophet of a new era (*The Coming Era of Nanotechnology* is the subtitle of his first book). His insight into the tremendous technological potentials provided by the direct rearrangement of atoms was fascinating to scientists and the public at large.

THE MACHINERY MODEL

For Drexler, the first benefit of referring to biology is a lesson of optimism. Biology shows that molecular machines can exist and build more molecular machines. There is no mystic stuff in living systems. And Drexler argues that nanoscientists will do much better than biology because they use more reliable materials.[7] Drexler thus rejuvenated the positivist crusade of nineteenth-century synthetic chemists like Marcellin Berthelot against the limits imposed by superstition or by the metaphysical belief in a vital force.[8] Drexler and his supporters combat the skeptics by insisting that the existence of life itself is the proof that nanomachines are feasible. For instance, Marvin Minsky from the MIT Media Lab and AI Lab has argued as follows: "It seems quite strange for anyone to argue that you cannot build powerful (but microscopic) machinery—considering that our very own cells prove that such machines can indeed exist. And then if you look inside your cells you will find smaller machines that cause disease. Most arguments against nanotechnologies are arguments against life itself."[9]

This quotation makes it clear that life provides more than just an invitation to build nanomachines; it constitutes an imperative to do so. For Kurzweil, evolution has created organisms that are able to overcome the limitations of the human brain by inventing computational technology. It is thus the evolution of life itself that presides over the building of nanobots. This vague notion of a process of hominization is all Kurzweil needs to establish himself as the prophet of a new era of spiritual machines. His argument rests on two postulates: (1) human technologies are the continuation of biological evolution; just as the flint chipper was an extension of the human hand, so the nanorobot extends the human brain; (2) exponential growth is the feature of any evolutionary process, of which technology is a primary example (Moore's law). The logical conclusion of this syllogism is this: the golden age of nanotechnology will come within a couple of decades as an unavoidable future. Because it is the continuation of the natural process of evolution, we have no choice in the matter. We must simply accept it and adapt our society to a world shared with nanobots.[10]

What is "life" for these prophets of nanotechnology? It is basically a mechanism defined by analogy with macroscopic machines. This is the second lesson Drexler derived from biology. From the publication of his very first article in 1981, Drexler deliberately chose to describe the components of biological cells by comparing them with the macroscopic components of machines.[11] He thus found struts, cables, fasteners, glue, motors, bearings, containers, pumps, and clamps in living material. His model is a machine made of a set of independent pieces—a few building blocks—mechanically assembled in the manner of a Lego construction. Molecules are rigid building blocks similar to the parts of Tinker Toys or Meccano construction sets. Molecular machinery thus clearly belongs to solid rather than to fluid mechanics. It is the extrapolation to the smallest scale of simple mechanical machines. The assembly process itself is described with the same metaphor of "mechanosynthesis," or "the use of mechanical control to guide the placement of molecules so as to build complex objects."[12] Nanosystems are like factories engaged in a rigid framework of controlled motions using the building blocks of matter as raw materials.

The functions performed by the various pieces of molecular machinery are also essentially mechanical. They position, move, transmit forces, carry, hold, store, and so forth. Drexler likes the metaphor of "molecular hands" manipulating nano-objects and placing them wherever they need to go to perform the desired function. He thus prefers the term molecular manufacturing (literally, "hand-making") to the word nanotechnology, which he sees as a buzzword overused by microcomputer engineers. The key term, however, is molecular assembler. This is the magic wand that binds together the pieces in an arrangement, allowing them to perform useful tasks. Molecular assemblers are "devices able to guide chemical reactions by positioning reactive molecules with atomic precision."[13] They are neither specific nor individual molecules. They are described as universal, all-purpose assemblers that can assemble all kinds of materials in the same way that ribosomes can assemble all kinds of proteins. They may be inspired by John von Neumann's self-replicating automata, which were able to grab components out of their location and put them together according to programmed instructions. Similarly, Drexler's assemblers would move atoms and selectively bind them. Although Drexler's citation of von Neumann in Engines of Creation is limited to his studies of self-replicating machines, he undoubtedly also borrowed the notion of "universal assemblers" from the same source. Drexler's program thus seems to combine two models of machines. On the one hand, his description of molecular manufacture rests on a classic mechanics requiring only space, matter, and motion. In this sense, his matter is like Boyle's uniform, catholic matter. Molecular machines, like clock mechanisms, require the hands and the brain of a clock-maker. On the other hand, Drexler implicitly refers to computational machines, but without facing the challenge of complexity that von Neumann clearly prophesied.[14]

The third possibility that biology opened up for Drexler is that of "a world in which digital data can be used to control general-purpose machines that will put the fundamental building blocks of matter in place to build almost everything."[15] What Drexler essentially retains from biological systems is that they operate under programmatic control. The DNA-RNA system provides the code and the instructions according to which the machine operates. Protein assembly works according to rigid instructions, in a clean and efficient manner. By comparison, Drexler explains, conventional chemical reactions are extraordinarily messy: "Whereas engineers ruling a chemical plant must work with vats of reacting chemicals (which often misarrange atoms and make noxious by-products), engineers working with bacteria can make them absorb chemicals, carefully rearrange the atoms, and store the product or release it into the fluid around them."[16] Chemistry looks so primitive and dirty when compared to protein machines that Drexler wonders how chemists, lacking the "molecular hands with which to put the parts where they want them," have managed to achieve such remarkable things.[17]

In living things, then, Drexler finds a precious guide to improving chemical technologies. Enzymes are his favorite model of assemblers. "[Enzymes] assemble large molecules," he explains, "by 'grabbing' small molecules from the water around them, holding them together so that a bond forms."[18] In this manner they assemble DNA, proteins, and many other biological items. It should therefore be possible to put them to work on metal ions or complex structures in order to produce molecules with the precision of programmed machines. But if enzymes and proteins show the way to build nanomachines, they do not provide a perfect model for nanotechnology. Drexler proposes to use protein machines only for the first generation of nanomachines because they present serious flaws as engineering materials. The amino acids of which they are composed are simply not tough enough for the construction of nanomachines. Drexler's ambition is to mimic life's devices working under genetic instructions in order to build machines more robust than organisms.

Kurzweil also conceives of nanotechnology in terms of the impression of information upon tiny pieces of solid material, but his goal is less to mimic living structures than to build intelligent machines capable of overcoming the limits of living creatures. "In my view," he testified before the U.S. House of Representatives, "the most significant implication of the development of nanotechnology and related advanced technologies of the twenty-first century will be the merger of biological and nonbiological intelligence."[19] Kurzweil thus joins the crowd of AI artisans who assert that nanotechnology is the means, but artificial intelligence is the end. Nanotechnology is only one aspect of the mainstream of twenty-first-century technology for Kurzweil; only in conjunction with existing trends in information technology and genetic engineering will it bring about a new era where nanorobots replace humans in most of their activities.

In summary, the propagandists of nanotechnology have both near-term and long-term expectations for their research. The immediate purpose is to design nanomachines able to perform the jobs nowadays done either by macro- or micromachines or by humans. But the ultimate aim is to create intelligent, self-replicating machines that can perform more tasks than the human brain. This grandiose program relies on two main actors: matter and algorithms. Matter is conceived of as passive, deprived of action and of qualities. Its architecture is made of ultimate building blocks that can be individually manipulated to serve useful purposes through powerful techniques such as STM. The laws of nature are seen as universal and uniformly applying everywhere, with no major problem at the quantum scale. Machines are thus a collection of independent pieces *partes extra partes*, operated by a program. Given this highly mechanistic conception of matter, the utility of biology for partisans of nanobots is largely restricted to the idea of algorithmic instructions. This bio-inspiration, preferably taken from genetic engineering, suggests an all-powerful code or program providing instructions for moving matter, similar to the mechanisms of simple automata. Interestingly, the machinery paradigm combines an old mechanistic model inherited from Cartesian mechanics and a more recent computational model of machines inherited from cybernetics. Both of these rest on the assumption of a blind mechanism operating without intentionality. In other words, they lead to nonhuman or posthuman technologies.

THE ORGANICIST MODEL

A quite different perspective is conveyed by a number of scientists—mainly materials chemists—who jumped on the bandwagon of nanotechnology. For most of them the age of nanotechnology is not exactly a radical break. After all, building molecular architectures is what chemistry has done for many centuries. Still, these scientists acknowledge that the idea of making nano-structures atom by atom does pose new challenges that call for various inventive solutions. The Harvard chemist George Whitesides, who frequently uses the term *art* in his papers on nanotechnology, epitomizes their approach to the field.[20] Nanostructures belong to "art" both in the Aristotelian sense of *technê*, or design for specific purposes, and in the sense of skill, since they require the invention of astute and unconventional methods of nanofabrication. Chemists took inspiration from living structures before the term *nanotechnology* became fashionable. In 1978, for example, bio-inspiration led to the creation of a new branch of chemistry—supramolecular chemistry—which aims to obtain molecular recognition without the help of the genetic code through chemical processes that mimic the selectivity of biological processes.[21] According to Jean-Marie Lehn, who coined the term *supramolecular chemistry*, "it is one of chemists' major motivations to see that biology successfully made highly complex properties on a molecular basis."[22]

Chemists whose work is bio-inspired are less concerned with genetic programs and genetic engineering than with the stuff of which living things are made. Their main purpose is to understand what is unique about biological materials both in their structure and in the dynamics of their development and morphogenesis.[23] Living organisms are models for nano-design primarily because they present materials adapted, by design, to a set of performances. In contrast with the structures usually designed by engineers at the macrolevel, however, biomaterials are built from bottom up. Life operates by bonding atoms or groups of atoms instead of by carving a structure from raw materials.

Like Drexler, materials scientists and engineers have shaped an artificial-ist view of nature. For them, biological evolution is a kind of engineer designing the most efficient systems. Unlike Drexler and Kurzweil, however, they assume that nature is an insuperable engineer. Nature is not so much a model of order as a model of ingeniosity (*ingenium*). For instance, Richard Smalley describes the works of nature in superlative and playful terms:

> Nature has played the game at this level [the nanoscale] for billions of years, building stuff with atomic precision. Every living thing is made of cells that are chock-full of nanomachines—proteins, DNA, RNA, etc.—each jiggling around in the water of the cell, rubbing up against other molecules, going about the business of life. Each one is perfect right down to the last atom. The workings are so exquisite that changing the location or identity of any atom would cause damage.[24]

In trying to understand the tricks used by nature to solve her "engineering problems," materials chemists draw three major lessons from biology. First, biomaterials are interesting because they are never chemically pure. Whereas engineered materials are usually processed for their one particular property, biomaterials are multifunctional, composite structures. The interest of material scientists, especially chemists working on high-performance composites, is to learn something about the art of associating heterogeneous structures from nature itself. In their effort to design composite structures at the molecular level, they either turned their attention to such familiar materials as wood, bone, or mucus, or they studied modest organisms such as algae, spiders, insects, or seashells. In the 1990s, they suddenly became interested in traditional topics of natural history: mollusk shells, insect cuticles, spider-silk, and so on. These composite structures—associating hard and soft, combining inorganic and organic components, and capable of high performance—appeared to be ideal models for human technology for three major reasons.

They are models of functional diversity, being optimally adapted for a variety of tasks, including growth, repair, and recycling. The key to success in living organisms does not lie within a single building block engineered so as to concentrate all the instructions or information needed to operate the machine. Rather, biology teaches us that success comes with improving the art of mixing heterogeneous components. Consequently the focus is less on the ultimate components of matter than on the relations between them. Interfaces and surfaces are crucial because they determine the properties of the components of composite materials and how they work together. In other words, organicist nanoscience distinguishes itself from the culture of purity characterized by high-vacuum chambers by advancing an impure process of composition and hybridization that mimics natural materials. Biology does not provide a model of highly concentrated information as suggested by Feynman's famous talk. It is a model of interaction and composition. Nature challenges nanomaterials scientists to design a composite that displays more properties than the sum of the properties of its components. In this case biology provides a model of emergence opposed to the traditional view that reduces living matter to physical and chemical phenomena.

The second major lesson of nature concerns the way in which the components of life are assembled. Since assemblers were also Drexler's major concern, the contrast between the two paradigms on this point is striking. Drexler saw in enzymes the model universal assemblers, a sort of molecular hand capable of moving parts to the right position for assembly. This assertion provokes the skepticism of chemists such as Smalley, for they believe that atoms are not independent and that it is therefore impossible to move them individually. Because he doubts that any assembler can manipulate individual atoms, Richard Smalley entered a debate with Drexler about the possibility of manipulating atoms. There is not that "much room at the bottom," Smalley objected. Molecular fingers would obviously take up too much space and prevent the closeness needed for reactions at the nanoscale (the "fat fingers" problem). Inevitably manipulators would adhere to the atom being moved, making it impossible to move a building block where you want it to go (the "sticky fingers" problem).[25]

Drexler replied to these objections in an open letter: "My proposal is, and always has been to guide molecular synthesis of complex structures by mechanically positioning reactive molecules, not by manipulating atoms. This proposal has been defended successfully again and again, in journal articles, in my MIT doctoral thesis."[26] He complained that Smalley attempted to undermine his scientific credentials and that for positioning reactive molecules no computer-controlled "Smalley fingers" are required. Smalley responded by asking, "So, if the assembler doesn't use fingers, what does it use?" If there is some kind of enzyme or ribosome in self-replicating nanorobots, he reasoned, then there should be water inside because enzymes and ribosomes can only work in water, where they find all the nutrients necessary for living systems. Since there is thus no possibility of fine chemistry without an aqueous milieu, Smalley denies that nanorobots are chemically plausible. Strangely enough, the debate about the feasibility of nanorobots seems to revive an old dispute between physicists' and chemists' views of material devices. As Philipp Ball noticed, "It is becoming increasingly clear that the debate about the ultimate scope and possibilities of nanotech revolves around questions of basic chemistry."[27]

For many materials chemists, Drexler's program to force chemical reactions by placing the reagents in the right position is useless. Their "art" consists in making heterogeneous components *spontaneously* converge in the right location and assemble into larger aggregates without any external intervention. In fact, neither manipulating the molecules nor programming the machines requires outside intervention because the components move by themselves. This fascinating perspective was opened up by George Whitesides in 1995:

> Our world is populated with machines, nonliving entities assembled by human beings from components that humankind has made. . . . In the 21st century, scientists will introduce a manufacturing strategy based on machines and materials

that virtually make themselves; what is called self-assembly is easiest to define by what it is not. A self-assembling process is one in which humans are not actively involved, in which atoms, molecules, aggregates of molecules and components arrange themselves into ordered, functioning entities without human intervention. . . . People may design the process, and they may launch it, but once underway it proceeds according to its own internal plan, either toward an energetically stable form or toward some system whose form and function are encoded in its parts.[28]

To be sure, Whitesides provides here only a negative definition of self-assembly, but this does not mean that it is an obscure process that chemists do not understand. Rather they take advantage of all possible resources of chemistry and thermodynamics in the competition to mobilize all sorts of interactions between atoms and molecules. Instead of using covalent bonds as traditional organic chemists do, they make use of weak interactions such as hydrogen bonds, Van der Waals and electrostatic interactions. They use microfluidics and surfactants in order to produce self-assembled monolayers which, in turn, permit the chemists to move from atomic- and molecular-level structure to macroscopic property.

Self-assembly presupposes that the instructions for assembly are integral to the material components themselves or that they are embedded in their relations. Matter can no longer be viewed as a passive receptacle upon which information is imprinted from the outside, because self-assembly rests on spontaneous reactions between materials. Molecules have an inherent activity, an intrinsic *dunamis* (potential) allowing the construction of a variety of geometrical shapes (helix, spiral, etc.). It is not an obscure and mysterious vital force. Over the past century, chemists have learned a lot from molecular biologists about the workings of biomaterials, from the mechanisms of molecular recognition to the process of morphogenesis. Ironically, it is the reductionist approach of molecular biology that eventually allowed chemists to develop emergentist views of molecular architectures.[29]

The third and final lesson that chemists learned from biology involves complexity. Here this term is taken in its general sense, referring to the study of nonlinear processes with high sensitivity to initial conditions. Complexity became a problem when chemists started to examine the behavior of single molecules and to ask how they act to produce the average properties and behavior of familiar macroscopic chemicals.[30] In fact, chemists had suspected that nanoparticles behaved differently than macroscopic chemical substances long before the coming of nanoscience.[31] Gold, usually characterized by its yellow color, becomes red when processed in nanospheres. More generally, the color of metal and semiconductor nanoparticles depends on their size, a property commonly used in the glass industry. Today, it is also used to design magnetic materials with iron/platinum colloids, an application that has rendered colloid synthesis a highly sophisticated and promising domain of nanochemistry.[32] Given this long-standing attention to size-sensitive properties, the discovery that carbon nanotubes behave more like metals or semiconductors than like sticks of graphite did not come as a revolution in theoretical chemistry. Chemists were prepared to admit that elements have special properties and behavior when processed at the nanoscale. Unlike computer scientists, who are eager to replicate conventional machines at the nanolevel, materials scientists focus mainly on size-sensitive properties. Their work comprehends the entire hierarchy of structures in life, from the large molecules that assemble at the nanoscale to form organelles to the cells, tissues, and organs that ultimately compose unique organisms. Therefore they cannot rely on a uniform view of nature as being the same at all scales. While it is true that the laws of nature are universal, chemists do not assume that they apply equally to all scales.

In brief, the design of nanomaterials seems to rely on a specific, underlying view of matter that revives a number of antimechanistic notions such as emergence; spontaneity, or *dunamis*; and, above all, complexity. Thus there are two diverging ways of taking inspiration from life. Certainly such a divergence was already inherent in biology, which embraces a wide variety of approaches, including biochemistry, molecular biology, genetic engineering, entomology, and ecology. However, the wide variety of models within life science cannot by itself account for the contrast between the mechanistic and organicist paradigms that I have outlined in this paper. The choice between reductionism and emergentism rests on a broader philosophical divide that calls into question the unity and the coherence of the field known as nanoscience. Whether the contrast here outlined is a simple rhetorical effect, an artifact of discourse analysis, or a real difference in research practices is an interesting question that remains to be explored. This would require a thorough study of experimental strategies as well as a survey of technical publications that exceed the limits of this paper.

It is important nevertheless to stress that the contrast outlined here seems to be fading into insignificance due to the increasing predominance of the mechanistic model in public presentations of nanotechnology. The main objective is to build nanomachines, do a better job, and bring prosperity or even immortality. Nanostructured materials are said to be just a step toward the building of nanobots, a means toward an end. For Chris Peterson, the president of Drexler's Foresight Institute, nanostructured materials are only "near term" technology while "advanced nanotechnology" will mean complete control of the physical structure of matter all the way down to the atomic level. The design of fully functional nanomachines is the ideal of nanotechnology. It is an ideal with such a fascinating power that materials physicists and supramolecular chemists are ready to redefine materials as machines as they design wheelbarrow molecules, cantilevers, springs, switches. . . . The machine metaphor is equally pervading the language of biologists. Each entity active in the cell is described as a machine: ribosomes are assembly lines, ATP synthases are motors, polymerases are copy machines, proteases and proteosomes are bulldozers, membranes are electric fences, and so on.[33]

Biological evolution itself is conceived of as a blind mechanism operating under the control of an all-powerful algorithm by those who advocate the creation of nanobots and posthumans as an organic necessity. The overwhelming prevalence of the machinery paradigm is one of the most visible symptoms of the convergence of nanotechnology, biotechnology, information technology, and cognitive science (NBIC). To be sure, the machinery paradigm has a heuristic power today; however, its epistemic relevance may be soon questionable. First, it may turn into an epistemic obstacle: as Blaise Pascal said a long time ago about mechanicist philosophy, "It is roughly correct" (C'est vrai en gros), which means that it may be essentially wrong as it ignores subtler agencies at work in the details. Second, in technology the machinery paradigm may lead to systemwide standardization and generate a technological imperative due a path-dependency phenomenon.[34] Finally, the whole project of biomimetics is undermined because it rests on a circular logic: biological objects are described as machines and subsequently utilized as models for making artifacts.[35] Erasing the boundary between nature and artifact is certainly not the best way to bridge the "Sistine gap."

NOTES

1. See, for instance, the report of the Interagency Working Group on Nanoscience, Engineering, and Technology (IWGN) chaired by M. C. Roco, at the Web site of Mihail Roco, National Nanotechnology Initiative, on the page entitled "Small Wonders: Exploring the Vast Potential of Nanoscience. A NSF Symposium": http://www.nsf.gov/nano. See also Hans Fogelberg and Hans Glimell, Bringing Visibility to the Invisible: Towards a Social Understanding of Nanotechnology (Göteborg, Sweden: Göteborgs Universitet, Science and Technology Studies Section, 2003).

2. Here the term *paradigm* refers not only to puzzle-solving techniques but also to a set of values and implicit assumptions about the world that both inspires and limits the questions addressed by the scientist.

3. Ray Kurzweil, *The Age of Intelligent Machines* (Cambridge, MA: MIT Press, 1998), and *The Age of Spiritual Machines:When Computers Exceed Human Intelligence* (New York: Viking, 1999).

4. Richard Smalley, "Discovering the Fullerenes," Nobel Lecture, December 7, 1996, available online at http://cnrst.rice.edu.

5. Richard Feynman, "There Is Plenty of Room at the Bottom," in *Miniaturization*, ed. H. D. Gilbert (New York: Reinhold, 1961), 282–96.

6. Interestingly, Christine Peterson, Drexler's early disciple and self-proclaimed historian of nanotechnology, emphasizes the location of the founding event: MIT's Department of Interdisciplinary Science in the late 1970s, where Eric Drexler studied the design of manufacturing systems for use in the novel environment of space; see her "Nanotechnology: Evolution of the Concept," in *Prospects in Nanotechnology*, ed. Markus Krummenacker and James Lewis (New York: John Wiley & Sons, 1995), 173–95 [Proceedings of the first general conference on nanotechnology: "Developments, Applications, and Opportunities," 11–14 November 1992, Palo Alto, CA].

7. Eric Drexler, *Engines of Creation* (Garden City, NY: Anchor Press/Doubleday, 1986), 6–14.

8. Ibid., 17: "One might doubt that artificial nanomachines could even equal the abilities of nanomachines in the cell, if there were reason to think that cells contained some special magic that makes them work. This is called vitalism. Biologists have abandoned it because they found chemical and physical explanations for every aspect of living cells yet studied, including their motion, growth, and reproduction."

9. Marvin Minsky, "Virtual Molecular Reality," in Krummenacker and Lewis, *Prospects in Nanotechnology*, 193. See also Edward A. Rietman, "Drexler Hypothesis of a Universal Assembler Is Supported Not by Theoretical Arguments Alone but by Existence Proof in the Form of Biological Life," in *Molecular Engineering of Nanosystems* (New York: Springer Verlag, 2001), 2.

10. This is the conclusion of Kurzweil's testimony, quoted in House Committee on Science, *Hearing to Examine the Societal Implications of Nanotechnology and Consider H.R. 766, the Nanotechnology Research and Development Act of 2003*, 108th Cong., 2nd sess., April 9, 2003, available online at http://www.kurzweilai.net/meme/frame.html?main=/articles/art0556.html. Technology has always been a double-edged sword, so we simply need to implement "defensive technologies" against self-replicating nanobots in the same way as our society is defending itself against computer viruses.

11. Eric Drexler, "Molecular Engineering: An Approach to the Development of General Capabilities for Molecular Manipulation," *Proceedings of the National Academy of Science, USA* 78, no. 9 (1981): 5275–78. Available online at the Web site of the Institute for Molecular Manufacturing: http://www.imm.org/PNAS.html.

12. Eric Drexler, "Introduction to Nanotechnology," in Krummenacker and Lewis, *Prospects in Nanotechnology*, 6.

13. Drexler, "An Open Letter to Richard Smalley," published 16 April 2003, on Kurzweil AI.net, at http://www.kurzweilai.net/meme/frame.html?main=/articles/art0560.html.

14. See Jean-Pierre Dupuy, *The Mechanization of the Mind* (Princeton, NJ: Princeton University Press, 2000). Drexler did his Ph.D. in Marvin Minski's laboratory and wrote his doctoral thesis under von Neumann.

15. Drexler, "Introduction to Nanotechnology," 17.

16. Drexler, *Engines of Creation*, 7; see also 13.

17. Drexler, "Introduction to Nanotechnology," 2.

18. Ibid., 12.

19. Testimony of Ray Kurzweil, in House Committee on Science, *Hearing to Examine the Societal Implications of Nanotechnology.*

20. See, for instance, George Whitesides, "Nanotechnology: Art of the Possible," *Technology, MIT Magazine of Innovation*, November–December 1998; George Whitesides and J. Christopher Love, "The Art of Building Small," *Scientific American*, September 2001, 38–47.

21. Jean-Marie Lehn, *Supramolecular Chemistry* (Veinheim: Verlag Chemie, 1995).

22. Jean-Marie Lehn, "Une chimie supramoléculaire foisonnante," *La lettre de l'Académie des sciences* 10 (2003): 12–13.

23. See, for instance, *Biomimetics: Design and Processing of Materials*, ed. Mehmet Sarikaya and Ilhan Aksay (Woodbury, NY: American Institute of Physics Press, 1995).

24. Richard Smalley, "Prepared Written Statement and Supplemental Material," Rice University, 22 June 1999, online at http://www.house.gov/science/smalley_062299.htm.

25. Richard E. Smalley, "Of Chemistry, Love, and Nanobots," *Scientific American*, September 2001, 76–77.

26. Eric Drexler, letter, *Chemical & Engineering News* 81, no. 48 (2003): 37–42.

27. Philipp Ball, "Nanotechnology in the Firing Line," available at the Web site of the online journal *Nanotechweb.org*, 23 December 2003: http://www.nanotechweb.org/articles/society/2/12/1/1.

28. George M. Whitesides, "Self-Assembling Materials," *Scientific American*, September 1995, 146–49.

29. Emergence here should be understood in thermodynamic terms as the production of higher order out of lower order, which, according to Norbert Wiener, is the major characteristic of machines and living organisms as well. Self-assembly is a process leading from less ordered to higher thermodynamically ordered ensembles of molecules or macromolecules. The resulting aggregates have new properties that could not have been predicted from the characteristics of individual components. A major difference between natural organisms and aggregates formed in a laboratory environment is that the latter are in a state of equilibrium, whereas in living beings most of them are out of equilibrium.

30. George M. Whitesides and Rustum F. Ismagilov, "Complexity in Chemistry," *Science* 284 (1999): 89–92.

31. This phenomenon had been observed in metal colloids or hydrosols by Michael Faraday in the mid-nineteenth century and became known as the Tyndall effect after Tyndall extended Faraday's earlier observations. Suspended particles that are small relative to the wavelength of visible light (with radii of approximately 20 nm) are brilliantly colored in red, green, and violet because the interaction with the incoming light is a combination

of absorption and scattering. See Hervé Arribart, "Les nanomatériaux autres que ceux des technologies de l'information et des communications (TICS)," in *Nanosciences, nanotechnologies: Rapport sur la science et la technologie n° 18*, ed. Académie des sciences, Académie des technologies (Paris: Éditions Tec & Doc, 2004), 361–82, on 363.

32. D. F. Evans and H. Wennerstrom, *The Colloidal Domain:Where Physics, Chemistry, Biology, and Technology Meet* (New York: John Wiley & Sons, 1999).

33. David Goodsell, *Bionanotechnology: Lessons from Nature* (New York: Wiley-Liss, 2004); Shuguang Zhang, "Fabrication of Novel Biomaterials Through Molecular Self-Assembly," *Nature Biotechnology* 21 (October 2003): 1171–78.

34. The notion of path-dependency is applied to cases of technological choices made long ago under the influence of a historical momentum that gradually become irreversible because of the advantages of standardization. The QWERTY keyboard is an example (see Paul David, "Understanding the Economics of QWERTY: The Necessity of History," in *Economic History and the Modern Economist* (Oxford: Blackwell, 1986). An example of an obstacle to alternative technologies is the electric vehicle; see David A. Kirsch, *The Electric Vehicle and the Burden of History* (New Brunswick, NJ: Rutgers University Press, 2000).

35. Because of the circularity of such a notion of biomimetics, Colin Milburn defines nanologic as a "cyborg logic, imploding the separation between the biological and the technological, the body and the machine"; see his "Nanotechnology in the Age of Posthuman Engineering: Science Fiction as Science," *Configurations* 10 (2002): 261–95.

12   Creating Insight:
Gestalt Theory and the Early Computer

DAVID BATES

INTRODUCTION

It would be a sad day if human beings, adjusting to the Computer Revolution, became so intellectually lazy that they lost the power of creative thinking. | *Martin Gardner*, Aha! Insight (1978)

Much recent work in cognitive science suggests that thinking is more fundamentally rhetorical than strictly logical. The mind, many argue, is less a processor of information than a creator and interpreter of associative relationships. A focus on symbolic logic has given way to an interest in how mental activities such as metaphor, narrative, and analogy can help reveal the essence of human reasoning.[1] After decades of work in artificial intelligence, however, it has become increasingly clear that it is precisely *these* forms of creative thought that have been most difficult to replicate in any serious fashion with digital computers. Despite rapid progress in the development of information-processing technologies, resulting in machines capable to-day of incredible logical sophistication, the field of Artificial Intelligence (AI) has repeatedly been frustrated when it comes to imitating those basic, rhetorical dimensions of human understanding that do not seem to reduce to hierarchical sets of rules nor involve clearly defined, sharply bounded conceptual representations.[2] And so the once powerful and productive anal-ogy between the mind and the computer has been radically transformed, largely because these new approaches to human thought have questioned the relationship between computational models and the essence of human cognition.[3] Human beings do not so much calculate, then, as intuitively *relate* ideas and experiences instantaneously and in flexible permutations, and it is difficult to imagine how any rule-governed logic machine could ever attain this kind of insight.

However, a new generation of AI work has abandoned the top-down approach altogether, replacing logical frameworks based on abstract symbolic representations with models and methods drawn in part from contemporary biology and the related discipline of Artificial Life.[4] In this research, complex behaviors of global systems are redescribed as emergent properties that arise from interacting, distributed parallel processes. These systems are "self-reproducing" and "self-organizing" because their order is a product of numerous operations performed by relatively autonomous individual agents.[5] From the perspective of the new wave in AI, we may view the human mind as a form of complexity that acts creatively and unpredictably, constantly transforming itself (through various feedback mechanisms) as it confronts new situations. So instead of seeing the mind as a form of logical "program" running on digital neural "hardware," we may now see the brain as a decentralized, flexible network of neural relations and connections that produces complex mental behaviors and even, perhaps, consciousness itself.[6]

These new models have certainly produced some impressive results in, for example, areas such as the learning and recognition of patterns. Still, it is also clear that simulated neural nets, as well as more elaborate and abstract parallel distributed processing (P-D-P) and connectionist programs, are still a long way from replicating the kind of focused, complex cognitive tasks that characterize human thinking of the highest order. Of course, many have faith that these new approaches will one day allow us to understand creative and insightful human thought,[7] though we should note that major players in traditional AI research made the same kind of claims for their work.[8]

What I want to do here is question the underlying assumption that grounds both traditional AI research, which is intimately tied to the theory and practice of computing logic, and newer forms of investigation into intelligence that are linked more to biological models of complexity and self-regulation. While classic AI assumes that the processes of thought can be broken down into sets of interacting rules, more recent "bottom-up" approaches deny that there are any "global" rules dictating mental behavior; however, they do suggest that thinking is the result of interacting "populations of simple programs or specifications."[9] We apparently must assume either that human thought, including its most subtle and creative dimensions, will succumb to *analysis* of some sort, or that thinking will be explained in terms of its *production*, as the "synthetic" result of constructed "aggregates" of simple systems, to use Christopher Langton's terms. Both approaches believe that thought will be replicated (theoretically at least) once these underlying processes are revealed.

My critique is historical. I focus on the rather surprising fact that ideas about the intrinsic, foundational unity of creative, insightful thought framed some important discussions of mind and brain at exactly the same time as the first electronic computers were being developed—and along with them, the first modern concepts of machine intelligence. Moving from the clash between cybernetics and Gestalt theory to John von Neumann's revolutionary automata theory, I argue that a rediscovery of the complex intersections drawn between minds, living organisms, and artificial intelligence at this origin point will alert us to some critical—if too often neglected—questions that still haunt the whole project of simulating creative thought today. This is not simply an issue in AI research. As we will see, the problem of creativity was also central to discussions about living bodies. If von Neumann's influential automata theory virtually laid out the research agenda for the new field of Artificial Life, we will see how he struggled from the start with the problem of how to conceptualize unprecedented moments of creativity within any organism, thinking or not.

BRAIN LOGIC

It is a noteworthy fact that the human mind and animal nervous systems, which are known to be capable of the work of a computation system, contain elements which are ideally suited to act as relays. These elements are the so-called *neurons* or nerve cells.

Norbert Wiener, Cybernetics or Control and Communication
in the Animal and the Machine (1948)

There are wholes, the behaviour of which is not determined by that of their individual elements, but where the part-processes are themselves determined by the intrinsic nature of the whole.

Max Wertheimer, "Gestalt Theory" (1924)

In 1951, the preeminent Gestalt psychologist Wolfgang Köhler published a review of Norbert Wiener's seminal work, *Cybernetics*, which included a famous chapter entitled "Computing Machines and the Nervous System." Though he praised certain features of the book, including the important theorization of feedback, Köhler had serious reservations about the idea that electronic calculators and other machines could serve as models for the human nervous system and thereby help to explain the origin and nature of human thought. As Köhler put it, this "now popular comparison" was entirely ungrounded—the kind of information processing carried on by these new computing machines was, he believed, "functionally" and "generically" different from human thinking. As Köhler pointed out—anticipating Dreyfus and other AI critics—it was obvious that only the "intelligent" decisions of human mathematicians gave any value to these blind mechanical processes; the machines could not really know anything, he concluded, "because among their functions there is none that can be compared with insight into the meaning of a problem."[10]

So Köhler would hardly have been surprised to find the influential cyberneticist and polymath Warren McCulloch admitting, a decade later, that "the problem of insight, or intuition, or invention—call it what you will—we do not understand."[11] On various occasions, McCulloch would acknowledge that imitating this kind of insight artificially with machines was quite problematic (though still possible, he believed).[12] That a leading cyberneticist like McCulloch would become so entangled in the question of insight and "productive thinking"—a question taken up by virtually all the leading Gestalt psychologists—was rather ironic. For it was in fact McCulloch's early work with his young colleague Walter Pitts, undertaken in the 1940s, that initially spurred the "popular" idea that human thinking could be compared to the work done by calculating machines, a central theme of cybernetic theories.[13] McCulloch and Pitts famously argued, in 1943, that it was possible to consider neuron structures as if they were in essence digital switching devices, and, armed with this idea, they went on to demonstrate (mathematically, that is) that networks of these "idealized" neurons were entirely capable of representing and calculating logical propositions and their consequences.

In effect—following here the revolutionary work of Kurt Gödel, Alonso Church, and Alan Turing in mathematical logic—McCulloch and Pitts argued that the brain, seen as a simplified, interconnected network of digital relays, was perfectly capable of fulfilling the functions of a Turing machine, an imaginary (at this point) form of digital computer that was capable of calculating logical sequences, by representing propositions and relations as a series of computable numbers. The implications of this brilliant and startling argument were obvious, and the two authors did not fail to make this explicit, closing their essay with this powerful claim: "Both the formal and the final [i.e., purposeful] aspects of that activity we are wont to call *mental* are rigorously deducible from present neurophysiology."[14] As Mc-Culloch would go on to explain, in a talk at the celebrated Hixon symposium on brain and behavior held at the California Institute of Technology in 1948, we did not need to postulate anything "extra" in thought to explain mental functioning. Indeed, we *should not* entertain the idea that something "new" could happen in the mental domain: "Our knowledge of the world, our conversation—yes, even our inventive thought—are," he stated emphatically, "limited by the law that information may not increase on going through brains, or computing machines."[15]

This seminal attempt to redescribe thinking as a kind of bio-logical mechanism may have fit well with the behaviorism that was dominant in American psychological research. And in an academic environment that valued "facts" and empirical research in these areas, the materialist implications of these arguments were hardly problematic. It was, however, precisely this approach that the German Gestalt psychologists criticized, especially as they arrived in America as exiles. They made their mark with a series of elegant empirical studies *demonstrating* the insufficiency of deterministic, behavioral models of cognition. Even the lowly chicken, Köhler once showed, possessed a kind of "insight" into new situations that could never be explained by the assumption that learning took place merely through mechanical association and feedback. When trained to release food by choosing one particular shade of gray that was lighter than another one presented to them, chickens would not, in new tests, choose the exact same shade of gray they had associated with the food, but rather a completely *new* shade that happened to be lighter than this original. The chicken *transposed* a relation, which itself could not be "located" in any specific experience or behavior.[16] Addressing such problems, Gestalt theorists were trying to understand how the mind (both animal and human) organized these relational perspectives that transcended the concreteness of perceptual phenomena. There was, it seemed to them, something radically "new" that entered even basic forms of thought, when the mind, for example, grasped the inner meaning of some unfamiliar situation.

The Gestalt psychologists' intense interest in this special phenomenon, which they called "insight" (*Einsicht*), did not, however, signal a rejection of empirical research into the relation between neurophysiological structures and psychological experiences. Just the opposite. Köhler, for example—who in fact participated in the Hixon symposium himself, delivering a paper on relational determination in perception—gave a penetrating critique of McCulloch's position on the brain as a logic machine. In the discussion period, Köhler said that while he admired "the courage with which Dr. McCulloch tries to relate his neurophysiology to facts in psychology," the early cybernetic position on this relationship was, Köhler thought, simplistic and thus confusing. He pointed out here that in order to explain the acknowledged complexity of certain rather basic psychological effects in the realm of perception, McCulloch had been forced to assume all kinds of arbitrary "histological" facts in his paper, despite the lack of any evidence for them. "Why then the elaborate constructions?" asked Köhler. "Most probably the reason is that the *atomistic* character of Dr. McCulloch's neurophysiology prevents any direct approach to *relationally* determined facts such as visual shapes."[17] If this was clear even at the level of perception, it was even more apparent at the higher levels of cognition, Köhler said. The experience of "having a goal," he observed, was an intrinsically "relational situation" that could hardly be represented by *specific* sets of nerve impulses. Ultimately, there was, according to Köhler, more complexity in the psychological terrain than in the neurological realm as McCulloch understood it—which would force us to admit either that the mental had some essentially "nonphysical" aspect, or that the physical model itself was substantially wrong.[18]

It is fair to say that Köhler (like many other Gestaltists) was open to both of these possibilities. In the postwar period, he was not only deeply engaged in research on neural current flows, but maintained his early interest in applying new theoretical advances in physics (particularly field theory) to work in neurophysiology and experimental psychology.[19] At the same time, Köhler was equally emphatic that the mental world had its own forms of organization that had to be unraveled, and for that reason he never tried to reduce that organization to some kind of foundational biological reality, invoking instead the rather slippery term *isomorphism* to explain connections between the nervous system and the mind. Köhler's rejection of Wiener's overly enthusiastic analogy between relays and neurons left open the possibility that the brain could be organized in some other way so as to make mental insight explicable. In effect, what Köhler and his colleagues were interested in was the structure of "insight" in *both* the mental and the physiological world. In order to approach this somewhat complex notion, we need to focus on the idea of "insight" itself.

INSIGHT THEORY

Insight is the *understanding* of the total situation, the *immediate perception* of the functional value of things in relation to the goal pursued. But how to explain the act of intelligence that is invoked by insight—a kind of *deus ex machina*?

E. Claparède, "La genèse de l'hypothèse: Etude expérimentale" (1934)

The starting point of Gestalt theory was the observation that human thought had a fundamental "organization" that went beyond the individual experiences and sensations that made it up. The famous origin of much of this research was the discovery by Max Wertheimer early in the twentieth century that humans perceive motion when two dots are shown quickly in succession in two different places—the mental process, in other words, was not *determined* by the specific (discrete) activations in the perceptual apparatus, but had an ordering logic all its own.[20] Many similar experiments proved the same point—we did not make judgments and "infer" qualities in experience (such as the "belonging together" of things, the continuity of visual figures, apparent motion, and so on) but rather saw them directly, without any awareness of the individual mechanisms involved in their production, and indeed in a way that demonstrably went *beyond* those mechanisms. Wertheimer proposed a law of *Prägnanz* to explain the fact that the mind always strove to overcome any concrete "gaps" in experience, to maintain coherent, unified phenomenal spaces.[21]

While perception and perceptual ordering remained an important topic for Gestalt researchers through the 1930s and 1940s, higher forms of cognition were also studied within the same basic framework. That is, in addition to "automatic" forms of phenomenal connectivity, Gestalt thinkers focused on the way the mind could actively "see" and understand—directly and intuitively—meaningful connections in the world of experience. Köhler's famous study of apes was groundbreaking in this respect. Undertaken during the First World War (when Köhler was stranded at his research station on an island off the coast of Africa), these experiments investigated the intelligence of ape subjects by looking at how well they could solve certain basic problems. Köhler swept aside the behaviorist model of animal psychology by devising ingenious tests that forced the animals to come up with *creative* responses to situations. By making food inaccessible, at least directly, Köhler showed how the apes were able to discover novel solutions to their various predicaments: for example, making tools to reach their goal, substituting found objects for these tools when they were taken away, finding detours around obstacles, building structures from available materials, and so on.[22]

Köhler called this ability "insight," for he wanted to stress that these solutions were never the product of mere blind trial and error, nor the mechanical application of established routines, but instead the often unpredictable result of a profound perceptual "shift." According to Köhler, these creative solutions could only appear if the inner "meaning" of the problem was understood by the subject. At this point, possible solutions, possible aids, and so forth were seen as meaningful in this context. When the direct "path" to a goal was blocked, Köhler explained, the animals had to find "alternate paths" by seeing the original problem in a broader perspective. Just as when we step back from an obstacle and open ourselves up to the discovery of a "roundabout path" that will take us to our goal (even though we may actually have to go *away* from this goal at certain points), Köhler suggested that the mind (both animal and human) could imagine *figurative* "paths" that would overcome less obvious, nonphysical obstacles, once we looked at our situation in a new light or from a new vantage point.[23]

The Gestalt theorists would delve more deeply into this insight phenomenon, developing ever more sophisticated models of how human beings in particular were able to solve problems in unfamiliar situations. The key assumption here was that the "unity" of our perceptual and cognitive world was not determined by the specifics of any empirical "reality." If we think of the famous Gestalt "switch" that takes place when we look at ambiguous figure-ground images, we can see how, at the higher cognitive levels, something similar could occur. Because the inner relations and fundamental perspectives that ordered our understanding of the world were essentially indeterminate, they could at times shift abruptly and radically, giving us new insight. The question was *how* and *why* such a reordering took place in our experience, and why a new organization that replaced a previous one was felt to be more appropriate or "right."

The young psychologist Karl Duncker, who conducted experiments on student subjects trying to solve mathematical and logical puzzles, argued in 1935 that the solution process could be broken down into phases.[24] First, the subject was able to "see into" the inner structure of the original problem. Then he or she could "restructure" this situation to open up a potential solution path. Finally, these more abstract solution possibilities were tested for practicality—that is, an attempt was made to discover or invent a concrete version of this solution idea. For Duncker, the problem situation could be defined by its internal "tension," and it was this problematic character that demanded attention—which is what Wertheimer's law of *Prägnanz* would suggest, since the "puzzle" was puzzling precisely because it did not cohere into a stable unity of any kind. The new order that emerged was one that *resolved* this tension in some way. "The decisive points in thought-processes, the 'Aha!,' of the new, are always at the same time moments in which a sudden restructuring of the thought-material takes place," said Duncker, and this new structure formed itself in thought precisely because it "altered" the original problem and provided a sense of closure.[25] This ability to restructure therefore required a transcendence of the immediate problem in some way. As Kurt Koffka would argue, it was the essential plasticity of the mind that allowed it to "disengage" from the immediacy of experience and then "reconfigure" it in creative ways.[26] Wertheimer would agree. In his own important posthumous work on *Productive Thinking*, he said that alongside "insight" into the "fundamental structural features" of a problem, the ability to "free" ourselves from specific situations to open up solution ideas was also essential for creative thought. And like Duncker before him, Wertheimer zeroed in on the importance of the intrinsic tension within problem situations, noting that solutions often followed a focus on what he called "trouble regions."[27]

But what made possible these new configurations? Like Köhler, Duncker would use the metaphor of the *path* to frame this problem, noting that "the man whose vision is not limited to the few feet just ahead, but who directly takes in the more distant possibilities as well, will surely and quickly find a practicable path through difficult terrain."[28] Psychologically speaking, these "distant possibilities" were generated from within—but with what logic? How did we know which ideas to bring to bear on radically new problems?

Again, the answer lay in the concept of "insight." Duncker, for example, explained that the mind could see what he called the "functional value" (and not just the surface characteristics) of a solution. This allowed us to see that particular solution as *adaptable* to entirely new situations—we could transpose past and present experiences because the specifics of both were not as important as the structural features expressed in the particular configurations. For example, we "see" not this particular object as an obstacle, but rather the object as an instantiation of "obstacle" as a generic category, and this allows the mind to apply previous detour strategies in this unfamiliar situation, if they can be practically found.[29] Or—as Köhler had noted in his ape studies—aspects of our present situation that seem irrelevant might take on functional value once we begin to see the current predicament as analogous to past situations that had concrete solutions. For instance, a longish, relatively straight object becomes the focus of attention when we are searching for something that might substitute for a stick used earlier that is no longer available.[30] Crucial, however, to the Gestalt theory was the idea that the *structural* insight came first—only then could the mind recognize meaningful connections and objects within specific situations. As Duncker phrased it, "The general or 'essential' properties of a solution genetically precede the specific properties; the latter are developed out of the former."[31]

For the Gestaltists, the mind organized its experience in terms of structural wholes, but since these organizations were not determined by the specific "facts" and immediate sensory inputs, nor by past associations, and not by any systematic "logic," new reconfigurations were always possible. But explaining the origin of these new insights was not easy. It was clear, though, that they were not generated by some mysterious mental force, which is why the Gestalt theorists always rejected any association with vitalist or mystical positions. Instead, they believed that the mental restructuring could be explained by the "forces" involved in the encounter between mind and environment. The mental world may have had an independent form of organization, yet it was not itself independent of the organism it served. This is why the Gestalt theorists often looked to the *physiological* realm as a source of understanding—not to reduce the mental to the physical, but to see more clearly the dynamic structures in play in both spaces of activity.[32] Psychological insight was not explained by physiological processes. Rather, the structure of insight helped explain the complex, nonmechanical behavior of living, acting organisms.

SOMATIC INSIGHT

The end to which each process normally tends, is determined by its significance for the essential tasks of the whole organism.

    Kurt *Goldstein*, Der Aufbau des Organismus: Einführing in die Biologie unter
    besonderer Berücksichtung der Ehfahrungen am kranken Menschen (1934)

Gestalt psychologists were interested in the neural dimensions of human thought, but not because they believed that the essential "insight" characteristic of cognitive functioning could be broken down and explained by deterministic physical systems. In fact, they believed that the physical systems involved in thought did not themselves obey a strictly deterministic logic. Köhler for one emphasized that some physical phenomena (just like psychological ones) were best approached by techniques that stressed relationship structures over discrete interactions. The intrinsic "wholeness" of these dynamic systems could not be understood as the mere product of atomistic "events" on a microscopic level. Köhler, who had studied physics with Max Planck, repeatedly invoked field theory to explain how "physical Gestalten" were configured (and reconfigured) on the basis of their inner unity as a system of forces, and not by sequences of local events.

Assuming that the nervous system functioned in this manner—dynamically, as a field of forces, that is, and not "geometrically" as a set of specific interactions—Köhler was able to postulate an "isomorphic" relation between fields of neural discharges and the psychological field of creative thought. That is, both the nervous system and the mental domain of thoughts acted like unitary "organisms" that responded to intervention and tension by reordering themselves to preserve some kind of complex equilibrium. Crucial to the Gestalt conception was the postulate, voiced here by Koffka, that "we accept order as a *real* characteristic, but we need no special agent to produce it, since order is a product of organization, and organization is the result of natural forces."[33] There was no priority, then, attached to either realm in practice. As Koffka noted at the end of his textbook on Gestalt psychology, "If a thought process that leads to a new logically valid insight has its isomorphic counterpart in physiological events, does it thereby lose its logical stringency and become just a mechanical process of nature, or does not the physiological process, by being isomorphic to the thought, have to be regarded as sharing the thought's intrinsic necessity?"[34]

Gestalt thinkers would repeatedly draw attention to the complex affinities between somatic and psychological processes. Alluding to the important early work of Hans Driesch on developmental biology, Köhler once observed that embryos were able to "compensate" for irregular conditions that threatened their existence, describing the process with language that recalled his own work on "insight" and ape intelligence. "Many embryos can be temporarily deformed (artificially) in ways utterly incongruous with their racial history and nevertheless regain by a *wholly other route* the same developmental stage reached by another, untouched embryo."[35] These radical "reorganizations" also occurred within the nervous system, which was, Köhler said, never restricted in its operation by specific pathways—the control and selection of these routes often changed over time according to *functional* requirements. He contrasted this form of flexible organization with what he called the "machine," in which the "form and distribution" of the process is not left to dynamics, but rather to precisely determined "external" constraint mechanisms. A dynamic process, he said, "distributes and regulates itself, determined by the actual situation in a whole field."[36] The proper approach to cognitive organization, as Koffka put it, "cannot be a machine theory based on a sum of independent sensory processes."[37] It is clear, then, why Köhler would later reject the idea that the computer could represent human (or even animal) thought, since, as he said in his review of Wiener, all "decisions are forced upon the machine from outside by devices in which a series of successively applied mechanical constraints prescribes the right sequence of operations."[38]

In this context it is interesting that at this historical moment the nervous system was being described as a *creative* entity of its own, with a particular form of "plasticity" that mimicked the mental structure of "insight." Kurt Goldstein, who had close links with the Gestalt school, made this explicit in his synthetic work on holistic organisms, published in 1934. Sudden "restructurings" were, he argued, characteristic of biological behaviors—behaviors that seemed to parallel, in a way, the forms of mental restructuring that the Gestaltists saw in intelligent cognitive work. Indeed, Goldstein described the organism as in essence a problem-solving entity, noting that "normal as well as abnormal reactions are only expressions of the organism's attempt to deal with certain demands of the environment. Symptoms are *answers, given by the modified organism, to definite demands*."[39] The organism, like the mind, dealt with these demands through a flexible process of restructuring that not only adapted the organism to new external conditions but also, at times, transformed the very operations of that entity. Suddenly, and without any training, injured organisms, Goldstein explained, could recover the performance of vital functions by alternative means—this is what he meant by modification. Frogs, for example, when one limb is amputated, will immediately substitute another to perform the wiping reflex, and some insects with missing legs will radically (but automatically) reconfigure their gait to maintain stable locomotion. Humans with their hands tied will unconsciously scratch themselves in some other way so as to relieve the itch.[40]

The point of these examples, for Goldstein, was to show that the organism was not a machine of isolated parts moving according to definite, predictable rules. The organism, confronted with obstacles to normal operation, will find the "shortest routes [*Wege*]," as he phrased it, to continue its work, by transposing functions if necessary. The internal order of the organism, he said, is "not confined to a definite anatomical structure, but represents a . . . Gestalt which can utilize, for its course, any available structure."[41] Goldstein adopted, in other words, Wertheimer's *Prägnanz* concept for biology, to explain the tendency of dynamic systems to seek "closure," that is, stability in some form of equilibrium. This stability, the foundational principle of its unity, did not entail any specific forms of organization, but only the most general tendency to "be ordered." Wholly new "norms" of existence were therefore possible for organisms, Goldstein claimed. A revolutionary turn could take place when, for example, an organism in crisis became "diseased" and had to establish a new norm of health to overcome the crisis.[42]

Despite the emphasis on bodily "performances," Goldstein—along with Köhler—was quick to point out that the "organism" had to be understood as a holistic entity that was not defined solely by the body and its functioning. The organism responded to its environment with both somatic and cognitive responses, depending on the situation; automatic processes took over in certain emergencies, for example, while focused, intelligent attention dominated in others.[43] The "mind-body" problem was, for Goldstein and the Gestalt thinkers, simply a nonproblem. The ultimate order of the organism was everywhere and nowhere, expressed through many different particular behaviors in any number of contexts—psychological, neurological, biochemical, and so on. The great plasticity of such a "transphenomenal entity" (Köhler's term) was the very source of its success and productivity.[44] "Insight," then, could be mapped isomorphically onto both psychological and somatic processes because the organism as a transcendent whole organized its own being with just this kind of creative understanding.

ARTIFICIAL INSIGHT: VON NEUMANN'S FLEXIBLE AUTOMATA

The most hopeful answer is that the human species has been subjected to similar tests before and seems to have a congenital ability to come through, after varying amounts of trouble. To ask in advance for a complete recipe would be unreasonable. We can specify only the human qualities required: patience, flexibility, and intelligence.

John von Neumann, "Can We Survive Technology?" (1955)

One of the reasons why the potentially fruitful encounter between Gestalt theory and cybernetics ended in antagonistic confrontation may well be that American intellectuals, especially scientists such as Norbert Wiener, found this German brand of "organicism" a foreign concept, difficult to accept or even understand.[45] While Gestalt thinkers were arguing that the mind-brain connection must be understood in relation to a "higher" form of unity that was expressed in both domains, Wiener was resisting what he saw as a turn to mysticism in such arguments. Clearly intending to rebuke Köhler, Wiener once dismissed the idea that one could adapt the techniques of field theory in physics to explain brain functioning. This was "nonsense," he said, and just showed ignorance of the physics involved—adding that the analogy between computer switching and brain switching was "valid and verifiable." In a withering comment, he then observed, "Finally, let me lay the ghost of another pseudo-scientific bogy: the bogy of 'wholism.' If a phenomena can only be grasped as a whole and is completely unresponsive to analysis, there is no suitable material for any scientific description of it; for the whole is never at our disposal."[46] The machine analogy, so important for cybernetic theory and the earliest forms of artificial intelligence theory, was therefore antithetical to these holistic approaches. Indeed, many critics of traditional AI have explicitly drawn their inspiration from Continental philosophical traditions compatible with (or even directly influenced by) Gestalt thinking, underlining the fundamental antagonism here.[47]

So it is perhaps surprising that John von Neumann—one of the key developers of the modern programmable digital computer, a leading figure in the cybernetic movement, and someone who ultimately tried to link machines and humans within a comprehensive theory of automata—actually shared Köhler's reservations concerning the powerful analogy that linked the structure of the brain to new electronic computers. True, in some early work on computing von Neumann had (with his colleague H. H. Goldstine) hinted that "human reasoning" might at some point be "more efficiently replaced by mechanisms."[48] And he did, to the puzzlement of his collaborators on the project, describe the EDVAC (the first stored-program computer) as a machine built from "idealized neurons."[49] But by the time he came to deliver his important series of lectures on the theory of automata (natural and artificial), von Neumann, like Köhler, questioned the value of any analogy between the brain and the computer: "It's terribly difficult to form any reasonable guess as to whether things which are as complex as the behavior of a human being can or cannot be administered by 10 billion switching organs. No one knows exactly what a human being is doing, and nobody has seen a switching organ of 10 billion units; therefore one would be comparing two unknown objects."[50]

One of the "complex" human behaviors von Neumann was interested in was intelligence—indeed, he often stressed the importance of insight in scientific and mathematical work in a way consistent with the Gestalt framework.[51] "In pure mathematics," he told his audience, "the really powerful methods are only effective when one already has some intuitive insight. . . . In this case one is already ahead of the game and suspects the direction in which the result lies." He did not claim that the computer would replace human insight. Instead, he merely suggested, pragmatically, that high-speed computers could help out in those areas, such as turbulent flows and shock analysis, where the problems facing physicists were so unique, so "singular," that very few "insights" were forthcoming, because the nonlinear equations involved were not analogous to the better-known linear domain and in fact "violated" all of our expectations. Here, the computer, with its astounding calculating ability, could serve as a kind of virtual space of experimentation—the machine could solve certain critical cases first isolated by the physicists as promising, and from analysis of these cases, we might gain some form of "intuition" (as he put it) as to where other critical cases might lie.[52]

This attention to insight helps, I think, to explain von Neumann's perhaps surprisingly harsh critique of McCulloch and Pitts's seminal work on computation and neurology in these lectures. Although he agreed with their mathematical conclusion that idealized neural networks could well operate as a biological Turing machine, von Neumann emphasized what was not demonstrated by the two collaborators—that anything like this system actually occurs in nature. That is, von Neumann resisted the logical simplification of a system that was, according to neurological research, almost overwhelming in its internal complexity.[53] But, more important, he was also acutely aware of the psychological complexities involved. The two realms were, obviously, intimately connected. As he noted in this suggestive illustration:

> Suppose you want to describe the fact that when you look at a triangle you realize that it's a triangle, and you realize this whether it's small or large. It's relatively simple to describe geometrically what is meant: a triangle is a group of three lines arranged in a certain manner. Well, that's fine, except that you also recognize as a triangle something whose sides are curved, and a situation where only the vertices are indicated, and something where the interior is shaded and exterior is not. You can recognize as triangle many different things, all of which have some indication of a triangle in them, but the more details you try to put in a description of it, the longer the description becomes.

Or, more precisely, the "logical" descriptions of such complex analogical recognitions may even prove impossible at a certain point. In a perhaps unconscious allusion to Gestalt theory, von Neumann noted that "with respect to the whole visual machinery of interpreting a picture, of putting something into a picture, we get into domains which you certainly cannot describe in those [i.e., logical] terms." There was no way of describing the "wholeness" we recognize despite distorted, partial, or ambiguous forms.[54]

At this point, von Neumann suggested an interesting "isomorphic" relation between these intuitive forms of thought and the complex organization of the nervous system. If the "visual brain" was responsible for such direct insights, he suggested, perhaps it was best to think of this neural network (consisting of two billion pathways) not as carrying out some formal logical computation to produce the insight but as itself physically organized as such a unity, according to its own vastly complicated logic—one that was at this point totally unknown and frankly beyond our comprehension. "I think that there is a good deal of reason to suspect that this is so with things which have this disagreeably vague and fluid impression (like 'What is a visual analogy?'), where one feels that one will never get to the end of the description." Neither the experience nor the neural processes could be reduced to simpler categories.[55]

Yet von Neumann went even further here, noting that not only were these processes complex, but they were not necessarily predetermined. After criticizing McCulloch and Pitts's theory of memory circuits with the observation that the brain's memory, whose existence is undeniable, seems to be located nowhere in particular, he went on to say that in general that "it is never very simple to locate anything in the brain, because the brain has an enormous ability to re-organize. Even when you have localized a function in a particular part of it, if you remove that part, you may discover that the brain has reorganized itself, reassigned its responsibilities, and the function is again being performed."[56] This is exactly what Kurt Goldstein had argued in his 1934 book on the holistic organism, which was based in part on the numerous studies Goldstein and his collaborators had done on brain-injured soldiers during the Great War.

For von Neumann, the creative plasticity of the nervous system served only to highlight the rather simplistic, and inferior, mechanical structure of the early computers, something he was of course well positioned to notice. This was no mere philosophical question at the time. Von Neumann, like many others involved in the origins of digital computing, was acutely aware of the problematic unreliability of these early machines. From the start, error loomed large in von Neumann's thinking on machine computing—and in fact, much of modern error analysis and correction techniques can arguably be traced to his seminal work in this area with H. H. Goldstine. But in his elaboration of his general automata theory, von Neumann was interested more in what was at stake conceptually with the problem of error. He could see that while the digital logic of computers had greatly increased precision because it eliminated the whole problem of "noise" that was so troublesome in analogical computing devices, the sequential logic of the digital systems was, practically speaking, much less reliable, since any single breakdown in the circuitry (a blown vacuum tube for instance) introduced potentially devastating errors that would quickly multiply in long and complex calculations.[57]

With this in mind, von Neumann made one important assumption in his automata theory: "In no practical way can we imagine an automaton which is really reliable."[58] This would have far-reaching implications. Von Neumann realized that any systematic form of error management, or any single, centralized control mechanism, would also have to be considered subject to failure.[59] Error, he once said, had to be considered "not as an extraneous and misdirected or misdirecting accident, but an essential part of the process under consideration—its importance in the synthesis of automata being fully comparable to that of the factor which is normally considered, the intended and correct logical structure."[60] Of course, it was here that the amazing plasticity of "problem-solving" biological organisms was so important, and von Neumann would invoke this nonlogical—or perhaps a-logical—plasticity as a model for any artificial automaton. "If you axiomatize an automaton by telling exactly what it will do in every completely defined situation, you are missing an important part of the problem," he stated. "It's very likely that on the basis of the philosophy that every error has to be caught, explained, and corrected, a system of the complexity of the living organism would not run for a millisecond. Such a system is so well-integrated that it can operate across errors."[61] In this context, von Neumann was fascinated by the fact that the nervous system did not seem to use digital forms of notation at all, operating instead with a statistical messaging technique that was less precise, to be sure, but much more reliable because it never relied on any single, specific message for its successful operation.[62]

Of course, the "system" could only operate across errors because its essential unity was not fixed in any one specific location. The system did not just spread its information over many neural connections, in case some failed. It also could actively bypass a whole series of performances for the good of the whole system when systematic error was detected. To accomplish this, the system had to possess a direct form of insight, von Neumann implied, and he used rather psychological language to describe this ability: "The system is sufficiently flexible and well-organized that as soon as an error shows up in any part of it, the system automatically senses whether this error matters or not. If it doesn't matter, the system continues to operate without paying attention to it. If the error seems to be important, the system blocks that region out, by-passes it, and proceeds along other channels."[63]

The "radically different" approach to error and organization in the physical nervous system raised a key question: What was the principle of organization that would explain this flexible form of self-regulation? What was this very active systematic order? At this point, the best von Neumann could do was suggest that "the ability of a natural organism to survive in spite of a high incidence of error . . . probably requires a very high flexibility and the ability of the automaton to watch itself and reorganize itself."[64] This entailed both an important autonomy of parts, where "there are several organs each capable of taking over in an emergency," but also a logic of systematic order that could in some way define these creative solutions to crisis. In other words, the parts could only take control if they had some inner connection to the principle structure of the whole organism. "The problem consists of understanding how these elements are organized into a whole, and how the functioning of the whole is expressed in terms of these elements," he pointed out—returning to one of the central tropes of Gestalt psychology.[65] Yet, until the end of his life, von Neumann admitted repeatedly that this logic was beyond any present understanding. His unfinished and undelivered Silliman lectures for Yale University, published as *The Computer and the Brain* after his death, in the end just detailed the fundamental differences between these two entities. His terse conclusion was that the logical structures involved in nervous system activity must "differ considerably" from the ones we are familiar with in logic and mathematics.[66]

CONCLUSION

Reading von Neumann's work from the perspective of other early twentieth-century concepts of mind and brain, we can see that he, like the Gestalt psychologists, was raising the important and difficult question of how thinking was related to what we might call the essential unity of the organismic structure—whether of a natural organism or some artificial being. This contextualization of von Neumann's automata theory, which is now being rediscovered by Artificial Life theorists, suggests that we may need to look more carefully at what are now familiar ideas concerning "self-organizing" and "self-regulating" systems. The plasticity and insight characteristic of both somatic and psychological processes was, it was thought, grounded by a real unity that nonetheless escaped any direct localization or representation. The Gestalt emphasis on problem-solving and "tension"; von Neumann's interest in error and unreliability; Kurt Goldstein's studies on brain pathologies and coping strategies—these perspectives all suggested that only with an understanding of this genuine unity would we ever begin to explain how a bounded, concrete entity could overcome unpredictable obstacles in new and challenging circumstances. The kind of unity these thinkers evoked in their work could never be *analyzed* as the mere product of a more fundamental logic, nor could it be understood as the end result of dynamically interacting simple processes, because it was thought to be present from the start—unity defined a systematic organization and was not merely some "emergent" property. With this in mind, we can predict that one of the challenges for AI in the twenty-first century will be to think clearly about whether this kind of elusive unity, in the mind *and* in the brain, will be taken seriously, whether it can be understood and replicated in any real way.[67] The least we can say at this point is that the historical rediscovery of an intellectual tradition that has been neglected by most contemporary AI researchers and cognitive scientists influenced by Artificial Life models could well provide some valuable new resources, at least conceptually.

NOTES

1. Some starting points would include Mark Turner and Gilles Fauconnier, *The Way We Think: Conceptual Blending and the Mind's Hidden Complexities* (New York: Basic Books, 2002); Douglas Hofstadter and the Fluid Analogies Research Group, *Fluid Concepts and Creative Analogies: Computer Models of the Fundamental Mechanisms of Thought* (New York: Basic Books, 1995); George Lakoff and Mark Johnson, *Philosophy in the Flesh: The Embodied Mind and Its Challenge to Western Thought* (New York: Basic Books, 1999); Mark Johnson, *The Body in the Mind: The Bodily Basis of Meaning, Imagination, and Reason* (Chicago: University of Chicago Press, 1987); Keith J. Holyoak and Paul Thagard, *Mental Leaps: Analogy in Creative Thought* (Cambridge, MA: MIT Press, 1996). More generally, see Howard Gardner, *The Mind's New Science: A History of the Cognitive Revolution* (New York: Basic Books, 1985).

2. Hubert L. Dreyfus, Stuart E. Dreyfus, and Tom Athanasiou, *Mind over Machine: The Power of Human Intuition and Expertise in the Era of the Computer* (New York: Free Press, 1986).

3. Hubert Dreyfus, *What Computers Still Can't Do: A Critique of Artificial Reason* (Cambridge, MA: MIT Press, 1993).

4. See Christopher G. Langton, ed., *Artificial Life: The Proceedings of an Interdisciplinary Workshop on the Synthesis and Simulation of Living Systems, Held September, 1987, in Los Alamos, New Mexico*, vol. 6 (Redwood City, CA: Addison-Wesley, 1989); and Luc Steels, "The Artificial Life Roots of Artificial Intelligence," *Artificial Life* 1 (1994): 75–110.

5. The seminal text is David Rumelhart and James L. McClelland, eds., *Parallel Distributed Processing: Explorations in the Microstructure of Cognition*, 2 vols. (Cambridge, MA: MIT Press, 1988). More recent overviews include E. Bonabeau, M. Dorigo, and G. Théraulaz, *Swarm Intelligence: From Natural to Artificial Systems* (New York: Oxford University Press, 1999); Steven Johnson,

Emergence: The Connected Lives of Ants, Brains, Cities and Software (New York: Scribner, 2001); Jacques Ferber, *Multi-Agent Systems: An Introduction to Distributed Artificial Intelligence* (Harlow: Addison-Wesley, 1999); a case study here would be Derek Partridge and Jon Rowe, *Computers and Creativity* (Oxford: Intellect, 1994).

6. See, for example, Daniel Dennett, *Consciousness Explained* (Boston: Little, Brown, 1991).

7. Douglas Hofstadter, for example, confidently claims that our creative intuitions and inventive discoveries are just the "unconscious" product of a "vast swarm of parallel activities" taking place in the brain. See his "Ambigrammatics and Creativity," in *Creativity in the Arts and Sciences*, ed. William R. Shea and Antonio Spadafora (Canton, MA: Science History Publications, 1990), 53. More generally, see Mark Bedau, "Philosophical Content and Method of Artificial Life," in *The Digital Phoenix: How Computers Are Changing Philosophy*, ed. T. W. Bynam and J. H. Moor (Oxford: Blackwell, 1998), sec. 2.

8. Herbert Simon and Allen Newell, for example, acknowledged that "isolated phenomena" such as human insight "must surely be included within the scope of a comprehensive theory" of intelligence—but they waited until the very end of a massive 1972 work on their ominous General Problem Solver program before just assuming, without any evidence or analysis, that this kind of thinking would eventually "yield" to the information-processing perspective. See Allen Newell and Herbert A Simon, *Human Problem Solving* (Englewood Cliffs, NJ: Prentice-Hall, 1972), 871–72.

9. This general statement of method appears in Christopher G. Langton, "Artificial Life," in Langton, *Artificial Life*, 3.

10. Wolfgang Köhler, review of *Cybernetics or Control and Communication in the Animal and the Machine*, by Norbert Wiener, *Social Research* 18 (1951): 128; my italics.

11. Warren McCulloch, "What Is a Number, That a Man May Know It, and a Man, That He May Know a Number?" in *Embodiments of Mind* (Cambridge, MA: MIT Press, 1965), 14.

12. McCulloch, "What's in the Brain That Ink May Character?" in McCulloch, *Embodiments*, esp. 389; and cf. "Where Is Fancy Bred?" another essay in the same volume that examines the insight problem in relation to AI.

13. Steve Joshua Heims, *The Cybernetic Group* (Cambridge, MA: MIT Press, 1991).

14. Warren McCulloch and Walter Pitts, "A Logical Calculus of the Ideas Immanent in Nervous Activity," in McCulloch, *Embodiments*, 38. Although Turing's iconic 1950 essay "Computing Machinery and Intelligence" is famous for introducing the "Turing test" for machine intelligence, one of the more disturbing (almost Nietzschean) undercurrents of the argument was the idea that human beings are no more than complex machines, whose "unpredictability" and "creativity" are just consequences of our own ignorance concerning the processes unraveling within.

15. Warren McCulloch, "Why the Mind Is in the Head," in McCulloch, *Embodiments*, 73.

16. Wolfgang Köhler, "Simple Structural Functions in the Chimpanzee and in the Chicken," in *A Sourcebook of Gestalt Psychology*, ed. Willis D. Ellis (London: K. Paul, 1938), 217–24; the experiments are also described in Wolfgang Köhler, *Gestalt Psychology* (New York: Liveright, 1947), 119.

17. Intervention by Köhler, in McCulloch, "Why the Mind Is in the Head," 95; my italics.

18. Ibid., 97–99.

19. See Wolfgang Köhler, *Die physischen Gestalten in Ruhe und im stationären Zustand: Eine naturphilosophie Untersuchung* (Braunschweig: Vieweg, 1920); and Wolfgang Köhler, William D. Neff, and Jonathan Wegner, "Currents of the Auditory Cortex in the Cat," *Journal of Cellular and Comparative Physiology* 45, supp. 1 (1955): 1–24; Wolfgang Köhler and Jonathan Wegner, "Currents of the Human Auditory Cortex," *Journal of Cellular and Comparative Physiology* 45, supp. 1 (1955): 25–54; Wolfgang Köhler and Donald Neil O'Connell, "Currents of the Visual Cortex in the Cat," *Journal of Cellular and Comparative Physiology* 49, supp. 2 (1957): 1–43.

20. Max Wertheimer, "Experimentelle Studien über das Sehen von Bewegung," *Zeitschrift für Psychologie* 61 (1912): 161–265.

21. See two excellent studies on Gestalt psychology and its varied contexts: Anne Harrington, *Reenchanted Science: Holism in German Culture from Wilhem II to Hitler* (Princeton, NJ: Princeton University Press, 1996), which takes a more cultural approach; and Mitchell Ash, *Gestalt Psychology in German Culture, 1890–1967: Holism and the Quest for Objectivity* (Cambridge: Cambridge University Press, 1995), which emphasizes institutional factors.

22. Wolfgang Köhler, *Intelligenzprüfungen an Menschenaffen* (Berlin: J. Springer, 1921). Translated into English as *The Mentality of Apes*, trans. Ella Winter (London: Kegan Paul, 1924).

23. See ibid., chap. 1, concerning *Umwege* (or "roundabout ways"), and 71–72 (103 in the English edition), on figurative *Umwege*.

24. Karl Duncker, *Zur Psychologie des produktiven Denkens* (Berlin: J. Springer, 1935). Translated into English as *On Problem-Solving*, trans. Lynne S. Lees, Psychological Monographs, no. 270 (Washington, DC: American Psychological Association, 1945).

25. Duncker, *Problem-Solving*, 4–6, 29.

26. Kurt Koffka, *The Growth of the Mind: An Introduction to Child Psychology* (New York: Harcourt, Brace, 1925), 125, 48.

27. Max Wertheimer, *Productive Thinking* (New York: Harper, 1945), 24, 34, 41.

28. Duncker, *Problem-Solving*, 39.

29. Ibid., 6–8.

30. Köhler, *Mentality of Apes*, 31–37, 103.

31. Duncker, *Problem-Solving*, 8.

32. Wolfgang Köhler, *The Place of Value in a World of Facts* (New York: Liveright, 1938), chap. 6, "On Isomorphism."

33. Kurt Koffka, *Principles of Gestalt Psychology* (New York: Harcourt, Brace, 1935), 175.

34. Ibid., 684.

35. Wolfgang Köhler, "Some Gestalt Problems," in Ellis, *A Sourcebook*, 66; my italics.

36. Ibid., 114, 193.

37. Koffka, *Gestalt Psychology*, 105.

38. Köhler, review of Wiener, 128.

39. Kurt Goldstein, *The Organism: A Holistic Approach to Biology Derived from Pathological Data in Man* (Boston: Beacon Press, 1963), 18. It is worth juxtaposing Georges Canguilhem's *The Normal and the Pathological*, trans. Carolyn R. Fawcett with Robert S. Cohen (New York: Zone Books, 1991), which explores a similar theory of disease as radical internal restructuring.

40. Goldstein, *Organism*, 52, 221–22, 229–31.

41. Ibid., 222, 232.

42. Ibid., 119, 429–38.

43. Ibid., 340.

44. Köhler, *Place of Value*, 187, and more generally, chap. 8: "A Discussion of Organic Fitness."

45. On the antagonism between Gestaltists and cyberneticists, see Jean-Pierre Dupuy, *Aux origines des sciences cognitives* (Paris: La Découverte, 1994); and Heims, *Cybernetic Group*. Anne Harrington, in *Reenchanted Science*, has noted the disjunction between a "German" intellectual tradition that informed the holistic approach of Gestalt psychology and the American academic environment in this period. Köhler himself reflected on the difficulty of "translating" the basic concepts of Gestalt theory for an American audience, using as his example the central idea of "insight." See Köhler, "The Scientists from Europe and Their New Environment," in *The Selected Papers of Wolfgang Köhler*, ed. Mary Henle (New York: Liveright, 1971), esp. 430–35.

46. Norbert Wiener, "Some Maxims for Biologists and Psychologists" (1950), in *Collected Works with Commentaries*, vol. 4: *Cybernetics, Science, and Society; Ethics, Aesthetics, and Literary Criticism; Book Reviews and Obituaries*, ed. P. Masini (Cambridge, MA: MIT Press, 1985), 454–56.

47. See, for example, the work of Hubert Dreyfus or Francisco Varela, both influenced by Heidegger and phenomenological thinkers such as Merleau-Ponty, who himself drew on Gestalt ideas.

48. See John von Neumann and H. H. Goldstine, "On the Principles of Large-Scale Computing Machines," in *Collected Works*, by John von Neumann, vol. 5: *Design of Computers, Theory of Automata, and Numerical Analysis* (Oxford: Pergamon Press, 1963), 2.

49. See William Asprey, "The Origins of John von Neumann's Theory of Automata," in *The Legacy of John von Neumann*, ed. James Glimm, John Impagliazzo, and Isadore Singer (Providence, RI: American Mathematical Society, 1990), 289–90.

50. John von Neumann, *Theory of Self-Reproducing Automata*, ed. Arthur W. Burks (Urbana: University of Illinois Press, 1966), 35.

51. A good example is John von Neumann, "The Mathematician," in *Collected Works*, vol. 1: *Logic, Theory of Sets, and Quantum Mechanics* (Oxford: Pergamon Press, 1961).

52. Von Neumann, *Self-Reproducing Automata*, 33–35. Cf. the editor's introduction to this text, esp. p. 3. For a broader perspective, see Peter Galison, "Computer Simulations and the Trading Zone," in *The Disunity of Science: Boundaries, Contexts, and Power*, ed. Peter Galison and David J. Stumpf (Stanford, CA: Stanford University Press, 1996).

53. Von Neumann, *Self-Reproducing Automata*, 45.

54. Ibid., 46.

55. Ibid., 47.

56. Ibid., 49.

57. On this point, see Claude E. Shannon, "Von Neumann's Contributions to Automata Theory," in "John von Neumann, 1903–1957," special issue of *Bulletin of the American Mathematical Society* (May 1958): 123–25.

58. Von Neumann, *Self-Reproducing Automata*, 57.

59. John von Neumann, "The General and Logical Theory of Automata," in *Collected Works*, 5:322. This point is made by von Neumann in the discussion period following this, his Hixon symposium paper.

60. John von Neumann, "Probabilistic Logics and the Synthesis of Reliable Organisms from Unreliable Components," in *Collected Works*, 5:329. This paper was edited from the notes for a 1952 lecture given at the California Institute of Technology.

61. Von Neumann, *Self-Reproducing Automata*, 57, 71–73.

62. Von Neumann, "General and Logical Theory," 307–8. Cf. von Neumann, *The Computer and the Brain* (New Haven, CT: Yale University Press, 1958), 76–79.

63. Von Neumann, *Self-Reproducing Automata*, 71; my italics.

64. Ibid., 73.

65. Von Neumann, "General and Logical Theory," 289.

66. Von Neumann, *Computer and the Brain*, 81–82.

67. For a move in this direction, see Susan Hurley, *Consciousness in Action* (Cambridge, MA: Harvard University Press, 1998). More to the point, see Susan Hurley and Alva Noë, "Neural Plasticity and Consciousness," *Biology and Philosophy* 18 (2003): 131–68. The emergence of a "dynamical systems theory" approach to cognition, as an alternative to computational or neural network models, recalls some of the early twentieth-century ideas presented here. See Timothy van Gelder, "What Might Cognition Be, If Not Computation?" in *Journal of Philosophy* 91 (1995): 345–81; and the essays collected in Robert F. Port and Timothy van Gelder, eds., *Mind as Motion: Explorations in the Dynamics of Cognition* (Cambridge, MA: MIT Press, 1995).

PART THREE

INTERACTIONS

13 Perpetual Devotion:
A Sixteenth-Century Machine
That Prays ELIZABETH KING

Lord, make me an instrument of your peace.
From the prayer of Saint Francis

In 1977 the Smithsonian Institution's National Museum of History and Technology purchased a small automaton believed to have been made in south Germany or Spain near the year 1560 (figures 13.1 and 13.2). The figure of a Franciscan monk, clad in the tunic, cowl, and sandals of the order, the automaton is sixteen inches high—about one-fourth life-size. The body is carved and turned linden wood, hollowed out to contain an animating mechanism whose many parts are made of hand-forged iron, all of which is hidden beneath the cloth of the habit. The visible hands, head, and feet are delicately carved, and polychromed. The paint, now cracked with age, was clearly once aglow with life: a faint flush still lingers on cheek and lip. An X-ray of the inside of the head reveals the machinery that controls the motions of the eyes and mouth (figure 13.3). The eyeballs themselves are made of iron, hammered round and painted. The monk grasps a small wooden crucifix and rosary in one hand, while the fingers and thumb of the other converge in the classic gesture of Latin emphasis (figure 13.4). Only the movement's mainspring and the monk's cloth garment are not original, and the cross and rosary have also been replaced in the intervening centuries. All else is remarkably well-preserved. The whole machine weighs just under five pounds.

In the history of European clock technology, the monk is an early and rare example of an automaton whose mechanism is self-contained and hidden within its body. Its performance takes place not on a mediating plinth or within an animated grotto (where there would be ample "backstage" room

FIGURE 13.1 | Automaton figure of a monk, south Germany or Spain, ca. 1560 (National Museum of American History, Smithsonian Institution, Washington, D.C.).

for concealing the drive hardware), but on a table or floor in the viewer's own world. Let us imagine it set in motion by a trusted steward, in a privileged setting. The mainspring has been wound ahead of time, out of sight of the assembled spectators, and the key withdrawn. The steward steps forward, holding the figure upright on the palm of one hand, and silently sets it down on a table, with a feint that discreetly releases the stop-work lever. Slowly the monk comes to life. He turns his head to single out one among the company. Left foot stepping forth from under the cassock hem, then right foot, the monk advances in the direction of his gaze, raising the crucifix and rosary before him as he walks. His eyes move: turning his head he looks to the

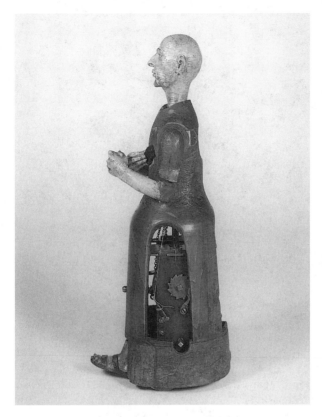

FIGURE 13.2 | Side view of automaton figure of a monk, south
Germany or Spain, ca. 1560 (National Museum of American
History, Smithsonian Institution, Washington, D.C.).

raised cross and then back to his subject. His mouth opens, then closes, af-
fording a glimpse of teeth and interior. He bends his right arm, and with the
gathered fingers of his hand, he strikes his breast. The small blow is audible.
And now he is lowering and turning his head as he walks: with elbow and
shoulder in synchronized motion he brings the cross higher, up to his lips,
and kisses it. Thirty seconds into the act, he has taken eight steps, beaten his
chest three times, kissed the cross, and traveled a distance of twenty inches.
At what seems like the last moment—for doubtless the subject of his attention has backed away from the table's edge—he looks away, arms still aloft,
executes a turn to his right, and makes a new appointment. He will make

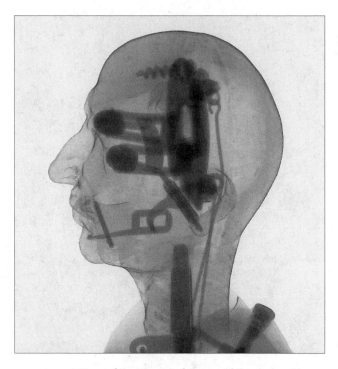

FIGURE 13.3 | X-ray of the interior of the monk's head, revealing the mechanism of the eyes, mouth, and neck (National Museum of American History, Smithsonian Institution, Washington, D.C.).

seven such turns and advances in his campaign if the mainspring has been fully wound. The uninterrupted repetition corresponds exactly to a trancelike performance of prayer, incantation.[1]

After more than four hundred years, the figure is still in good working order. Tradition connects its manufacture to Juanelo Turriano, engineer and mechanician to Emperor Charles V and then to his son King Philip II of Spain. How it came to be made is a tale a curious viewer today might hear if he or she presses anyone at the museum. King Philip, the legend goes, praying at the bedside of a dying son of his own, promised a miracle for a miracle, if his child be spared. And when the young prince did indeed recover, Philip kept his bargain by having Turriano construct a penitent clockwork homunculus.

Historian Otto Mayr, then curator at the Smithsonian's National Museum of History and Technology, wrote to his acquisition board before the

FIGURE 13.4 | The monk's hands. Photograph © Rosamond
Purcell, 2002.

purchase, "For the history of technology, this automaton is important as
one of the earliest program-controlled self-acting machines that have survived to the present. It is a direct predecessor of the eighteenth-century automata of Vaucanson and Jaquet-Droz, of the Jacquard loom and Babbage's
computer." Mayr equally understood its significance to philosophy and to
the literature of artificial life, for few of the miraculous automata attributed
to inventors and thinkers from Archimedes to Descartes have materialized
beyond legend. Fate is reversed in the case of the monk: it arrives intact
from the tumult of history with no letter of introduction. Yet among the
documents in the accession archives, there are some credible recent records
proposing a Spanish origin in connection with the court of Philip II.

A MIRACLE AND A PROMISE

History records the illness of Don Carlos, Philip's son and Spain's crown prince, in 1562. We know Carlos today as the hero of Verdi's opera *Don Carlo*, the story of a defiant prince destroyed by his father in a tale of empire, doomed love, and Inquisition. The historical Spanish prince, king, and court are a different story, though remnant contention yet clings to the subject of Don Carlos's short and strange life. He was seventeen years old in 1562 when, perhaps on an illicit errand, he took a fall down a flight of little-used stairs in his royal lodgings in the university town of Alcalá de Henares and struck his head against a closed door in the passageway below.

William Prescott, in his 1874 *History of the Reign of Philip the Second*, describes what happened to Don Carlos after the fall:

> He was taken up senseless, and removed to his chamber, where his physicians were instantly summoned. . . . At first it seemed only a simple contusion on the head. . . . But soon the symptoms became more alarming. Fever set in. He was attacked by erysipelas; his head swelled to an enormous size; he became totally blind; and this was followed by delirium. It now appeared that the skull was fractured. . . . Meanwhile, the greatest alarm spread through the country at the prospect of losing the heir-apparent. Processions were everywhere made to the churches, prayers were put up, pilgrimages were vowed, and the discipline was unsparingly administered by the fanatical multitude, who hoped by self-inflicted penance to avert the wrath of Heaven from the land.[2]

A recent study by historian L. J. Andrew Villalon, "Putting Don Carlos Together Again: Treatment of a Head Injury in Sixteenth-Century Spain," takes us through day by appalling day of the ordeal, quoting from the accounts of the prince's surgeon.[3] On the afternoon of May 9, in the aftermath of an unsuccessful attempt by physicians to trepan the patient's skull, the townspeople of Alcalá gathered at the Church of Saint Francis. "With Franciscan friars in the lead, they marched toward the palace, carrying with them the remains of a fifteenth-century member of the order, Diego de Alcalá, for whom they had long hoped to win sainthood."[4] Prescott transmits the story a little differently, having the king himself and his court fetching "the mouldering remains of the good father, still sweet to the nostrils, as we are told." Laying the corpse on the prince's bed, they removed the cloth that wrapped the dead man's head and placed it on Carlos's forehead.[5] But by evening so imminent was Carlos's death that the king took his closest advisors and departed before midnight, riding toward the Jeronymite monastery outside Madrid, to await the final news. Yet that night Carlos slept peacefully for the first time in weeks. The next day commenced a sudden, extraordinary recovery. The patient regained his sight a week later, his fever disappeared soon thereafter, and within a month, he was completely healed.

In the aftermath of this recovery, there was controversy as to how the cure was effected, but soon attention focused on Fray Diego de Alcalá, with a groundswell of feeling that here was the agent of a miracle. We can look back today and see two miracles, the second being the patient's survival of the physicians' purging and bleeding, and the daily probing of the wound with nonsterile instruments. ("They went on placing upon the exposed portion of the skull a powder made of iris and birthwort, and on the lips of the wound a mixture of turpentine and egg yolk. Following this, they cleaned the wound with honey of roses and applied a plaster of betony.")[6] The most dramatic evidence came from Don Carlos himself. When he regained his senses, he reported that on the night of May 9, a figure dressed in Franciscan habit and carrying a small wooden cross had entered his room and spoken to him. Certain that the apparition had been Brother Diego himself, Don Carlos vowed, together with his father, to bring this miracle before the Pope.[7] In Counter-Reformation Spain, even Andreas Vesalius, among the physicians laboring at the bedside, would risk heresy to contest such testimony to divine intervention.

In the end, it took Philip twenty-six years of respectful petitions to four consecutive popes to bring about the institutional confirmation of the miracle of Don Carlos's cure. The documents his scribes amassed are available today in the archives of the Escorial; they are the source of the close detail we have of the illness. Diego de Alcalá was canonized as a saint by the Roman Catholic Church in 1588. He was the first Counter-Reformation saint, commemorated in paintings by Zurbarán (figure 13.5), Murillo, and Carracci. A mission in his name was established in the new world in 1769, which later grew into San Diego, California.

The cure itself was not the first wonder performed by this thaumaturgical corpse. Through the hundred years before its appearance on the royal stage, Diego's body had drawn a steady procession of ailing pilgrims come to touch the holy flesh of an illiterate Franciscan lay-brother (figure 13.6).[8] He had lived a life of uninterrupted poverty, famous for the asceticism of his practice, but his death in 1463 was only the beginning of his holy career. The first miracle was the perfect absence of death's mark on the body: the monastery guardian apparently had second thoughts after the burial and had the body disinterred, and one visitor in the first month tried to take its

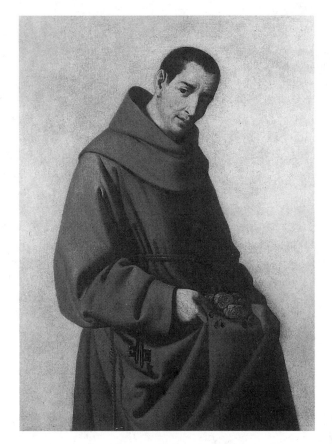

FIGURE 13.5 | Francisco de Zurbarán, *Portrait of Diego de Alcalá*
(Museo de la Fundación Lázaro Galdiano, Madrid).

pulse.[9] His remains, 540 years later, are still in Alcalá de Henares; the body
has been moved now to the Iglesia Magistral. Over the small tomb, an inscription appears on the wall:

DIEGO DE ALCALÁ
CUERPO INCORRUPTO

The name of Juanelo Turriano, celebrated Italian-born clockmaker (1500?–
1585), has been connected not just to the monk, but to several other automata that have survived to the present day. Court mechanician and engineer to Philip II, and before Philip to his father Charles V, Turriano worked in
Madrid and Toledo for the better part of his life. "The prince of the masters

FIGURE 13.6 | Cornelius Galle, *Diego de Alcalá* (vita), engraving, first published in 1614 (Museo Francescano, Istituto Storico dei Cappuccini, Rome).

of clockmaking," the emperor called him. To this day, a street in the heart of Toledo bears his memory: Calle del Hombre de Palo—"street of the wooden man"—so named for a fabled automaton he built that walked daily to the archbishop's palace and returned "laden with an allowance of bread and meat, after doing ceremonious obeisance to the donor."[10] Turriano would certainly have known the young prince Don Carlos, for his friend Ambrosio de Morales, court annalist to Philip II, had been appointed to oversee the preparation of documents on the life of Diego de Alcalá, submitted in petition to the Vatican. The same Ambrosio de Morales, in his encyclopedic volume *Las antigüedades de las ciudades de España*—a descriptive inventory of Spain's material history, commissioned by Philip and published in Madrid in 1575—provides us with an eyewitness account of the works of Turriano. Inventions large and small—but especially his clocks, how he made them and what he had to say about them—were chronicled by Morales in eloquent Old Spanish. Among all these marvels is a small automaton, "la dama que tañe y dança":

> Juanelo as a diversion also wanted to create anew the ancient statues which moved and, on that account, were called automata by the Greeks. He made a lady more than one *tercia* high who, placed on a table, dances all over it to the sound of a drum which she meanwhile beats herself, and goes round in circles, returning to where she started. Though it is a toy and fit for mirth, it is nevertheless a great proof of his high intelligence.[11]

A *tercia* is a Spanish linear measure of approximately eleven inches. In 1934, an automaton turned up in Vienna that so closely resembled this description that the horological historian Ernst von Bassermann-Jordan cited Morales to suggest an attribution to "Master Juanelo."[12] The figure plays a lute rather than a drum, but her dancing steps and circular tour and her small size certainly fit the slipper. The material and design of her clockwork, like that of the monk, indicate a mid-sixteenth-century pedigree. She is now in the collection of Vienna's Kunsthistorisches Museum. Turriano's name, once connected to her, moves through her in turn to circle the monk; we see at least a schematic similarity between the two iron movements. But tantalizing as the Morales passage is, we have no further words from him in reference to any other automaton.[13]

Yet, in the archives of the Smithsonian, we learn that the monk, purchased through a private broker in Geneva in 1977, arrived in Washington with documents that included a letter written in 1975 by Servus Gieben, director of the Istituto Storico dei Cappuccini in Rome. Solicited by the broker for his opinion of the automaton, which by then had been dated materially and stylistically to the sixteenth century, Gieben confirmed its Franciscan iconography and hypothesized that the monk was a portrait of Diego de Alcalá. He based his opinion on a constellation of factors: his strong impression that the figure was Spanish; the fact that Diego's saintly attributes are precisely the small cross and rosary; the observation that the automaton is given the shaved head of a lay brother (as Diego was) and not the tonsure of an ordained priest; the opinion that the face even bore a likeness to other portraits of the saint (Gieben had enclosed photocopies of two engravings from the collections of the Istituto's Museo Francescano, one of which I reproduce here as figure 13.6); and the reasoning that the illness and miraculous cure of prince Don Carlos provided the occasion for a votive portrait:

> It is in this climate of religious exaltation—processions, public prayers, pilgrimages for obtaining the prince's cure—that one must locate the fabrication of the automaton, as a kind of ex-voto and—why not?—as an exhortation to the young prince to a more serious life (he was very capricious). The date should be 1562 or a little later, and the author Juanelo Turriano (or de la Torre, dead in 1585), who was chief engineer to Philip II.[14]

In 1999, twenty-four long years after this letter was written, I was lucky to contact Servus Gieben in Rome, where today he is once again, at age eighty-two, head of the Istituto Storico dei Cappuccini and director of its Museo Francescano. A scholar of medieval theology and a specialist in the lives of the saints as depicted in the history of art, he responded to my letter in fluent English (one of his six languages), writing, "I certainly will be pleased to read your paper on the 'Mechanical Monk,' of which I preserve nice photographs and curious memories." My first question for him: Had the Geneva broker mentioned either Turriano's name or the precedent of the Vienna automaton? No, he replied, no link whatsoever had been made to another automaton, and in fact he learned of such a connection only from reading the draft of the essay I had just sent him.[15] At most, he recalled the suggestion of a sixteenth-century origin for the piece. His opinion that the monk represented a Spanish figure was based on the physiognomy of the carved face, a judgment confirmed by his search through the Museo Francescano print archives and the match, not just with the person of San Diego, but with the saint's symbols, lay status, and gestures.

Servus Gieben's hypothesis comes not from the history of technology, but from the history of theology, for he arrived at his conclusion without an awareness of the attribution to Juanelo Turriano of any similar automaton figure. It is remarkable that two separate tracks converge on so close an explanation of the monk's origin. Nonetheless, the attribution can of course remain only an educated guess.[16] The clockmaker himself evades us, for none of the machines Juanelo Turriano made with his own hands have survived in a form we can surely identify. His greatest astronomical clock, the work of a lifetime, he signed with an engraved inscription in Latin: "QVI. SIM. SCIES. SI. PAR. OPVS. FACERE. CONABERIS," which can be very roughly translated as "You will know who I am if you try and make this."[17]

TWO QUESTIONS

Years ago, when I first saw the automaton monk perform on a table in the museum conservation lab in Washington, I wondered how this object was seen in the age in which it was made. Could this lifelike thing have been perceived as alive? Could it even momentarily have been perceived as alive, just for the space of a shudder? In pursuing the source of the legend, a second question took hold. *What* was perceived to be transferred from the corpse to the prince? With death so close, it could almost be said that one dead body was resurrected by another. Could my two questions in any way be useful, one to the other, in the attempt to glimpse something of the emotional and imaginative undertow beneath declared understandings of the substance of living matter—judging the quick from the dead—in mid-sixteenth-century Europe? After all, the automaton, in good running order after four and a half centuries, comes to life in animated defiance of time and decay. Like the saint's body, it still "delivers" after all these years. Whether or not the automaton commemorates the cure, just the proximity of the machine and the corpse on the historical stage is worth considering (figure 13.7).

The automaton monk may be small and portable, but to consider it a toy—*juguete* is the word Morales uses for the dancing lady—presents an awkward dilemma for the interpretation of its performance.[18] So drilled are we against "presentism" in interpreting historical events, we hardly know what to do with our own emotional response to the objects we have inherited. Everyone who sees the monk in action agrees that it is intimidating—when the redesign of the museum's Hall of Timekeeping began in 1997 and the monk was taken off display, some of the staff were actually relieved.[19] I can offer my own testimony that when this machine heads in my direction on a table, my animal flight urge stirs. The bare mechanics of the gear motion are straightforward compared to other automata of the age. But look at the delicacy of the carved face in the X-ray (figure 13.3)! Seeing past the skin of cracked and discolored paint, we can make out the sculpture itself. The

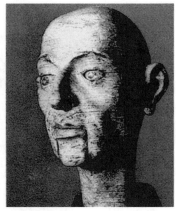

FIGURE 13.7 | Comparison of the automaton's head with an engraved portrait of San Diego de Alcalá. From the auction catalog *Très importante collection de tableaux Espagnols du XIIIe au XVIIIe siècle* [...] (Paris: Pierre Cornette de Saint-Cyr, 1976), 84.

character of the image, together with its head-on motion, makes an object that is impossible to regard with objective remove. And is this not precisely how works of art connect us most intimately to the past? As David Freedberg asks in *The Power of Images: Studies in the History and Theory of Response*, how can we hope to understand how representational objects were seen in history if we repress our own responses to these works?[20]

Freedberg's book tracks the history of human response to many classes of objects. The votive image, the effigies of sorcery, the wax museum, the religious relic and object of pilgrimage, acheiropoietic images (those believed not made by human hands), images that bled, or wept real tears, or whose eyes moved—all these as well as works from the high canon of Western painting and sculpture come under his scrutiny. The book itself is a master's response to Roland Barthes's call for a "History of Looking."[21] Although he does not focus on automata per se, his broad theme of figural verisimilitude and his discussion of objects associated with miracles provide correlative insights.

A strong devotional image may invoke what Freedberg calls an answering act of *reconstitution* performed by the beholder. Using the terminology of semiotics, he proposes that the gap between sign and signified may remain unacknowledged in the devout response. The statue is not just a metaphor or stand-in for the saint. It is a metonym: a piece of the real cloth, God Himself in the image. The power of an image, together with the empathy and desire of the beholder, collapses the distinction between representation and presence:

> The sign has become the living embodiment of what it signifies. Perhaps it will be suggested that a strong or ingrown belief in the powers of the Virgin easily inclines the believer to see her present, disencumbered of everything that makes her dead representation. Perhaps the suggestion will run that one cannot believe that the Virgin is in the picture—or is the picture—unless one believes, to begin with, in the Virgin. Then, wanting her to be there, to exist (because of the love we bear her), we willingly concentrate on the image, and what is represented on it becomes present again. She is, quite literally, re-presented. The slip from representation to presentation is crucial, from seeing a token of the Virgin to seeing her there. What happens? How do we proceed without engaging in the analysis of the propositional status of statements about belief, or the relations between the nature of beliefs and reports about them?[22]

This slip from representation to presentation, Freedberg argues, is something we experience every time we respond emotionally to an image, and such a slip accounts for the unacknowledged traces of animism in our own time: "not necessarily 'animism' in the nineteenth-century ethnographic sense of the transference of spirits to inanimate objects, but rather in the sense of the degree of life or liveliness believed to inhere in an image."[23] Just as the body of Christ inheres in the host, the divine spirit (the Holy Ghost) inheres in the corpse, and the saint inheres in the image, in a fusion effected by will and belief.

Artists have long understood the dangers of an overdetermined verisimilitude, for too much realism (or too visible an effort to achieve it) can halt the imaginative motion of the viewer. Michael Baxandall speaks of the asymmetry of relations between the image and the beholder in the devotional setting, offering a glimpse of the burden of expectation and desire we address to images.[24] The fusion Freedberg proposes between an image and what it represents may paradoxically be induced precisely by what the image *does not* do. That it is finally only a picture, for example, lets the viewer meet the Virgin more than halfway. But what if the *image* is doing the moving? If the motion is slight, uncertainty can ignite perception. In the case of the monk, whose activities are anything but slight, there are two conditions that check a potential excess of "signal." One is the possibility that it may only have been seen for the four minutes of its performance—that is, it was removed from the table and the room at the end of the play. The other is its small size.

Size is always a matter of weight and material: the nature of a thing is often a consequence of pure conditional imperative. The iron, spring-driven table clock had to be small enough so its motive force could accommodate its weight. An automaton was a virtual concert of clocks in one piece, with compound gear trains and multiple axes of rotation to accomplish both sequenced and simultaneous motions levered into space. To conceal the full iron mechanism within the body of a wooden figure, and an ambulatory figure at that, the maker had to precisely calculate its net mass. In fact, at sixteen inches and five pounds, the monk does so much *because* he is so small. Could the monk's size, in relation to his lifelike figure and performance, be urgent to his effect? The combination of "small" and "moving" holds something of an animal, even primal anxiety for us: the monk can play to only so large an audience, but his relation to a spectator is *personal*.

PETITION AND REPETITION

When his performance is over and I pull myself together, I look at the object itself, now at rest (the luxury of this!) and marvel at the almost seamless elegance by which the sculpture and the machine are combined. Here is a remarkably characterized figure carved in wood, and inside it is a functional clockwork of forged iron that has well withstood the test of time. How was it made? Inside out, first the mechanism, then the wooden parts carved and pieced together around it? Or outside in? The X-ray of the wooden head reveals a crisp, vertical, transverse seam just behind the ears: the head could have been carved and then sliced in half and hollowed out. Or perhaps a wooden blank was sliced first, roughly hollowed, and then temporarily clamped back together to be carved. Already the sculptor is operating on both sides of the divide. I am convinced that the chassis assembly supporting the eyes and jaw, together with the linkages for the movement of these parts, were made before the final carving of the face. The interface between wood and mechanism is precise—it must be, to permit the rotational and reciprocating motions of eyes and mouth to take place without loss of registration. Either or both the wood and the iron had to be cut and teased and adjusted for the fit. The automaton was made from the inside *and* the outside. Back and forth, we can look at the motions of the arms and feet, and read the same story: the artful negotiations between the very separate requirements of image and movement.

The question of how the monk was made is as challenging as that of how it was seen. How did the maker regard his task? In tracing the rise and perfection of the gear-based clock in the period between 1550 and 1650, Otto Mayr has illustrated the parallel rise of the machine metaphor that would come to be paradigmatic in describing not only the animate human body but the animating principles of state, world, and cosmos.[25] In a clockwork universe, concepts of harmony, hierarchy, regulation, period, authority, balance, system, division of labor, and predictable cause and effect were rendered visible in the mechanism of the timepiece: coiled spring as originating force, fusee as gatekeeper, verge and foliot as balance and regulator, gear train as transfer and distribution, cam and lever as memory, and the multitude of linkages as fibers to locate the motion in space, be it the limbs of an automaton, the hands of a clock, or the heavenly bodies of the astrarium.[26] The clockmaker who fashioned these parts and assembled them to make the internal mechanism of the monk was producing a coherent model of the motive force and coordinate will of the human anatomy.

We will see the reach of the mechanical/rational paradigm in Enlightenment philosophy and engineering. But among the early clockwork automata of the late fifteenth and the sixteenth centuries, the monk to my mind is the example par excellence of the birth of the machine in the Age of Faith. Here is nothing less than a machine that prays. As Carlos Fuentes once wrote, "'El movimiento se demuestra andando,' you demonstrate movement by moving."[27] Among the ecclesiastical automata we know, I am not aware of another independently moving clockwork figure that so explicitly performs the authorized and orthodox gestures of Catholic prayer.[28] Moreover, considering that the rosary is an instrument guiding the cumulative repetition of a spoken prayer—and repetition is the key to efficacy—why not build a machine to make the petition perpetual?

THE AUTOMATON AND THE HOMUNCULUS

The sixteenth century, ignited by Reformation iconoclasm and Counter-Reformation Inquisition, also saw the rise of Paracelsian alchemy. The monk's diminutive realism and functional autonomy warrants comparison with another small, willful, artificial being: namely, the alchemist's homunculus. For here is a competing paradigm for understanding and manufacturing the live body from the inside out. If the automaton's internal mechanism was forged in the smith's fire, the fabled homunculus—the alchemist's little test-tube man—was made in the heat of the flask. To us these are two very different models for the creation of artificial life: we might say that one is technological, the other mystical. But William Newman has argued that the alchemists of the late medieval period, in their profound attention to the world of matter and change, made one of the earliest and most articulate defenses of technology in the history of Latin Christendom.[29]

There are two points of contact between the mechanical and the chymical arts for the context of this study; one pertains to practice, the other to product.[30] The metallurgy of iron in this period—especially the literature on the transformation of iron to steel and steel to temper—shares common territory and history with the occult literature of alchemical transmutation, and both in turn are ancestors to experimental chemistry. With the rise of printing and literacy, by the sixteenth century such a book as Giambattista Della Porta's *Magia naturalis*, to take one example, became a best seller among schooled and lay readers alike. First published in 1558, it was re-edited and expanded by Della Porta over the next twenty-five years, in spite of the censure, publishing bans, and arrests of the author by the Inquisition. Ultimately comprised of twenty *libri* gathered into one volume, its pages offered prescriptions and experiments in everything from the medical arts to alchemy, optics, pneumatics, magnetism, the technology of fire, the secrets of generating new kinds of animals and plants, the marvels of putrefaction, the distillation of elixirs, perfuming, invisible ink, and the arts of disguise—all presented, as William Eamon says, in *sprezzatura* style.[31] The forbidden, the practical, the technically innovative, the scientifically advanced, and the apocryphal flourish side by side. Of the twenty *libri*, the fifth, "*Which treateth of Alchymy; ∫hewing how Metals may be altered and transformed, one into another*," together with the thirteenth, "*Of tempering Steel*," give us a glimpse in turn of the alchemist in his laboratory and the smith in his shop, each at work at his coal-fired forge with its bellows, surrounded alike by mortars, anvils, crucibles, cupels, ladles, hammers, alembics, and by plasters, liquors, fluxes, and quenches. If we squint through the smoke, they are brothers.

But how can we compare their handiwork? While the making of a homunculus is not exactly standard practice in the alchemical program at large, neither is the making of an automaton within the broader sphere of the metalcrafts, which included blacksmiths, lock and gunsmiths, smiths for the precious metals, many kinds of toolmakers, instrument makers, nail makers, and so on, along with the makers of large tower clocks and small table clocks. And, as we may have suspected by now, it is all but certain that the monk was made by more than one master. A clockmaker, but also a smith, a sculptor or wood carver, perhaps a painter, and certainly a tailor are all likely to have contributed their skills to the task. So I embark on a rarefied comparison, and between incommensurates: a recipe and a machine. But, contemporary to the second half of the sixteenth century, both required a master's (or adept's) proprietary skill with nature's materials. Both entailed the creation of an artificial being with the power to perform tasks to aid or excel human agency. Both were associated with magic and sorcery. And both were small.

While Della Porta and many other writers extolled the fundamentals of "putrefaction" and its role in the production of living creatures both imperfect and perfect,[32] the now classic text on the homunculus is found in De natura rerum, supposedly written in 1537 and published as the work of Paracelsus in 1572 by the iatrochemist and physician Adam von Bodenstein. Some scholars, including William Newman, have questioned the authorship of this text, although Newman suggests that it may be reworked material from a genuine Paracelsian script. Referring to its author provisionally as "pseudo-Paracelsus," Newman has made a close study of the work in the context of alchemy's pursuit of artificial life.[33] The famous recipe for the making of the homunculus appears in De natura rerum after a lengthy discussion of the transmutational virtues of putrefaction and the powers of human art. Here is Newman's translation from the original Latin:

> We must now by no means forget the generation of homunculi. For there is something to it, although it has been kept in great secrecy and kept hidden up to now, and there was not a little doubt and question among the old philosophers whether it even be possible to nature and art that a man can be born outside the female body and [without] a natural mother. I give this answer—that it is by no means opposed to the spagyric art and to nature, but that it is indeed possible. But how this should happen and proceed—its process is thus—that the sperm of a man be putrefied by itself in a cucurbit for forty days with the highest degree of putrefaction in a horse's womb, or at least so long that it comes to life and moves itself, and stirs, which is easily observed. After this time, it will look somewhat like a man, but transparent, without a body. If, after this, it be fed wisely with the arcanum of human blood and be nourished for up to forty weeks, and be kept in the even heat of the horse's womb, a living human child grows therefrom, with all its members like another child, which is born of a woman, but much smaller.[34]

Newman examines the recipe in relation to other passages on the homunculus from Paracelsus's oeuvre, to earlier sources from the Hermetic tradition, and—as in his essay "The Homunculus and the Mandrake" in this volume—in relation to the narratives of artificial generation in popular folk legend. This last comparison Paracelsus himself makes, and Newman translates the following passage from the *De vita longa libri quinque*: "The homunculus, which the necromancers falsely call *alreona* and the natural philosophers *mandragora*, has become a topic of common error, on account of the chaos in which they have obscured its true use. Its origin is sperm, for through the great digestion that occurs in a *venter equinus*, the homunculus is generated, like [a man] in all things, body and blood, with principal and lesser members."[35]

The legends of the *alraun*, or mandrake, indeed provide a remarkably analogous narrative to that of the homunculus. The *alraun*, according to German folk legend, was the root—in the shape of a tiny man—of a plant that sprouted in the earth beneath the gallows from the sperm of a hanged criminal. Such a man-root had powerful magic properties if one could get hold of it, but legend warned that anyone who tried to pull the *alraun* out of the ground would perish from its lethal scream. It had to be harvested by secret means. A lively trade in roots carved into human shape and sold as *alraun* naturally flourished in the sixteenth and seventeenth centuries, and Newman details some memorable cases of peddlers arrested for fraud. "You little extract of a man!" cries a thief, in another folk legend originating in the sixteenth century, the story of Tom Thumb.[36]

In surveying the myths and burlesques of the mandrake in search of the homunculus's own roots (Paracelsus's "homuncular ruminations," as he calls them), Newman emphasizes the alchemist's claim to improve on nature, not imitate her, and this distinction between the perfective and the merely mimetic is central to the difference not only between the homunculus and the *alraun*, but broadly between alchemy itself and the pictorial or plastic arts.[37] The alchemist distills and extracts the pure essences of matter, accomplishing by natural means what nature in her imperfection cannot. The homunculus was the promise of an incorrupt man, whose powers would be concentrated and increased. One may assume that an automaton would have been considered a thin imitation by the alchemist's standard since, as in painting or sculpture, it is made of "found" and not "like" materials.[38] In the homunculus as a made thing, Freedberg's concept of *reconstitution* in the context of works of art becomes literal and technical, a generative phenomenon occurring not in the beholder's ardent response to a hypnotic signal, but instead entirely in the laboratory of the maker—with elemental matter, from scratch. The resulting creature becomes agent to its own powers in a way that short-circuits the orthodox channel through which Freedberg's icon comes to life.

"MORE ALIVE THAN LIFE ITSELF"

Renaissance high realism in the visual arts was underwritten by a rhetoric built on biological metaphor. Fredrika Jacobs has made a study of the language that was used in the Italian sphere to define likeness, lifelikeness, and liveness in works of art, both painting and sculpture.[39] In regard to portraiture in particular, one finds a wealth of commentary rendering images as alive in their own right. Consider Giorgio Vasari's words of praise for Raphael's great 1518 painting of Pope Leo X: "più vivo . . . che la vivacità," by which he meant "more alive than life" or perhaps even, given his own inimitable *sprezzatura*, "more alive than life itself."[40] The word *risuscitare* (to resuscitate, to bring back to life) was widely used to describe the power of a commemorative portrait; and a sculpture might be declared "a speaking likeness," for example. Further, the classicizing impulse to correct nature—to make perfect—in reproducing the human physiognomy meant that the artist, like the alchemist, was in the business of perfecting nature. The creative process itself was exalted as a form of begetting, one that mixed the artist's own *ingenium*—his genius, his *spiritus*—with what he saw in order to bring the work of art to life.[41] The biological metaphor in art for both process and product extends back into linguistic time; the word *engineer* also takes its meaning from the Latin *ingenium*. Implicit in the metaphor is precisely the feat of infusing inanimate materials with life, through the breath and issue of the artist's own soul.

Across all of these models—from the devotional image and icon to the automaton, the homunculus, the magic *alraun*, and the cult of the artist—we can notice an implicit shift in the way images were seen. The concept of an inherent or originating divinity is replaced in varying degrees by a chronicle of human agency. Artist, engineer, and alchemist are introduced as masters *sine qua non* of bringing ingredient matter to life. While the charisma of the icon is a form of grace vouchsafed by God alone—for to "animate" means, literally, to "give a soul to"—the difference between an icon and a homunculus is nothing less than a gnostic shift from Faith to Art in a bid to claim the means of implanting a soul in senseless stuff. No wonder the homunculus was kept such a secret. Is it possible the monk survives in such good condition because he, too, was carefully hidden? The Reformation made even the holy icon itself dangerous, for if an image is so powerful that God himself is seen as present in it, then the image will be worshipped and it becomes an idol. And an idol laid a seductive and deceiving path away from the true God.

In the next century, the whole glittering array of artificial beings is swept aside in a passage that William Newman takes from the Calvinist divine John Edwards. Edwards, in his *Demonstration of the Existence and Providence of God* published in 1696, extols the irreducible vitalism of the living human body:

> This is no Workmanship of Humane Skill, here is no *Automaton* made by Art, no *Daedalus's walking Venus*, no *Archytas's Dove*, no *Regiomontanus's Eagle and Fly*. Here is none of *Albertus magnus* or *Frier Bacon's speaking head*, or *Paracelsus's Artificial Homuncle*. Here is nothing but what proceeds from a divine Principle and Art, and therefore cannot be reckoned among those mechanical Inventions which have an external Shew of Sensation and Life for a time, but are destitute of a vital Spring.[42]

As Newman declares, Edwards turns on its head the alchemist's claim of superiority over nature and adds it to the list as just one more feeble human art.

THE MACHINE IN THE AGE OF FAITH

Vitalism, in its various guises since the time of Aristotle, insists on the presence of a life-granting force within the body that is not reducible to material explanation. But what about the wounded body, "the thousand natural shocks that flesh is heir to?" Perhaps the deeper human motivation in the effort to control nature has less to do with power and procreation and more to do with sheer repair. The practice of medicine in the sixteenth century was itself a manifold art, woven from responses to the mysteries of nature and divinity. For example, what are we to make of the miraculous absence of putrefaction in Fray Diego's disinterred body, and the life-inducing marvels of the same putrefaction in the alchemist's glass? It would seem that decomposition was as promising in the laboratory as its absence was marvelous in the monastery. Don Carlos's suppurating head wound, the physicians' pharmacopoeia for the treatment of infection, the ascetic Franciscan and his uncanny corpse (still sweet to the nostrils), the apparition of the midnight visit, the immaculate cure—all this forms the subject of another kind of study of sixteenth-century substance. What is living stuff made of? All the players here are busy with ideas about the hidden matter within the *corpus*.

If, as legend has suggested, the automaton monk is a portrait of San Diego, its status as a votive object invokes the broad and age-old tradition of promising a form of thanks that would commemorate the providential gift for all time, a form of thanks that would outlast the supplicant, the broker—that would continue after his death. To return to the two questions I asked at the beginning of this study—might the automaton monk have been taken as live? and what resuscitating essence was divined to be transferred from the corpse to the prince?—it would seem that the very impulse to reach beyond death affords a clue. The automaton may be a votive object for the giving of thanks, a prophylactic talisman, a promise, a petition, or a phantom, but like the homunculus it is also a maiden work of biotechnology. As for the corpse: in 1589, the year following Fray Diego's secure entrance into the pantheon of saints, his tomb was formally opened and the bones of his lower legs removed (still bearing flesh) to provide relics for the king.[43] Veneration—but health insurance too.

In 1974 (three years before he purchased the monk for the Smithsonian), Otto Mayr published an essay in German entitled "Automaton Legends in the Late Renaissance."[44] It focuses on the interpretation of precisely the Antique and Medieval legends of artificial life that Edwards lists, together with the efforts of philosophers from Cardano to Descartes to separate myth from legend from fact. As Mayr himself negotiates the uneven border between rational mechanism and legendary magic, he speaks of a double world: "Medieval literature certainly displays no lack of interest in the fantastic and the supernatural. But at the same time the official philosophy—scholasticism—was, in its way, strictly rational. Such contradictory schools of thought exist side by side without being reconciled."[45]

This double world is with us still. Surveying the historical record, I feel a distinct companionship with the uninitiated devout who were tricked by statues made to speak and move by hidden means.[46] Chapter 62 of Don Quixote, entitled "Which Deals with the Adventure of the Enchanted Head, Together with Other Trifling Things That Cannot Be Left Untold" (from part 2 of the novel, 1615), presents an Iberian version of Albertus Magnus's brazen head complete with a company of the fooled whose hair stands on end when the head speaks. Its owner must smash it "lest word of this should reach the alert ears of those reverend gentlemen, the sentinels of our faith."[47] Here was a machine meant to deceive the viewer, and Cervantes gives us an antic satire of the response. One must tread carefully in the business of simulating miracles. A different and final class of objects to which the monk might be compared can be found within the Spanish tradition of penitential processions during Holy Week. On these occasions, elaborately dressed wooden figures of patron saints—some with hidden mechanisms for animating the arms and heads—were carried through the streets in a commingling of theater and devotional passion. These articulated sculptures, *imágenes de vestir* as they are called (images with clothes), were seen throughout the year in static repose in the churches. But at Easter out they came, live from the altar as it were, and suddenly at large.[48] The full community participated in the ritual enactment of a wooden figure coming to life. There is no deception, but the peoples' hair *still* stands on end.

Such processional objects reenact the moment when the hand of providence touches and repairs the broken world. The Christian mysteries of resurrection and transubstantiation—not to mention the persuasions of the Inquisition's faith police—would surely have vivified an iconic, ecclesiastical automaton—or ghosted it. David Freedberg's less feverish term *reconstitution* carries the sense of this for the rest of us. The story of San Diego, even if no more than contemporary to the monk, helps us see how invested this machine is with the burden of faith.

FIGURE 13.8 | Photograph © Rosamond Purcell 2002.

NOTES

This essay is a study for the final chapter of a forthcoming monograph on the monk, *A Machine, a Ghost, and a Prayer: The Story of a Sixteenth-Century Mechanical Monk*, written with W. David Todd and with a foreword by Silvio Bedini and photographs by Rosamond Purcell.

I want to thank Jessica Riskin, William Newman, Fredrika Jacobs, Leslee Asch, and Joan Baixas, who together represented the worlds of science, art, and theater on a panel entitled "The Automaton, the Homunculus, and Other Artificial Beings: Toward an Interdisciplinary History" that David Todd and I organized for the College Art Association Conference in Philadelphia in 2002. The published works of these individuals have been essential to my study.

To Father Servus Gieben, president of the Istituto Storico dei Cappuccini in Rome, and director of its Museo Francescano, whose 1975 research on the automaton monk was pivotal in suggesting a connection to the Spanish court, and whose wisdom has guided my own search, I owe the deepest respect and gratitude.

1. This description of the monk's performance is based on the detailed conservation work of W. David Todd, clockmaker, associate curator emeritus, and former conservator of timekeeping at the Smithsonian Institution.

2. William H. Prescott, *History of the Reign of Philip the Second, King of Spain*, ed. John Foster Kirk, 3 vols. (Philadelphia: J. B. Lippincott, 1874), 2:467–68. (Compare this publishing date with the 1867 premiere performance of Verdi's opera in Paris.) *Erysipelas*, also called St. Anthony's fire, was a spreading inflammation of the skin and subcutaneous tissues, now understood as caused by a streptococcus.

3. L. J. Andrew Villalon, "Putting Don Carlos Together Again: Treatment of a Head Injury in Sixteenth-Century Spain," *Sixteenth Century Journal* 26, no. 2 (1995): 347–65. Chief among the primary texts Villalon consults is the document written by Dionisio Daza Chacon, the prince's surgeon: "Relación Verdadera de la herida de cabeza del Serenísimo Príncipe D. Carlos nuestro Señor, de gloriosa memoria, la cual se acabo en fin de julio del año de 1562," printed in *Colección de Documentos Inéditos para la Historia de España*, 112 vols. (Madrid: Academia de la Historia, 1842–95), 18:537–63.

4. Villalon, "Putting Don Carlos Together," 356.

5. Prescott, *Philip the Second*, 469.

6. Villalon, "Putting Don Carlos Together," 354 (betony or lousewort: a plant native to Eurasia).

7. For eyewitness accounts of Don Carlos's report, see Fr. Lucio M. Núñez, "Documentos sobre la curación del príncipe D. Carlos y la canonización de San Diego de Alcalá," *Archivo Ibero-Americano*, año 1, núm. 4 (Julio–Agosto 1914): 424–46. This four-part article series continues to año 2, núm. 9 (Julio–Agosto 1915): 374–87; año 3, núm. 13 (Enero–Febrero 1916): 107–26; año 4, núm. 19 (Enero–Febrero 1917): 421–31.

8. L. J. Andrew Villalon, "The Miracle Book of San Diego de Alcalá; or, The Fifteenth-Century Failure to Canonize the First Counter-Reformation Saint," *Mediterranean Studies* 10 (2001): 9–35.

9. Ibid., 26.

10. Sir William Stirling-Maxwell, *The Cloister Life of the Emperor Charles V*, 4th ed. (London: John C. Nimmo, 1891), 444. Comparison here with the Jewish Golem is irresistible.

11. "Tambien ha querido Ianelo por regozijo renouar las eftatuas antiguas, que fe mouian, y por eſſo las llamauan los Griegos Automatas. Hizo vna dama de mas de vna tercia en alto, que puefta fobre vna mefa dança por toda ella al fon de vn atambor, que ella mifma va tocando, y da fus bueltas, tornando a donde partio. Y aunque es juguete y cofa de rifa, todauia tiene mucho de aquel alto ingenio." Ambrosio de Morales, *Las antigüedades de las ciudades de España* (Madrid: 1575), 93v. The English translation here is from José A. García-Diego, *Juanelo Turriano, Charles V's Clockmaker: The Man and His Legend*, trans. Charles David Ley (Sussex: Antiquarian Horological Society; Madrid: Editorial Castalia, 1986), 101.

12. Ernst von Bassermann-Jordan, *Alte Uhren und Ihre Meister* (Leipzig: Wilhelm Diebener, 1926), 58, 66–69. Bassermann-Jordan had seen the figure in a private collection sometime before 1922.

13. I have elsewhere traced out the larger web of primary and secondary sources that may shed light on the automaton monk, both in relation to the story of Don Carlos, and to the workshop of Juanelo Turriano. See Elizabeth King, "Clockwork Prayer: A Sixteenth-Century Mechanical Monk," published in *Blackbird: An Online Journal of Literature and the Arts* 1, no. 1 (Spring 2002), at http://www.blackbird.vcu.edu/v1n1/nonfiction/king_e/king_e.htm, and in *Breaking the Disciplines: Reconceptions in Knowledge, Art and Culture*, ed. Martin L. Davies and Marsha Meskimmon (London: I. B. Tauris, 2003), 84–128. This essay includes a close comparison of the monk with the several kindred tabletop figures that survive from the same period, together with a discussion of the likely court scenarios in which they were viewed. The online version includes a streaming media video of the monk in action.

14. Excerpt, letter written in French, dated 16 November 1975, to Georges Sedlmajer in Geneva from Servus Gieben in Rome. Photocopy in the Registrar's Office of the Smithsonian's National Museum of American History. The English translation is my own.

15. This was an early draft of "Clockwork Prayer" (see n. 13 above).

16. There is a competing argument for a south German guild scenario for the production of both monk and lady—particularly of the internal iron clockwork. A forthcoming monograph on the monk, *A Machine, a Ghost, and a Prayer: The Story of a Sixteenth-Century Mechanical Monk*, which I coauthored with W. David Todd, explores this alternative.

17. Morales, *Antigüedades de las ciudades de España*, 93v.

18. The possibility that the monk is a species of toy begs consideration of what is and was meant by *toy* and what kinds of work toys do in the world. In his important essay "Automata and the Origins of Mechanism and Mechanistic Philosophy," *Technology and Culture* 5, no. 1 (1964): 9–23, Derek J. de Solla Price describes the interdependent rise of biological and astronomical simulacra in antiquity: "Perhaps it is not altogether fanciful to see the astronomical zodiac as the first primitive coming together of a cosmic model and a set of animal models" (12). In ancient Greece, "these two great varieties of automata go hand-in-hand and are indissolubly wedded in all their subsequent developments" (15). More tellingly: "Amongst historians of technology there seems always to have been private,

somewhat peevish discontent because the most ingenious mechanical devices of antiquity were not useful machines but trivial toys. Only slowly do the machines of everyday life take up the scientific advances and basic principles used long before in the despicable playthings and overly ingenious, impracticable scientific models and instruments" (15). Daniel Tiffany, in his recent book *Toy Medium: Materialism and Modern Lyric* (Berkeley: University of California Press, 2000), weaves a philosophically rich narrative around Price's thesis in developing his own "discourse of automata"; see the chapter entitled "The Natural Philosophy of Toys." We continually remap the border between useful things and frivolous ones, regardless of the complexity of the objects themselves. Among classes of automata unique to the sixteenth century—figure clocks, musical tableaux, dancing animals, animated figures of pagan myth (and often the figures of play: Diana, Cupid, or Bacchus)—one finds a considerably less ludic subgroup of pieces representing biblical figures and Christian themes. Klaus Maurice, in an essay entitled "*Propagatio fidei per scientias*: Jesuit Gifts to the Chinese Court," published in the catalogue for the exhibition he curated with Otto Mayr, *The Clockwork Universe: German Clocks and Automata, 1550–1650*, ed. Klaus Maurice and Otto Mayr (Washington, DC: Smithsonian Institution; New York: Neale Watson Academic Publications, 1980), touches on the role such objects played in the Jesuit mission.

19. In 1980, three years after the monk entered the Smithsonian collections, the name of the Museum of History and Technology was changed to the National Museum of American History. In December 1997, the monk and other rare instruments and clocks that once graced the museum's great Hall of Timekeeping were removed from display, and the galleries were redesigned for the new exhibition "On Time," which examined timekeeping from an exclusively American perspective. As I write, in the summer of 2006, that exhibition in turn is being dismantled as the museum itself closes down for a two-year renovation.

20. David Freedberg, *The Power of Images: Studies in the History and Theory of Response* (Chicago: University of Chicago Press, 1989): "[When our study] is too strictly based on reclamation of context [it] resolutely refuses to allow the integration of ourselves into the past, or—to put it less sentimentally—to allow the lessons we learn from our own responses to inform our judgments about the past" (431).

21. Roland Barthes, *Camera Lucida*, trans. Richard Howard (New York: Hill and Wang, 1981), 12.

22. Freedberg, *Power of Images*, 28.

23. Ibid., 32.

24. Michael Baxandall, *The Limewood Sculptors of Renaissance Germany* (New Haven, CT: Yale University Press, 1980): "The disposition to infer character and feeling from a representation of a human figure is both strong and deep, and certainly one of the few constants in the older European art criticism. Indeed it seems a natural enough transference from our normal social interest, for character and feeling are things we want and need to know about in persons we address, and we are all very skilled in interpreting visual appearance to this end: posture, gesture, glance, the fixed lineaments of the body and the face. In particular we are sensitive to what all these imply of an attitude towards ourselves. The devotional image is often a special case because acts of devotion involve urgent and complicated kinds

of expectation and desire; to pray to an image—even, as the theory of images would have it, through an image—with a view to a spiritual or material return, or to meditate on an image as exemplification of spiritual quality or as simulacrum of the divine, is to enter an encounter of a testing kind, interpersonal in its general form but abnormally asymmetrical. For the image this means both that quite small cues of either appearance or context are open to more than usual attention, and that the beholder may project into a figure intimations of character and feeling not so much initiated as admitted by it" (153).

25. Otto Mayr, "A Mechanical Symbol for an Authoritarian World," in Maurice and Mayr, Clockwork Universe, 1–8. See also Otto Mayr, Authority, Liberty, and Automatic Machinery in Early Modern Europe (Baltimore, MD: Johns Hopkins University Press, 1986).

26. Thanks to W. David Todd for help with this morphology. See also n. 18 above.

27. Carlos Fuentes, "Velázquez, Plato's Cave and Bette Davis: When Narration Is a Visual Art," New York Times, 15 March 1987, Arts and Leisure sec.: "Where does true reality reside? On canvas, on film or in the imagination?" (38).

28. For my argument on this, see King, "Clockwork Prayer," under the heading "Part 3: A Sixteenth-Century Mechanical Masterpiece" in the online version (see n. 13 above), or in Davies, Breaking the Disciplines, 104–6.

29. See Pamela O. Long, Openness, Secrecy, Authorship: Technical Arts and the Culture of Knowledge from Antiquity to the Renaissance (Baltimore, MD: Johns Hopkins University Press, 2001), 146. Long quotes William Newman, "Technology and Alchemical Debate in the Late Middle Ages," Isis 80, no. 303 (September 1989): "The alchemists and their supporters gave a conscious and articulate defense of technology, indeed, one of the earliest and most thorough to be found in Latin Christendom" (427).

30. A detailed comparison of the works of the mechanician and the alchemist in the creation of artificial life in the sixteenth century would make a welcome and fascinating study. Newman's essay cited above and his more recent essay "Alchemy, Assaying, and Experiment," in Instruments and Experimentation in the History of Chemistry, ed. Frederic L. Holmes and Trevor H. Levere (Cambridge, MA: MIT Press, 2000), 35–54, make it clear how intertwined were the practices of alchemist and metalsmith in the pursuit of an operational knowledge of materials. Earlier works addressing the links between the alchemy and technology (as well as the folklore) of artificial beings include Robert Plank's short but pioneering essay "The Golem and the Robot," Literature and Psychology 15, no. 1 (Winter 1965): 12–28; John Cohen's indispensable Human Robots in Myth and Science (London: Allen & Unwin, 1966); and Jean-Claude Beaune's "The Classical Age of Automata: An Impressionistic Survey from the Sixteenth to the Nineteenth Century," in Fragments for a History of the Human Body, Part One, ed. Michel Feher (Cambridge, MA: MIT Press/Zone, 1989). Plank, Cohen, and Beaune offer a comparison of the products of mechanical and alchemical enterprises but not the practices themselves.

31. William Eamon, Science and the Secrets of Nature: Books of Secrets in Medieval and Early Modern Culture (Princeton, NJ: Princeton University Press, 1994). Eamon has illuminated the medieval origins of the libri secretorum, the rich genre of manuscripts and texts passed hand to hand that formed the nascent beginnings of experimental science and recorded craft tech-

nology. Chapter 6, "Natural Magic and the Secrets of Nature" (194–233), presents a fine, detailed portrait of Della Porta, his work, and his fortunes; see especially 217–21, "The Magician as Artisan." The English translation of *Magia naturalis* that Eamon cites is John Baptista Porta, *Natural Magick*, ed. Derek J. de Solla Price (New York: Basic Books, 1957), originally published in London by Thomas Young and Samuel Speed, 1658.

32. Della Porta, *Natural Magick*; see "The Second Book of Natural Magick: Shewing how living Creatures of divers kinds, may be mingled and coupled together, that from them, new, and yet profitable kinds of living Creatures may be generated," chapter 1: "the firſt Chapter treateth of Putrefaction, and of a ſtrange manner of producing living Creatures" (26–27).

33. William R. Newman, "The Homunculus and His Forebears: Wonders of Art and Nature," in *Natural Particulars: Nature and the Disciplines in Renaissance Europe*, ed. Anthony Grafton and Nancy Siraisi (Cambridge, MA: MIT Press, 1999), 321–45; "Alchemy, Domination, and Gender," in *A House Built on Sand: Exposing Postmodernist Myths about Science*, ed. Noretta Koertge (New York: Oxford University Press, 1998), 216–26; and most recently, "Artificial Life and the Homunculus" in Newman's book *Promethean Ambitions: Alchemy and the Quest to Perfect Nature* (Chicago: University of Chicago Press, 2004), 164–237.

34. Newman, *Promethean Ambitions*, 203–4, translated from "[Pseudo?-] Paracelsus," *De natura rerum*, in *Theophrastus von Hohenheim, genannt Paracelsus, Sämtliche Werke*, I. Abteilung, ed. Karl Sudhoff, 14 vols. (Munich: Oldenbourg, 1922–33), 11:316–17. A "horse's womb"—*venter equinus* in Latin—refers to decaying equine dung used as a source of heat. For the perceived homology between womb and dung, and for the context of this recipe in the broader history of beliefs on the procreative roles of male versus female, see Newman, "Artificial Life and the Homunculus," 164–237.

35. See Newman, "The Homunculus and the Mandrake," in this volume. See also Newman, *Promethean Ambitions*, 210–15: translated from Paracelsus, *De vita longa libri quinque*, in Sudhoff, *Theophrastus von Hohenheim*, 3:274.

36. "Hop-O'-My-Thumb," in *Grimm's Goblins* (London: George Vickers, ca. 1865–69), 155. But see W. Carew Hazlitt's reprint, *Tom Thumbe, his Life and Death:Wherein is declared many Maruailous Acts of Manhood, full of wonder, and ſtrange merriments:Which little Knight liued in King Arthurs time, and [was] famous in the Court of Great Brittaine* (London: John Wright, 1630), in *Remains of the Early Popular Poetry of England; Collected and Edited, with Introductions and Notes*, ed. W. Carew Hazlitt (London: John Russell Smith, 1866), 167–250. In the Grimms' version, the infant Thumb comes into the world by natural birth, "well formed in all its limbs, but no bigger nor higher than [a] thumb." In lyric form, the passage seems to come directly from the homunculus, "with all its members like another child, which is born of a woman, but much smaller." But in the original tale, as Hazlitt gives it, none other than Merlin himself produces the little man.

37. See Newman, "The Homunculus and the Mandrake," in this volume. Newman pursues the distinction in detail in *Promethean Ambitions*, chap. 3, "The Visual Arts and Alchemy," 115–63.

38. It is worth noting here that "image magic," or *envoûtement*, as Freedberg presents it in his chapter on effigies and witchcraft in *Power of Images* ("Infamy, Justice, and Witchcraft: Explanation, Sympathy, and Magic," 246–82), is widely based on the material principle of

like producing like. Freedberg offers a passage from James George Frazer's 1913 work on the history of magic, *The Golden Bough*, to identify this principle, and in that passage we find a second precept that resonates with the themes of the present essay:

> If we analyze the principles of thought on which magic is based, they will probably be found to resolve themselves into two: first, that like produces like, or that an effect resembles its cause; and second, that things which have once been in contact with each other continue to act on each other at a distance after the physical contact has been severed. The former principle may be called the Law of Similarity, the latter the Law of Contact or Contagion. From the first of these principles, namely the Law of Similarity, the magician infers that he can produce any effect he desires merely by imitating it. (J. G. Frazer, *The Golden Bough* [London: n.p., 1913], 1:52; quoted in Freedberg, *Power of Images*, 272)

The question of what was perceived to be transferred from the holy corpse to the ailing prince in the story of Don Carlos's cure has an echo here in the Law of Contact, from the world of sympathetic magic. (As for the ancient concept of "action at a distance," let us remember that this is now a leading field in contemporary quantum physics.)

39. Fredrika Jacobs, *The Living Image in Renaissance Art* (New York: Cambridge University Press, 2005). My grateful thanks to Professor Jacobs for permitting me to read final drafts of this book.

40. Giorgio Vasari, *Le opere di Giorgio Vasari*, ed. Gaetano Milanesi, 9 vols. (Florence: G. C. Sansoni, 1906), 4:352. Again, my thanks to Fredrika Jacobs for providing this quotation. Raphael's painting is *Pope Leo X with Cardinals Giulio de' Medici and Luigi de' Rossi*, panel, 60½ × 47 inches, Uffizi Gallery, Florence.

41. I closely paraphrase both Jacobs, *Living Image*, 7, and, in turn, her reference, David Summers, "*Aria* II: The Union of Image and Artist as an Aesthetic Ideal in Renaissance Art," *Artibus et Historiae* 20 (1989). Summers writes, "Petrarch closely follows Seneca, who wrote that, in effect, painting was fully capable of imitation. That is, it did not simply and literally duplicate; rather, the painter mixed his own *ingenium*, his own talent and vision, his own *spiritus*, with what he saw in order to make it seem alive, and to make it true" (26).

42. Newman, *Promethean Ambitions*, 226, quoting John Edwards, *A Demonstration of the Existence and Providence of God, from the Contemplation of the Visible Structure of the Greater and the Lesser World*, part 2 (London: Jonathan Robinson, 1696), 124.

43. Núñez, "Documentos," año 3, núm. 13 (Enero–Febrero 1916): 119–22.

44. Otto Mayr, "Automatenlegenden in der Spätrenaissance," *Technikgeschichte: Beiträge zur Geschichte der Technik und Industrie* 41, no. 1 (1974): 20–32. My thanks go to Henning Bauer for a fine translation.

45. Ibid., 30.

46. See Freedberg, *Power of Images*, chap. 11, "Live Images: The Worth of Visions and Tales," 283–316.

47. Miguel de Cervantes, *Don Quixote de la Mancha*, trans. Samuel Putnam (New York: Viking, 1949), 1090–91. Cervantes was a native of Alcalá, born just two years after Don Carlos. Part 1 of *Don Quixote* was published in 1605; part 2 in 1615.

48. Susan Verdi Webster, *Art and Ritual in Golden-Age Spain* (Princeton, NJ: Princeton University Press, 1998):

> The spatial and temporal status of the sculptures in procession significantly enhance their mimetic effects, and their unique kinesthetic character allows them dramatic entrance into the realm of human experience. They are able to move both physically (through articulated limbs) and spatially (through the streets of the city). Furthermore, the incorporation of sculpture within a processional context acts to change a most fundamental aspect: the sense of time. No longer the static, eternal images of altars and *retablos*, their temporal state is extended so that they merge with the spectators' own experience of space and time. (167)

14 Motions and Passions: Music-Playing Women Automata and the Culture of Affect in Late Eighteenth-Century Germany

ADELHEID VOSKUHL

In the long and diverse history of attempts to create artificial humans, stretching all the way from antiquity to the early twenty-first century, the eighteenth century stands out as a period that was unusually productive in the manufacture of android automata—mechanical artifacts that look and move like human beings and perform cultural techniques, such as writing, drawing, or music-making.[1] Particularly remarkable is the perfection with which android automata from this period replicated humans and human activities with basically nothing but extremely sophisticated clockwork, which was well-hidden in the respective bodies of the automata. On the basis of contemporary android automata, Diderot's *Encyclopédie* from 1751 defined *automaton* as a "device that moves by itself, or a machine that carries in itself the principle of its motion," and the encyclopedia defined *android* as an "automaton with a human figure, equipped through the means of a certain spring drive to act like and conduct functions which apparently resemble those of humans."[2] Between 1738 and 1810, the artistic and mechanical project of creating android automata culminated in a set of extraordinarily beautiful and perfected artifacts.[3]

The period not only provided the cultural and the economic conditions for the tremendous efforts that went into making mechanical humans such awe-inspiring spectacles, but also spawned literary commentary on automata, arising from the intersection of contemporary social and cultural concerns with these artifacts. In this essay I analyze the intersection of automaton-making and literary and cultural commentary by investigating the context of two eighteenth-century android automata that both represent women playing a keyboard instrument. At the center of my analysis—apart from the two automata—is a short, complex satire entitled "Humans Are Machines of the Angels," which was written in 1785 by Jean Paul, one of Germany's prominent writers of the Late Enlightenment. The satire is one of the few contemporary literary texts that explicitly mentions music-playing women automata. Jean Paul uses the motif to illustrate mockingly an observation he makes about women's (and women automata's) musical performance, drawing a parallel between the musicians' moving bodies and contemporary efforts to develop rules and codes for the communication of affects in musical performance through carefully coordinated bodily motions. These codes were part of larger concerns to establish conventions and boundaries of sentimental selfhood, social interaction, and bodily comportment in the newly emerging bourgeois culture in the German lands.[4]

To investigate the intersection of music-making automata, literary reflection on "artificial humans," and theoretical and pedagogical literature on musical performance in the eighteenth century, I first briefly describe the environments in which the two music-making women automata were built. Following that, I deal with a few aspects of Jean Paul's work in the context of late eighteenth-century literary Germany and then undertake a close reading of his satire. I trace the terms and allusions appearing in this text to corresponding passages in two influential contemporary works from musical theory and pedagogy, one on playing the flute, by Johann Joachim Quantz, the flute teacher and court composer of Frederick the Great, and the other on playing the piano, by Carl Philipp Emanuel Bach, one of the sons of Johann Sebastian Bach. The historical connections between these two types of texts, the music-playing automata, and their respective cultural contexts suggest that Jean Paul's commentary on music-playing automata was not primarily concerned with the contradiction of "life" and "mechanism" ostensibly inherent in the android automata of the eighteenth century. Rather, his commentary was an illustration and a mockery of the characteristically deliberate, self-conscious, and at times unintentionally funny ways in which the movers and shakers of the German Enlightenment organized their efforts to constitute and establish techniques and practices for social interaction in the newly emerging spaces of bourgeois culture.

AUTOMATA

The two music-playing women automata considered here were among the most important and most famous android automata of the eighteenth century. The first one, called *La musicienne* (The woman musician), was built by the clockmakers Pierre and Henri-Louis Jaquet-Droz in La Chaux-de-Fonds in the French-speaking part of Switzerland and presented to the public in 1774 (figure 14.1). The other one, called *La joueuse de tympanon* (The dulcimer player), was made by the cabinetmaker David Roentgen and the clockmaker

FIGURE 14.1 | *La musicienne*, by Pierre and Henri-Louis Jaquet-Droz from La Chaux-de-Fonds (Switzerland); introduced to the public in 1774. Reproduced by kind permission of the Musée d'art et d'histoire, Neuchâtel (Suisse).

FIGURE 14.2 | *La joueuse de tympanon*, by David Roentgen and Pierre Kinzing from Neuwied (Germany). Presented as a gift to Marie Antoinette in 1785. Reproduced by kind permission of the Conservatoire National des Arts et Métiers, Paris (France). © Musée des arts et métiers-CNAM, Paris. Photo: P. Fallgot/Seventh Square.

Pierre Kinzing, who ran a manufacture in Neuwied, a small principality about fifty kilometers south of Cologne (figure 14.2). Roentgen and Kinzing presented the dulcimer player as a gift to Marie Antoinette in 1785.[5]

Both the Jaquet-Droz workshop and the Roentgen/Kinzing manufacture were unusually successful and productive sites of artisanship in their time. Complex sets of social, religious, and political factors had brought each firm to the height of its economic and artistic success in the late eighteenth century. Pierre Jaquet-Droz and his son Henri-Louis were unusually well-educated and gifted clockmakers. They came from an affluent and well-connected family in La Chaux-de-Fonds, and they both received university educations. Pierre Jaquet-Droz was originally supposed to embark on a clerical position—a typical choice of profession for the son of an aspiring middle-class family at the time. He matriculated as a student in philosophy at the University of Basel in the late 1730s, where he supposedly studied, among other things, mechanics and mathematics with Johann and Daniel Bernoulli. Following his training in Basel, he enrolled in a theological seminary in Neuchâtel, near his home. Upon his return to La Chaux-de-Fonds from his studies in the 1740s, however, he decided to settle on the art of clockmaking.[6] In the later 1760s, he sent his son Henri-Louis to the University of Nancy in France, where he studied music and drawing with, among others, the abbé de Servan. Connections to the local political elites in La Chaux-de-Fonds gave Pierre Jaquet-Droz the opportunity to pay a visit to the Spanish king in Madrid in the year 1759 and to present his clocks and other mechanical artifacts to him. The patronage and financial support resulting from this visit eventually enabled Pierre and Henri-Louis Jaquet-Droz to devote several years to the construction of their spectacular automata in the late 1760s and early 1770s.[7]

La Chaux-de-Fonds, a small town in the rural Jura mountains, was a culturally and economically remarkable place. In the course of the eighteenth century, formerly agricultural communities in this area had developed into prospering, "proto-industrial" cultures of mechanical artisanship.[8] In La Chaux-de-Fonds and in the neighboring Le Locle, for example, every third adult was in one way or another working in the business of clockmaking after 1750.[9] Karl Marx uses the clockwork industry in this region as an example to illustrate his concept of the "manufacture" in the famous twelfth chapter of his *Capital*: in a footnote, he describes the city of La Chaux-de-Fonds as "one big clock manufacture."[10] Because of its political, religious, and economical peculiarities, Switzerland as a whole had always occupied a unique place in the economic history of early modern Europe. Among the factors contributing to its particular economic and cultural development in the early modern period were the strong and distinguished influence of the Swiss Reformation on virtually all other matters of the country's history, the country's geographic location, the frequent influx of religious refugees from the fifteenth to the eighteenth century, and the characteristically liberal culture in economic and political matters in the Swiss Confederation.[11]

La joueuse de tympanon, too, is the product of a very specific economic and cultural environment. The cabinetmakers Abraham and David Roentgen—also father and son—and their works were at least as distinguished in their time as those of the Jaquet-Droz family. In the second half of the eighteenth century, their furniture manufacture was the biggest and most productive of its kind in the entire Old Reich. Between 1742 and 1793, they produced more than two thousand pieces of every conceivable type of furniture and delivered them to the most prominent and influential individuals and courts on the European continent. Among their clients were Katharine II of Russia, Louis XV of France, and Frederick the Great. The Roentgens' distinction and success as producers of furniture was largely the result of the contemporary transition in artisanship and production from the traditional guild order in estate society and court culture in the Old Reich to industrial production and entrepreneurship in civil society. David Roentgen in particular, who ran the manufacture from 1767 to 1793, took the "best of both worlds" from the cultural and economic conditions of his time, relying both on the accomplishments of traditional artisanship and on the principles of industrial production. He served the demands and social norms of the court culture with his luxury commodities while at the same time tapping into the resources of modern marketing and trade.[12]

Furthermore, the Roentgens were members of a Protestant-Pietist sect, the Herrnhut Brethren. This religious community was a dense and widespread network consisting of mobile capital, trade contacts, skilled labor, and an elaborate work ethic. Because they were members of this sect, the Roentgens were not organized in a guild, which meant that they were not allowed to sell any furniture within their town. A viable alternative set of clients for them were the major political, cultural, and religious players, courts, and other institutions on the European continent. Having been expelled from Oberhessen under the rule of the prince of Isenburg-Büdingen in the 1740s, the Roentgens and their Herrnhut community subsequently settled in Neuwied, where Count Friedrich Alexander aimed to rejuvenate his county, which had been devastated by the Thirty Years' War. He granted the Herrnhut Brethren right of settlement, freedom of worship, and exemption from the strict guild rules. These conditions enabled the Roentgens to build up their flourishing furniture business in the decades to come.[13]

JEAN PAUL: AUTOMATA AND MOTIONS THAT "BETRAY AFFECTIVE INVOLVEMENT"

Among the few literary writers of the time to deal explicitly with music-making women automata was Jean Paul, a contemporary and interlocutor of the Weimar classicists Schiller and Goethe as well as of Herder, Karl Philipp Moritz, and Friedrich Nicolai.[14] Of the many satires Jean Paul wrote in his youth in the 1780s, five are concerned with "machine-men" or related types of "artificial humans."[15] He was very familiar with the famous automata built in the course of the eighteenth century, and many of the contemporary spectacles—as well as other, fictitious, ones—figure in these satires. Depending on the respective contexts in which they appear, these automata write, speak, pray, are servants, make meteorological observations, conduct musical orchestras, or preach sermons in church.[16] Jean Paul's work is deeply entangled in virtually all other major philosophical and literary debates of his time as well. His texts are particularly famous as original and highly imaginative microcosms of the complex and numerous debates in the late eighteenth century. At the same time, they are notoriously fragmentary and enigmatic collages, and they contain a dense variety of motifs, metaphors, and references. The satires about artificial humans are no exception in this respect.[17]

The particular form and content that Jean Paul's texts exhibit in this period can be partially explained by his characteristic research and writing techniques, which were based on extensive copying and excerpting from contemporary books, magazines, and journals. As a teenager, Jean Paul began to keep elaborate notebooks with meticulous and comprehensive excerpts from the major works and journals of the time. A local pastor had originally suggested that he excerpt books in preparation for his university education in theology, and this enabled young Jean Paul to develop in just a few years a profound and comprehensive understanding of the heated political and philosophical debates of the Late Enlightenment in Germany. In the course of the following decades—not least because he was too poor even in his adulthood to afford a personal library—Jean Paul created sixty such notebooks. This collection of text fragments is a remarkable encyclopedic index to the intellectual landscape of late eighteenth-century Germany. When Jean Paul quit his university studies and became a freelance writer in the early 1780s, his excerpt-writing turned into a literary method: the excerpts provided him with an enormous collection of details and comments on virtually all relevant contemporary topics, from which he quoted extensively.[18]

Jean Paul took up the theme of piano-playing women in a short, satirical text entitled "Humans Are Machines of the Angels," which he wrote in 1785.[19] The text opens with the narrator's claim that we had to be enlightened for a long time to come to realize that the world does not exist because of us; and that it will take even more Enlightenment to make us realize that we actually live here because of higher beings that we call angels; they are the true inhabitants of the earth; we are just furniture. The narrator offends the reader further by stating that, of our activities here on earth, so few seem to contribute to our well-being that we should have had doubts a long time ago about whether our activities actually serve our own purposes. Indeed, to him it seems "obvious [that] our industriousness, which works against our happiness, is conducive to other beings' happiness, whose hands conduct ours as their tools." The narrator concludes that it is therefore not a poetic saying but the "bleak, naked truth" that we humans are "mere *machines*" that serve higher beings who were first chosen to inhabit the earth.[20]

What kind of machines are humans, then, according to the logic of this text? And how did they come into being? The narrator explains that "when the angels first entered the earth, they did not yet have the numerous human-machines for which they can now congratulate themselves. Rather, it was only step by step that [they] invented machines, or, as we call them, humans." Humans are thus machines that the angels built for various purposes, until, so the narrator says, there were enough machines on the earth to serve "all needs" of the angels.[21]

In the following passage, the narrator introduces an example. He talks about an angel who built chess-playing machines, for the sake of "curiosity and pleasure" rather than for the sake of "utility."[22] Jean Paul's choice of chess-playing machines here is not accidental; it has an important historical correlative. The Hungarian nobleman Wolfgang von Kempelen built a chess-playing automaton in the late 1760s that quickly became famous and remained a topic of discussion in the years and decades to follow. It depicted an almost life-size Turkish man, dressed in a cloak and a turban, who was sitting behind a wooden cabinet and playing chess. The cabinet contained a clockwork mechanism that supposedly enabled the automaton to play chess against human players. The Kempelen chess automaton inspired a number of myths and legends, including those according to which it played chess (and won) against Frederick the Great, Napoleon, Benjamin Franklin, and Charles Babbage. The automaton's performance, however, was based on a hoax: a human chess-player was hidden in the cabinet and conducted the moves of the game.[23]

The example of chess-playing machines, though not explicated in great detail in the satire, has important consequences for the scenario that the text is building up. The chess-playing automata raise the significant issue that the machines of the angels (that is, humans) in their turn also build machines, and that they mimic the angels in their machine-building: humans build machines that "look like" or "pass as" humans and perform human tasks. While the passage on chess-playing automata thus imparts crucial information about the ways in which humans and machines relate to each other in the text's logic, the passage is also—both in form and in content—a well-crafted source of profound confusion for the reader. The narrator artfully exploits in the following sentences the two kinds of machine-building by talking alternately—in his typically cryptic and casual-ironic way—about angel-made chess-machines (that is, chess-playing humans) and human-made chess-machines (such as the Turk by von Kempelen) with consistently insufficient distinction between the two kinds. When talking about chess-playing humans, the text says, for example, "All my readers must have seen creatures of this kind, which play chess without any help of an angel, just through a mechanism in their head; they move their right arm by themselves, and they even shake their head . . . upon a wrong move of their opponent; and once the king is checkmate, they won't make another move under any circumstances."[24]

Not only are the readers invited here to remember the occasions on which they were watching other humans play chess, but the description of chess-playing humans is also followed by the suggestion that readers will recognize easily how similar these (human) chess-machines are to the "well-known chess-machine invented by Mr. v. Kempelen."[25] The text thus comes full circle in its deliberate confusion of chess-playing humans and chess-playing machines. The juxtaposition of angels' and humans' machine-building, the similarity between the products of this activity, and the consistently unclear references in the text leave the reader incessantly bewildered about the status of the story's actors (as well as the reader's own status) as machine or as human. The narrator continuously proposes to read the actors in the text as machines, when they are in fact alternately humans or machines.

The doubling-up of machine-building in the story and its confusing consequences eventually lead the text to the fundamental question about the difference between humans and machines. In the closing sentence of the passage about the chess-playing machines, the narrator makes the first brief reference to this theme. After describing von Kempelen's chess-machine as a "copy" of human chess-players, he says, "Notwithstanding all this, there will still always remain a tremendous difference among the two kinds of machines, and the work of the angels will always stand out by far against the work of a human."[26]

This is an unusually clear statement on the part of the narrator on this topic. While in the introductory parts, the narrator introduced humans as machines—point-blank with little further qualification—he is here willing to admit the superiority of humans over machines. The claim, however, still appears in the context of profoundly ambiguous passages, and the ostensibly serious tone remains conspicuous, not least against the background of the very mocking tone at the beginning of the text, which anticipated the continuous offending of the reader as an organizing principle of the entire text.[27] The last sentence of the chess-playing passage remains the only remark on the topic for the time being. The narrator does not comment any further on the question of the difference between humans and machines until the very end of the text.

In the final sentence, the narrator picks up this theme again, concluding the text with the following words:

> The machines of the earth must almost always step back behind the machines of the angels, and one does not really offend the former in claiming that they are . . . mere replicas and weak copies of those machines which the angels contrived: this woman, for example, who plays the piano is at most a fortunate copy of those female machines who play the piano and who accompany the music with bodily motions, which obviously seem to betray affective involvement.[28]

At this very prominent place—right at the end—the text thus turns the question about the difference between humans and machines into an intriguing collage in which "female machines," bodily motions, and the "betrayal" of affective involvement—the German word is *Rührung*—are related to each other.[29]

The bewildering writing technique encountered in previous passages reappears in this final sentence. While the collection of motifs in the second half of the sentence—female machines and motions that "seem to betray affective involvement"—pretends to serve as an example for the argument in the first half of the sentence (the claim that machines of the earth have to "step back" behind the machines of the angels), there is in fact much more confusion than illustration going on in this passage. The alleged illustration of the distinction between real piano-playing women and piano-playing automata is so convoluted that it is impossible to extract the crucial references and referents within the sentence. Neither in content nor in form does the sentence provide grounds for the reader to clarify the difference between humans and machines. As a result, the sentence itself becomes an illustration of the confusion and the mistaking of humans for machines and vice versa.

While the distinction between humans and machines remains utterly unclear, and the sentence fails to serve its purported function to demonstrate humans' superiority over machines, Jean Paul does refer in this passage in more detail than anywhere else in his machine-men satires specifically to music-making women and music-making women automata. And aside from being a beautiful manifestation of confusing humans with machines, the sentence does something else. It uses the rich and suggestive motif of piano-playing women automata to link the classic human-machine question to a specific concern over bodily motions in musical recital that "betray" affective involvement. This concern—which may seem odd and out of place at first sight, or, in fact, just another instance of Jean Paul's ironic and offensive chatter—has a concrete historical correlate in the contemporary literature on the theory and practice of music-playing.[30] To trace further this connection between Jean Paul's choice of motif and concerns in contemporary musicology, I now reconstruct briefly the textual conduits through which music-playing women automata entered his work and then investigate, on the basis of two examples, the debates in contemporary musicology on the role of bodily motion and affective involvement in musical recital.

JEAN PAUL'S WORK AND THE CONTEMPORARY LITERARY WORLD

One of Jean Paul's excerpt notebooks from the year 1785 contains a short entry from a German newspaper (the *Dessauische Zeitung*) about the dulcimer player built by Roentgen and Kinzing.[31] Jean Paul excerpted a passage that describes how the automaton plays the dulcimer with two sticks, and how it moves its eyes rapidly or slowly, according to the character of the musical piece it is playing. Yet Jean Paul never mentions the dulcimer player or Kinzing and Roentgen anywhere in his literary texts. Exactly the opposite is true for the Jaquet-Droz automata: they do not appear anywhere in Jean Paul's notebooks, but in another satire on machine-men, entitled "The Machine-Man and His Properties," Jean Paul mentions the Jaquet-Droz family by name.[32]

The relationship between the mention of a theme in Jean Paul's notebooks and this theme's relevance in his life and in the larger context of contemporary debates is not straightforward. Many books that were of fundamental importance for him (and for the larger intellectual context) and that he said he knew inside out—he called these books "*Schoos-Bücher*" (lap-books)—do not appear at all in his excerpts, probably because they were too obvious to be mentioned and too prominent in his life.[33] While the collection of Jean Paul's excerpts is an important documentation of contemporary intellectual and textual production, it is not a mirror image of his literary universe. Jean Paul mentions in his satires those automata that never appeared in his excerpts, and he never mentions in his satires the automaton about which he wrote an excerpt.

Although Jean Paul was quite familiar with the automata spectacles of his time, we cannot know if he ever got the chance to see an automaton "in action." Throughout his life, he was too poor even to pay his bills.[34] He received his information about automata from mentions and short accounts about them that were published, like similar stories and anecdotes on other spectacles and curiosities, in the typical mass media of the time—newspapers, almanacs, pocket calendars, and periodicals. Short mentions or descriptions of the Jaquet-Droz automata appeared, for example, in the very popular calendars *Almanac de Gotha* and the *Göttinger Taschenkalender* (edited by Georg Christoph Lichtenberg), in the French *Journal de politique et de littérature*, and in a variety of travel reports about Switzerland, such as William Coxe's *Travels in Switzerland* and Johann Bernoulli's *Bericht einer Reise durch Deutschland und die Schweiz*.[35] The media that disseminated the information about automata were the same ones that Jean Paul used for his elaborate system of excerpts and that constituted his literary universe.

Jean Paul's excerpting procedures and the short accounts of automata were both products of a profound transformation in public communication that was happening in the late eighteenth century in the German-speaking lands. This transformation included rapidly increasing literacy; a rapidly growing market for the production and consumption of books, journals, and periodicals; the increasing use of public libraries and reading circles; and a change in the economic conditions of writing itself as well as in the economic and political self-understanding of writers. The changes in the literary and media landscape and the political and epistemological revolutions at the time were two aspects of the large process of social and cultural change occurring in the societies of the European continent in the late eighteenth century.[36] The rapid increase in the production and consumption of print media is often noted as one of the characteristics—and even the causes—of the emergence of a specifically Enlightened literary and political public in late eighteenth-century continental Europe.[37] The intersections between these media and Jean Paul's research and writing techniques are thus more than historical contingencies: both processes and their products were part of a media industry that constituted the emerging civil society, and they were part of a changing understanding of the production and reception of texts.

A key characteristic of accounts of music-making women automata in newspaper articles and calendars is their detailed description of how the automata move their bodies while playing their instruments. The reader learns, for example, that the mechanism of *La musicienne* makes the automaton turn its head, lean forward and backward, and lift and lower its chest while it is playing, to suggest that it is breathing. Descriptions of *La joueuse de tympanon* relate that the automaton pays a compliment to the audience before it starts playing, and that its eyes move slowly with the *adagio* and quickly with the *allegro*.[38] The reports emphasize in general the enchanting and deceptive grace with which the mechanical women move their bodies.[39] These descriptions and their ramifications are key characteristics marking the gendering both of eighteenth-century music-making automata and of the corresponding discussion around musicians' bodily motions during recital. The descriptions of the automata in these accounts do not refer explicitly to the gender of the automata. However, the discussion concerning the expression of passions through bodily motions in musical recital took on, as did many other cultural debates, a distinctly gendered edge when it came to the limits, and potential dangers, of arousing affects in social interaction: generating and calming sentiments in a controlled way was a key task a musician (or a speaker) had to perform, while making sure that these sentiments remained within the boundaries of bourgeois moderation and self-determination. The resulting paradoxes were often (although not exclusively) expressed in terms of female musical performers—an issue I return to in the final part when discussing eighteenth-century pedagogical literature on how to play musical instruments.

Jean Paul reproduces this same emphasis on bodily motions in his own mention of music-playing women automata in the satire on humans, machines, and angels. In proposing that it is these motions that "seem to betray" the musicians' affective involvement in the music, he underlines, furthermore, the connection between bodily motions and the expression and communication of passions in a musical recital. These relationships between the bodily motions of music-making humans (or machines), the expression of their affective involvement, and the content of a musical piece were also the subject of elaborate pedagogical and theoretical efforts on the part of musicians and music theorists in the second half of the eighteenth century. As with practically any other contemporary debate, Jean Paul was also part of these discussions. The common musical textbooks of the time as well as daily practice on the piano were part of his life.[40] Furthermore, he was interested both in questions of musical aesthetics as such and in discussions about the relationship between musical art and literary art. Wilhelm Dilthey in fact named Jean Paul the "musical poet of this age."[41] Taking seriously his convoluted phrase at the end of his satire on humans, machines, and angels—which brings automata together with women, music, motion, and passion—I now examine two exemplary contemporary texts in musical theory and pedagogy to investigate the historical connection between this literature and Jean Paul's commentary on machine-men.

EMPFINDSAMKEIT AND THE MUSIC-MAKING BODY

Many aspects of the cultural and literary activity of the late eighteenth century in Europe—including philosophy, natural philosophy, and music—were characterized by the concept and ideology of "sensibility" (Empfindsamkeit). For the German-speaking areas, the term Empfindsamkeit describes specifically the growing attention that was being paid to the "inner life" of a subject and to the individuation of feeling in the social, literary, and ethical discourses of the mid-eighteenth century.[42] The ideology of Empfindsamkeit was not a countermovement or reaction to the rationalism of the Enlightenment. Rather, it was a moral conception of virtue and of "sensible" or "reasonable" habits, along the lines of which new forms of bourgeois intersubjectivity were formulated and practiced.[43]

In the theory and practice of music, there was talk about a period of Empfindsamkeit already before 1750. The generation of affects was first formulated as a principle of musical composition and performance in the early eighteenth century, and it was a principle that music shared with literature. Modeled after the tradition of rhetorical theory, music theorists conceived of a theory of affects in music-making as a theory of musical forms in which specific techniques of composition and performance were correlated with specific affects to be evoked in the audience during a recital.[44]

Eighteenth-century textbooks on how to play musical instruments document the attempts of musicians and pedagogues to conceptualize and systematize this process of generating affects in musical recitals and to impart other principles of musical performance to the emerging bourgeois classes interested in newly developing cultural practices. Among the most prominent theorists of musical form and performance of the time were Johann Joachim Quantz and Carl Philipp Emanuel Bach.[45] Quantz was the flute teacher of Frederick the Great and composer of many concerts for his master, as well as the author of an important treatise on flute-playing entitled Essay of an Instruction to Play the Transverse Flute, published in 1752. Carl Philipp Emanuel Bach was one of Johann Sebastian Bach's sons and a leading musician and composer in the eighteenth century. The two volumes of his Essay on the True Manner of Playing the Piano were published in 1759 and 1762, respectively.

Quantz's Essay breaks down into twenty-eight chapters that deal with issues such as breathing, posture, lipping, intonation, fingering, scales, trills, and appoggiaturas. Regarding the present concern with affects in musical performance, chapter 11 is of particular interest. It deals specifically with performance and is entitled "The Good Recital in Singing and Playing in General."[46] In this chapter, Quantz famously relies on the theory of forms from the rhetorical tradition to build his theory of the principles of musical performance. The chapter's often-cited first paragraph formulates this close relation between rhetoric and music: "The musical recital can be compared with the performance of a speaker. A speaker and a musician have one and the same intention, both with respect to the content of their performance and with respect to the recital itself, namely: to win the hearts, to excite and calm the passions, and to set the listeners now into this, then into that, affect."[47]

Following that, Quantz defines criteria for a successful performance: he asserts that the intonation must be correct and clear, that the notes must be presented in an easy, flowing way, that the performance must be varied and multifarious, and that "light and shadow should be permanently entertained."[48] Part of a good performance, however, is also the generation of affects. About how to do so in an appropriate manner, Quantz says, "The recital must be *expressive and in accordance with each passion which appears*. . . . The performer must aim to put himself into those primary and secondary passions which he is meant to express."[49] In Quantz's system, the musician must first provoke, according to the piece's musical program, the respective affects in himself and then communicate them to the audience.[50]

Quantz also illustrates how these processes are supposed to operate in a timely manner and how the musician should be keen to exercise precise control over the sequence of affects to be generated in himself and in the audience in the course of a musical recital. From the musician's ability to exercise this control, Quantz eventually derives another criterion for a successful performance: "One would have to set oneself, so to speak, into a different affect with each bar, in order to be now sad, then merry, then grave, and so on. . . . He who is able to master this art appropriately will not easily miss out on the listener's approval, and his performance will thus always be *affectively moving* [*rührend*]."[51] A successful recital, according to Quantz, then, is a recital that provokes affective involvement—Rührung—in the audience, initiated and communicated by the musician's own Rührung. Furthermore, in a successful recital, these processes happen in a precise and well-timed sequence, according to the pattern suggested by the musical piece.

The scenario in the final passage of Jean Paul's satire discussed above, in which the bodily motions of music-making machines "betray" affective involvement during musical performance, thus has intriguing parallels to contemporary textbook instruction for musicians on the how to generate, communicate, and control affects in musical performance. Furthermore, one of the key terms in Jean Paul's mocking observation on the relationship between the bodily motions of music-playing women (and women automata) and affective involvement—Rührung—appears at an equally central point in Quantz's work, in the concluding passage on the relationship between a musician's and an audience's affective involvement. Corresponding passages in the other central textbook on music-playing of the time, the work by Carl Philipp Emanuel Bach on playing the piano, not only support this finding, but also make explicit the connection between the musician's communication of affects and his or her bodily motions.

C. P. E. Bach's treatise *Essay on the True Manner of Playing the Piano* consists of two parts. The first has three main sections and deals with fingering, manners, and the public performance of musical pieces. These three things, so C. P. E. Bach says in the introduction, belong to the "true manner of playing the piano."[52] The second part deals with the *basso continuo* and the art of accompaniment. For my purposes, the third section of the first part is relevant, which—similar to Quantz's section—is entitled "On Recital" (*Vom Vortrage*). Again, like Quantz, Bach begins this part with an explanation of the criteria for good (and bad) recitals. According to him, a good recital consists of "nothing else but the ability to make musical thoughts transparent and sensitive in their true content and affect to the ear, either in singing or playing." The definition of a bad recital is worth citing for its rhetorical qualities: "The subjects of a performance are the strengths and weaknesses of the tunes, their pressure, their flicking, pulling, pushing, their thrust, their breaking, holding, their dragging and persistence, and their moving forward. He who uses these things either not at all or at the wrong time has a bad recital."[53] The almost comical listing of the tunes' "activities"—it comes across even more powerfully in German—illustrates expressively the dynamics of shaping musical tunes in a musical performance.

The communication of the musician's affects to the audience plays as important a role in Bach's textbook as it does in Quantz's. Where Quantz says that the musician has to "set [himself], so to speak, into a different affect with each bar," Bach emphasizes that the musician has to generate the affects in himself first in order to communicate them to the audience: "And because a musician cannot move others [affectively] unless he is moved himself, he thus has to put himself into all the affects which he intends to arouse in his audience; he conveys his sentiments to them, and this is the best way to make them feel the same with him." Bach explains this further by illustrating that the musician, accordingly, is dull and sad during dull and sad passages, and he is merry and fierce during merry and fierce passages. The dynamics of this process over time work like this: "Once the musician has calmed one affect, he arouses another one. He thus constantly alternates between the different passions." Bach finally links the musician's ability to control the processes of arousing and calming in a timely manner the appropriate affects—or, as he puts it, the musician's ability to "appropriate his listeners' hearts"—with the musician's bodily motions. He says shortly thereafter: "That all this could happen without any gestures and motions could only be denied by one whose insensitivity makes him sit before the instrument like a piece of carved wood."[54]

Both Quantz and Bach make it clear that, in the course of a musical performance, a good musician is supposed to arouse the correct set of affects in the audience in the order suggested by the musical piece. The musician does this by first generating these affects in himself and then communicating them precisely and effectively to the listeners. The bodily motions, which Jean Paul considers to be "betraying affective involvement," are taken to serve that same function in Quantz's and Bach's pedagogical works. Furthermore, when explaining the dynamics of the complex process of arousing affects in musical recital, Bach and Quantz use interesting rhetorical techniques, which make this process sound at times as if the musician is supposed to "switch on and off," like a machine, the affects in himself and in the audience, and to use his moving body for the transmission of this on-and-off switching. The dynamics of switching on and off various affects over the time span of a musical piece and the musician's ability to control this dynamic in himself and the audience add a peculiarly "technical" taste to Bach's and Quantz's instructions on performing musical pieces in public. Despite Bach's use of the metaphor of a "piece of carved wood" as an anti-example, musical performance is described as a rather mechanical business in crucial passages of Quantz's and Bach's pedagogical works.

We thus find in these works of Bach, Quantz, and Jean Paul remarkable and diverse juxtapositions and confusions, intentional and unintentional, of human and technical music-making bodies. Both musical pedagogues—Bach and Quantz—emphasize in their treatises the significance of controlling affects in performing music, and they are keen to distinguish creative and affective music-making from "mechanical" or "impassive" activity. While Bach uses the metaphor of a "piece of carved wood" to make this point, Quantz insists in the preface to his treatise that he is "trying not only to educate a mechanical flute player, but also an apt musical expert," and that his goal is "not only to form [the musician's] lips, tongue, and fingers, but also to form his taste and to sharpen his sense of judgment."[55] When elaborating the criteria of a good musical performance, however, both authors are comfortable with describing the arousal of affects as a calculated, controlled, and pointed effort and with describing the human body in this process as almost machinelike.

Jean Paul, in his turn, exploits very effectively this ambivalence residing in late eighteenth-century efforts among bourgeois circles to develop codes and taxonomies for cultivated bodily comportment and "sensible" (*empfindsam*) forms of interaction: he takes the machinelike character of bodily comportment in music-making, which he finds expressed both in contemporary automata and in contemporary pedagogical literature, and sketches a scenario of human and mechanical women piano-players who communicate affective involvement as an alleged example in his argument for the superiority of humans over machines. He thus caricatures both the efforts and activities involved with "practicing" bourgeois culture and the question of the distinction between humans and machines itself.

The gendered aspect of this complex of themes becomes most explicit not in Jean Paul's satire or in the pedagogical literature, but in those instances where musicians and other cultural critics express concern over the possibility and the propriety of subjecting a musician's audience to the arousing and calming of passions: the potential paradoxes and tensions arising from attempts to cultivate passions in a context of bourgeois discipline, moderation, and rationality were formulated in restrictions for women's bodily comportment in musical recital and in restrictions as to what instruments women were allowed to play.[56] Related types of contemporary literature on rules of social communication—such as manuals on table manners or on gestures—display similar tensions in regulating appropriate comportment of the gendered and embodied bourgeois subject. This is not least a reflection of the fact that "sensible" social interaction was both the very instantiation of bourgeois culture and a dangerously open space in which bad influences could become threats to the very bourgeois virtues that were supposed to be established and exercised.[57]

CONCLUSION

The gendered, music-making human body, the mechanical renditions of this body, the question about the role of affects in music-making, and the literary and musicological commentary of the time form a complex intersection of topics and concerns in late eighteenth-century Europe. Two contemporary music-making automata and a satire about humans and machines by Jean Paul have served as a starting point in this essay to explore some of the diverse contexts in which the spectacular android automata of the time played a role. Music-making women automata functioned as a motif in Jean Paul's commentary on the ramifications of contemporary efforts to understand the affective "inner self" of individuals and to codify its manifestations. He takes prominent contemporary habits and social conventions, creates a scenario in which this particular type of sociability matters, and casts it in terms of machine behavior. In doing so, he relies heavily on contemporary efforts to systematize and codify social interaction and intersubjectivity on both rational and "sensible" grounds. Among the most compelling examples for this codification were the rules and taxonomies developed for performing and listening to music, and Jean Paul elaborates on this in the story about humans, machines, and angels, relating it ingeniously to the classic question about the difference between humans and machines.

The problem of the constitution of rational, autonomous, and bourgeois selfhood in the beginnings of modern society in the late eighteenth century spawned a variety of famously extensive political and philosophical discussions. As the examples discussed here have shown, some aspects were negotiated in the context of music-making, others were negotiated in the context of "machine-men." These two discourses coincided in the two android automata and the commentary on them considered here. Both technical and literary manifestations of "machine-men" were historically very specific products. In the late eighteenth century, automata and their textual elaborations were products of the last blossoming of an early modern artisan culture and of the emerging media industry of the time, and they were reflections and embodiments of the intensely debated problem of rational and autonomous selfhood in civil society. Machine and affect—motion and passion—were often closer to each other in the respective arguments than the polarizing dichotomy of "mechanism" and "life" would suggest. While the pedagogical commentary on this issue displayed a concern with the "mechanical" performance of music, a closer textual analysis indicated that the textbook instructions on how to play a musical instrument themselves conveyed a rather "technical" or "mechanical" connotation. The satirical commentary supports this finding, deploying the ambiguous character of contemporary instruction about bodily comportment to convey sentiments in musical recitals and taking it to an extreme.

The automata by Jaquet-Droz and by Roentgen and Kinzing as well as the early satires by Jean Paul document the rich and productive technical and textual work on automata in this period. Both types of work contributed simultaneously to the confusion, distinction, and identification of humans with machines. Both the stunning and wonderfully deceptive beauty and perfection of the artifacts, and Jean Paul's artful satirical play on angels, humans, and machines—as well as his convoluted articulation of the connection between bodily motion, deception, and *Rührung*—indicate that sentiment and mechanism cannot, and perhaps should not, be distinguished in reflections on the mechanical nature of the human self.

NOTES

This essay is a revised and extended English version of Adelheid Voskuhl, "'Bewegung' und 'Rührung': Musik spielende Androide und ihre kulturellen Bedeutungen im späten 18. Jahrhundert," in *Artifizielle Körper—Lebendige Technik: Technische Modellierungen des Körpers in historischer Perspektive*, ed. Barbara Orland (Zürich: Chronos Verlag, 2005), 87–103.

1. Among the earliest and still the most comprehensive and authoritative accounts of mechanical automata are the works by Alfred Chapuis, who in the first half of the twentieth century compiled massive amounts of data on mechanical devices depicting humans and animals from antiquity to the twentieth century. His works are the foundation for many scholars working mainly or tangentially on automata. See Alfred Chapuis and Edouard Gelis, *Le monde des automates: Étude historique et technique*, 2 vols. (Paris: Chez les auteurs, 1928); and Alfred Chapuis and Edmond Droz, *Automata: A Historical and Technological Study* (Neuchâtel, Switzerland: Éditions du Griffon, 1958).

2. *Automate* is defined as an "engin qui se meut de lui-même, ou machine qui porte en elle le principe de son mouvement" (1:448). *Androide* is defined as an "automate ayant figure humaine & qui, par le moyen de certains ressorts, &c. bien disposés, agit & fait d'autres fonctions extérieurement semblables à celles de l'homme" (1:448). (The article refers mostly to the automata by Jacques de Vaucanson.) More recently, two criteria have been established among historians of technology and art historians to define "android" automata. First, the mechanism driving and conducting the automaton's motions must be entirely contained within the automaton's body; second, android automata performing cultural techniques must do so just as humans would: a piano-playing automaton, for example, would move its hands across a real keyboard and press the appropriate key; a writing automaton would use real ink and a quill to compose a text; and a flautist would produce a melody by blowing air from the lips into the instrument and moving fingers on the flute. Both criteria imply profound challenges for an android automaton's design and construction, and they make it notoriously difficult to build android automata compared to building other mechanical devices representing human activity. Important counterexamples to android automata are the devices created before and after the eighteenth century: sixteenth- and seventeenth-century machines often display automata whose principles of motion do not resemble those of a human. Their music-making capacities are, for example, located outside their bodies, their "walking" mechanisms are wheels, or their motions are in general not as sophisticated (e.g., they can only move one limb). In the nineteenth century, automata became mechanical dolls that were mass-produced and no longer android automata. See, for example, Christian Bailly, *Automata: The Golden Age* (London: P. Wilson, for Sotheby's, 1987); and Annette Beyer, *Faszinierende Welt der Automaten: Uhren, Puppen, Spielereien* (München: Callwey, 1983), 133–200.

3. About ten artifacts comprise the most important and most characteristic android automata of the eighteenth century. The three automata made by Jacques de Vaucanson in the 1730s—a flute player, a drummer, and a *galoubet* (an early modern French shepherd's pipe, whose player accompanies himself on a drum) player—provoked a great deal of discussion among natural philosophers and served as models for artisans in the following decades. Pierre and Henri-Louis Jaquet-Droz built a writer and a draughtsman in addition to their woman musician, and they presented this trio to the public in 1774. Another writing automaton (although not strictly an android) was the *Allesschreibende Wundermaschine* by Friedrich von Knaus, who was a mechanic at the court of Maria Theresia in Vienna. He presented the writing machine to her in 1760. The instrument-making Kaufmann family from Dresden presented a life-size trumpeter to the public in 1810. It played fanfares and melodies in two voices. Illustrative survey histories of automata include Beyer, *Faszinierende Welt der Automaten*, and Herbert Heckmann, *Die andere Schöpfung: Geschichte der frühen Automaten in Wirklichkeit und Dichtung* (Frankfurt: Umschau Verlag Breidenstein, 1982). For thorough investigations of the epistemic and political roles of automata in the eighteenth century, see Jessica Riskin, "The Defecating Duck; or, The Ambiguous Origins of Artificial Life," *Critical Inquiry* 29 (2003): 599–633, and "Eighteenth-Century Wetware," *Representations* 83 (2003): 97–125; and Simon Schaffer, "Enlightened Automata," in *The Sciences in Enlightened Europe*, ed. W. Clark, J. Golinski, and S. Schaffer (Chicago: University of Chicago Press, 1999), 126–65.

4. Examples for other types of social interaction and bodily comportment include manners in conversation and table manners. An insightful analysis of manuals about bodily comportment and manners written for bourgeois audiences of the time is Kirsten O. Frieling, *Ausdruck macht Eindruck: Bürgerliche Körperpraktiken in sozialer Kommunikation um 1800* (Frankfurt: Lang, 2003). Freia Hoffmann, in *Instrument und Körper: Die musizierende Frau in der bürgerlichen Kultur* (Frankfurt: Insel Verlag, 1991), is specifically concerned with the gendering of bodily comportment in musical performance. Norbert Elias contributed groundbreaking studies on the larger sociohistorical background of bodily comportment in eighteenth-century bourgeois and aristocratic society in his *Die höfische Gesellschaft. Untersuchungen zur Soziologie des Königtums und der höfischen Aristokratie, mit einer Einleitung "Soziologie und Geschichtswissenschaft"* (Neuwied: Luchterhand, 1969), and *Über den Prozess der Zivilisation: Soziogenetische und psychogenetische Untersuchungen* (Frankfurt am Main: Suhrkamp, 1978).

5. Both automata are still existent and functioning. *La musicienne* is on display in the Musée d'Art et d'Histoire in Neuchâtel in Switzerland, and *La joueuse de tympanon* is on display in the Conservatoire Nationale des Arts et Métiers in Paris. Demonstrations of their performance are staged regularly several times a month in both museums.

6. See, for example, Charles Perregaux and F.-Louis Perrot, *Les Jaquet-Droz et Leschot* (Neuchâtel, Switzerland: Attinger Frères, 1916); Charles Perregaux, *Les Jaquet-Droz et leurs automates* (Neuchâtel, Switzerland: Wolfrath & Sperlé, 1906); Roland Carrera, Dominique Loiseau, and Olivier Roux, *Androiden: Die Automaten von Jaquet-Droz* (Lausanne: Scriptar, 1979); and F. Faessler, S. Guye, and Edmond Droz, *Pierre Jaquet-Droz et son temps* (La Chaux-de-Fonds: Courvoisier, 1971).

7. Together with the *musicienne*, the Jaquet-Droz built two other automata, a writer (l'écrivain) and a draughtsman (le dessinateur), that were a bit smaller. Both represented boys

of about four years of age. The trio was designed and built as a unit, and the automata were presented together when they were shown around in the 1770s and 1780s throughout Europe. Many aspects of the biographies and background of the Jaquet-Droz family are well-documented. Among the most prominent works are Michel Schlup, ed., *Biographies Neuchâteloises*, vol. 1, *De saint Guillaume à la fin des Lumières* (Neuchâtel: Editions Gilles Attinger, 1996), and vol. 2, *Des Lumières à la Révolution* (Neuchâtel: Editions Gilles Attinger, 1996); Car-rera, Loiseau, and Roux, *Androiden*; Alfred Chapuis and Edmond Droz, *Les Automates des Jaquet-Droz* (Booklet of the Musée d'Art et d'Histoire in Neuchâtel, Switzerland: n.p., n.d.); and Perregaux and Perrot, *Les Jaquet-Droz et Leschot*. On the specific artistic and mechanical training that Henri-Louis received during his stay at Nancy, see, for example, Marc Vanden Berghe, "Henri-Louis Jaquet-Droz: Horloger mécanicien," in *Biographies Neuchâteloises*, vol.1, *De saint Guillaume à la fin des Lumières*, 149–53; and Jacques François Lanier, *L'abbé Michel Servan; ou, De Servan: Prêtre, historien, ingénieur, Romans 1746–Lyon 1837* (Valence, France: J. F. Lanier, 2000). The patron who arranged for Pierre Jaquet-Droz to visit the king of Spain was Lord Keith, a Scottish aristocrat, who was then governor for the Prussian king in the Principality of Neu-châtel. For a detailed report of Pierre Jaquet-Droz's trip to Spain, see André Tissot, *Voyage de Pierre Jaquet-Droz à la cour du roi d'Espagne, 1758–1759, d'après le journal d'Abraham-Louis Sandoz son beau-père* (Neuchâtel, Switzerland: Editions de la Baconnière, 1982). For the political situation of Neuchâtel under Prussian rule, see, for example, Henry Favre, *Neuenburgs Union mit Preußen und seine Zugehörigkeit zur Eidgenossenschaft. Ein Beitrag zur Verfassungsgeschichte von Neuenburg bis zu seinem Aufgehen in der Eidgenossenschaft* (Leipzig: Theodor Weicher, 1932), 51–70.

8. The term *proto-industrialism* was coined by Franklin Mendels, Peter Kriedte, Hans Med-ick, and Jürgen Schlumbohm in the 1970s and has developed since then into a key con-cept in the historiography of technology, artisanship, and economics in seventeenth- and eighteenth-century Europe. See Franklin Mendels, "Proto-industrialization: The First Phase of the Process of Industrialization," *Journal of Economic History* 32 (1972): 241–61; and Peter Kriedte, Hans Medick, and Jürgen Schlumbohm, *Industrialisierung vor der Industrialisierung: Gewer-bliche Warenproduktion auf dem Land in der Formationsperiode des Kapitalismus* (Göttingen: Vandenhoeck und Ruprecht, 1977). Maxine Berg discusses how the term was developed as a critique of Marx's influential phrase "the manufacture period," which he introduced to characterize and periodize the modes of production in Europe from the mid-sixteenth century to the last third of the eighteenth century. See Maxine Berg, *The Age of Manufactures, 1700–1820: Indus-try, Innovation, and Work in Britain*, 2nd ed. (London: Routledge, 1994), 66–72.

9. See, for example, Rudolf Braun, *Das ausgehende Ancien Régime in der Schweiz* (Göttingen: Vandenhoeck und Ruprecht, 1984), 110–21; B. M. Biucchi, "The Industrial Revolution in Switzerland," in *The Fontana Economic History of Europe*, vol. 4, *The Emergence of Industrial Socie-ties*, ed. Carlo M. Cipolla (London: Collins/Fontana, 1972), 639–47; and Albert Hauser, *Schweizerische Wirtschafts- und Sozialgeschichte* (Zurich: Eugen Rentsch Verlag, 1961), chaps. 2, 3, and 4.

10. Karl Marx, *Das Kapital* (Berlin: Dietz Verlag, 1970), 1:363n32.

11. Jean-François Bergier, *Die Wirtschaftsgeschichte der Schweiz* (Zurich: Benziger, 1983), 59–60; Wolfgang von Wartburg, *Geschichte der Schweiz* (München: Oldenbourg, 1951), 121–25.

12. The history of the Roentgen-Kinzing manufacture is well-researched and documented, in particular from the viewpoints of art history, the history of technology and economics, and cultural history. See especially Dietrich Fabian, *Kinzing und Roentgen: Uhren aus Neuwied* (Bad Neustadt/Saale: Internationale Akademie der Kulturwissenschaften, 1983) and *Abraham und David Roentgen: Das noch aufgefundene Gesamtwerk ihrer Möbel- und Uhrenkunst in Verbindung mit der Uhrmacherfamilie Kinzing in Neuwied* (Bad Neustadt/Saale: Internationale Akademie der Kulturwissenschaften, 1996); Stürmer, *Handwerk und höfische Kultur*; and Michael Stürmer, *Luxus, Leistung und die Liebe zu Gott* (München: Bayerische Vereinsbank, 1993).

13. Stürmer, *Luxus, Leistung und die Liebe zu Gott*, 41–50; Michael Stürmer, "Die Roentgen-Manufaktur in Neuwied (1)," *Kunst und Antiquitäten* 5 (1979): 24–36.

14. Even though he spent a few formative years at Weimar, Jean Paul never became a close member of the circles around Goethe and Schiller, whose political and aesthetic views differed in important aspects from his. However, other influential philosophers, poets, and publishers, such as Herder, Karl Philipp Moritz, and Friedrich Nicolai, supported his work. For an illuminating biography, see Gert Ueding, *Jean Paul* (München: C. H. Beck, 1993).

15. The other contemporary author to mention piano-playing women automata is E. T. A. Hoffmann (1776–1822), in his tale *Die Automate* (1813). Hoffmann mentions such an automaton in the context of a tragic encounter between the student Ferdinand and a mysterious professor who puts on shows of automatic music-making machines. These were so-called "orchestrions," fashionable automatic orchestras, which were built in Germany from the 1790s onwards. Unlike Jean Paul's biting satires, E. T. A. Hoffmann's texts are famous for developing literary expressions of the dark, tragic, and uncanny dimensions of encounters between humans and their artificial counterparts. His tale *Der Sandmann*, written in 1817, describes the tragedy of a young artist who falls in love with the female automaton Olimpia and famously inspired Freud's psychoanalytic conception of the "uncanny." See Sigmund Freud, "Das Unheimliche," in *Gesammelte Werke*, Bd. 12, ed. Anna Freud (Frankfurt am Main: Suhrkamp, 1947).

16. I quote Jean Paul's works from the edition by Norbert Miller, which is in two parts, of which the first consists of seven volumes and the second of three volumes. See Jean Paul, *Werke*, ed. Norbert Miller and Wilhelm Schmidt-Biggemann (München: Carl Hanser, 1976). The five relevant texts in the present context are entitled "The Machine-Man and His Properties," pt. 2, 2:446; "Simple but Well-Disposed Biography of a New, Pleasant Woman Made of Pure Wood Whom I Invented and Have Married Long Since," pt. 2, 2:393; "Most Humble Introduction of Us, of All Players and Talking Women in Europe, against the Establishment of Von Kempelen's Playing and Speaking Machines," pt. 2, 2:167; "Humans Are Machines of the Angels," pt. 2, 1:1028 (these were all written in 1789); and "Personal Data on the Servant and Machine Man," pt. 1, 4:901 (written in 1798 as a revision of the "Machine-Man").

17. Jean Paul's writing style and the conditions under which it developed are also relevant against the backdrop of the heated and convoluted nature of the debates in which he intervened. Wilhelm Schmidt-Biggemann, for example, calls German literature in the late Enlightenment a "syndrome" of motifs, metaphors, and forms of argumentation and takes

Jean Paul's earlier works to be "proto-typical" for this syndrome; see Wilhelm Schmidt-Biggemann, *Maschine und Teufel: Jean Pauls Jugendsatiren nach ihrer Modellgeschichte* (Freiburg: Karl Alber, 1975), 15. Feelings of confusion and alienation about Jean Paul's writing are also expressed by Jean Paul's contemporaries; see Wolfgang Pross, *Jean Pauls geschichtliche Stellung* (Tübingen: Max Niemeyer, 1975), 2. While the "strangeness" of Jean Paul's style is debated among German literary historians, there is also resistance against stereotyping him and subsuming him, together with Kleist and Hölderlin, into a heterogeneous triad of the "Weimar antipodes." See, for example, Pross, *Jean Pauls geschichtliche Stellung*, 8.

18. From about 1782 onward, Jean Paul produced excerpts as a foundation for his satires. For a partial edition of Jean Paul's excerpts that also comments on the reasons for and details of those excerpts, see Götz Müller, *Jean Pauls Exzerpte* (Würzburg: Könighausen und Neumann, 1988), esp. 9–13. On the relationship of Jean Paul's excerpts to his satires, see Birgit Sick, "Jean Pauls unveröffentlichte Satiren- und Ironienhefte (1782–1803)," *Jahrbuch der Jean-Paul-Gesellschaft* 35/36 (2001): 205–17. On the relationship between Jean Paul's reading and writing, see Hans-Walter Schmidt-Hannisa, "Lesarten: Autorschaft und Leserschaft bei Jean Paul," *Jahrbuch der Jean-Paul-Gesellschaft* 37 (2002): 35–52. On writing techniques historically related to Jean Paul's excerpts, see, for example, Ann Blair, "Humanist Methods in Natural Philosophy: The Commonplace Book," *Journal of the History of Ideas* 53 (1992): 541–51. On the interrelations between the material and the epistemological dimensions of early modern and modern practices of reading, see Marina Frasca-Spada and Nick Jardine, eds., *Books and the Sciences in History* (Cambridge: Cambridge University Press, 2000).

19. The original title reads: "Menschen sind Maschinen der Engel" (Jean Paul, *Werke*, pt. 2, 1:1028–30). The editor Eduard Berend most likely gave the story this title in his edition of Jean Paul's complete works in 1927.

20. Jean Paul, *Werke*, pt. 2, 1:1028.

21. Ibid., 1029.

22. Ibid.

23. The history of the chess-automaton is well researched from a variety of perspectives. See, for example, Herbert Heckmann, *Die andere Schöpfung: Geschichte der frühen Automaten in Wirklichkeit und Dichtung* (Frankfurt: Umschau Verlag Breidenstein, 1982), 219–30, 258–62; Simon Schaffer, "Enlightened Automata," in *The Sciences in Enlightened Europe*, ed. W. Clark, J. Golinski, and S. Schaffer (Chicago: University of Chicago Press, 1999), 126–65; and Tom Standage, *The Mechanical Turk: The True Story of the Chess-Playing Machine That Fooled the World* (London: Allen Lane, 2002).

24. Jean Paul, *Werke*, pt. 2, 1:1029.

25. Ibid.

26. Ibid., 1030.

27. Hans-Walter Schmidt-Hannisa has investigated in depth the relationship between author and reader in Jean Paul's work. Addressing the reader and engaging in a conversation with him/her is a key feature of Jean Paul's writing in general. Furthermore, ironic ambiguities coupled with offenses against the reader are fundamental aspects of Jean Paul's

work and of the ways he conceives of and construes the relationship between author and reader. See Schmidt-Hannisa, "Lesarten," esp. 51.

28. Jean Paul, *Werke*, pt. 2, 1:1031.

29. The term *Rührung* and the verb *rühren* in German connote both a person's affective involvement as well as the very act of being touched and moved (affectively) by something or someone else. A musician's performance, inner emotions, bodily motions, and ability to communicate and to "pass on" these affects to an audience are thus drawn together in this term and its connotations. One suggestive parallel between the French and the German cultures of sentiment in the eighteenth century is that, in both cases, sensibility and sentiment, apart form their moral and epistemological significance, also describe an affective movement in response to a sensation. See Jessica Riskin, *Science in the Age of Sensibility: The Sentimental Empiricists of the French Enlightenment* (Chicago: University of Chicago Press, 2002). On the gendered aspect of "the fainting woman and the mechanical man," see Anne Vincent-Buffault, *The History of Tears: Sensibility and Sentimentality in France* (London: Macmillan, 1991), 42.

30. Hans-Walter Schmidt-Hannisa has also investigated in depth the relationship between reading and writing in Jean Paul's work; see Schmidt-Hannisa, "Lesarten." One aspect of this relationship is the way in which Jean Paul "compiles" references, facts, and details in his texts. For one thing, Jean Paul's texts do not require a common hermeneutical technique that would assume a subject bestowing coherence upon a set of details (42). Schmidt-Hannisa argues accordingly that, for Jean Paul, data processing was a priority over understanding (41). Also, the heterogeneity of Jean Paul's texts is a product of a "mechanical" process of combining things, rather than a synthesis of coherence. Schmidt-Hannisa points out that Hegel criticized also the unreadability of Jean Paul's texts and the systematic overburdening of the reader (42).

31. Müller, *Jean Pauls Exzerpte*, 159.

32. The context here is a concert organized by the "machine-man." In this concert, neither the conductor nor the composer nor the musicians are alive, and the narrator mentions "Jaquet-Droz and his son" as those who fabricated all the instruments (Jean Paul, *Werke*, pt. 2, 2:447).

33. The musicologist Julia Cloot has demonstrated this for Jean Paul's reading and excerpting of literature on music; see her *Geheime Texte: Jean Paul und die Musik* (Berlin: Walter de Gruyter, 2001), 20–45.

34. Ueding, *Jean Paul*, 109.

35. See *Almanac de Gotha, contenant diverses connoissances curieuses et utiles* (Gotha: C. G. Ettinger, 1789), 90–93; *Taschenbuch zum Nutzen und Vergnügen fürs Jahr 1780*, ed. Georg Christoph Lichtenberg (Göttingen: Johann Christian Dieterich, 1780), 66–73; *Journal de politique et de littérature, contenant les principaux Evènemens de toutes les Cours; les Nouvelles de la République des Lettres*, &c., no. 3 (25 January), Brussels, 1775, 99–100; and William Coxe, *Travels in Switzerland in a Series of Letters to William Melmoth, Esq.*, vol. 2 (London: T. Cadell, 1789), 106–13.

36. Hans-Ulrich Wehler suggests, for example, that these processes taken together are "elements of one grand process of socio-cultural mobilisation" that changed irreversibly

the political conditions in Europe; see his *Deutsche Gesellschaftsgeschichte*, vol. 1, *Vom Feudalismus des Alten Reiches bis zur Defensiven Modernisierung der Reformära, 1700–1815* (München: C. H. Beck, 1987), 303. Even though this was a pan-European phenomenon, this mobilization process holds in particular for the German-speaking lands.

37. The groundbreaking work in which the term public sphere first appeared in the early 1960s is Jürgen Habermas, *Strukturwandel der Öffentlichkeit* (Frankfurt: Suhrkamp, 1990 [Neuauflage mit Vorwort]). Hans-Ulrich Wehler summarizes important aspects of this process in *Deutsche Gesellschaftsgeschichte*, 295–332. For a closer examination of the relationship between the public and literature, see, for example, Falko Schneider, *Öffentlichkeit und Diskurs: Studien zu Entstehung, Struktur und Form der Öffentlichkeit im 18. Jahrhundert* (Bielefeld: Aisthesis Verlag, 1992). Falko Schneider, Nikolaus Wegmann, and York-Gothard Mix investigate individual aspects of this process, such as the notion of "public sphere," the notion of "sentiment" in social and literary interaction, and the cultural and economic relevance of pocket books and pocket calendars. See Schneider, *Öffentlichkeit und Diskurs*; Nikolaus Wegmann, *Diskurse der Empfindsamkeit: Zur Geschichte eines Gefühls in der Literatur des 18. Jahrhunderts* (Stuttgart: Metzler, 1988); and York-Gothard Mix, *Almanach- und Taschenbuchkultur des 18. und 19. Jahrhunderts* (Wiesbaden: Harrassowitz, 1996).

38. "Compliment" in this period and context refers to a bow; see Frieling, *Ausdruck macht Eindruck*.

39. In the 1780 edition of the *Göttingen Taschenkalender* (see n. 35 above) 69–70, for example, we read the following short account of the woman musician:

> The third piece is a girl of about 12 years, who has a piano before herself, also seated on a *tabouret*. In completely natural motions of the eyes, hands and fingers, she plays a few pieces on the piano with utmost precision, turning her head and eyes now to the music, then to her fingers, sometimes also to the audience; yes, occasionally she even leans forward with her whole body to be able to read the music, and the whole time, her chest rises and sinks for breathing.

Freia Hoffmann has analyzed in detail the relevance of gender in eighteenth-century perception and criticism of musical performance and aesthetics; see Hoffmann, *Instrument und Körper*. Kirsten Frieling investigates contemporary differentiation according to gender in manual of cultivated bodily comportment; see Frieling, *Ausdruck macht Eindruck*, 77–109.

40. On the details of Jean Paul's musical training, see Cloot, *Geheime Texte*, 36–40.

41. The most recent, very thorough, analysis of this relationship is Cloot, *Geheime Texte*. She quotes Wilhelm Dilthey on p. 13.

42. Peter Hohendahl uses these phrases in his definition of the term in *Der europäische Roman der Empfindsamkeit* (Wiesbaden: Athenaion, 1977), 1.

43. Gerhard Sauder, "Die empfindsamen Tendenzen in der Musikkultur nach 1750," in *Carl Philipp Emanuel Bach und die europäische Musikkultur des mittleren 18. Jahrhunderts*, ed. Hans Joachim Marx, (Göttingen: Vandenhoeck und Ruprecht, 1990), 41–42; Hohendahl, *Der europäische Roman der Empfindsamkeit*, 2. While the respective connotations and aspects of *Empfindsamkeit*, or sensibility, varied, the movement of sensibility was a European phenomenon. Jessica Riskin

illustrates beautifully, for example, the relationship between natural philosophy, empiricism, and sentiment in the French Enlightenment, investigating the "sentimental youth of scientific empiricism" in its epistemological, moral, and political relevance; see her *Science in the Age of Sensibility*. Whereas the French tradition revolved around the connection between sentiment and empiricist epistemology, the German culture of *Empfindsamkeit* was an ideology about subject-formation and the formation of intersubjectivity, and it was located in literary traditions and the ethics of social interaction.

44. For helpful analyses of these matters, see Sauder, "Die empfindsamen Tendenzen"; Bellamy Hosler, *Changing Aesthetic Views of Instrumental Music in Eighteenth-Century Germany* (Ann Arbor: University of Michigan Research Press, 1978), chap. 3; Mark Evan Bonds, *Wordless Rhetoric: Musical Form and the Metaphor of the Oration* (Cambridge, MA: Harvard University Press, 1991), chap. 2; John Neubauer, *The Emancipation of Music from Language: Departure from Mimesis in Eighteenth-Century Aesthetics* (New Haven, CT: Yale University Press, 1986); and Hugo Goldschmidt, *Die Musikästhetik des 18. Jahrhunderts und ihre Beziehungen zu seinem Kunstschaffen* (Zurich: von Rascher, 1915). Johann Joachim Quantz makes the correlation between musical techniques and affects explicit in several places in *Versuch einer Anweisung, die flute traversière zu spielen*, 3rd ed. (Breslau: Johann Friedrich Korn, 1789). In a chapter on appoggiaturas, for example, he describes their function and how they should be played, explaining that certain ornaments, depending on the character of the piece, serve the purpose of cheering up and producing happiness. Others are means to create sentiments of mollification and sadness (81). Quantz also explains that "the funny" is expressed through short notes that move in small and large intervals. The glorious is expressed both through long notes accompanied by fast movements in the other voices, as well as with dotted notes. Flattery is expressed both through drawn out notes that go up and down in small steps and through syncopation (116). In his chapter on cadences (chap. 15), Quantz explains that the purpose of a cadence is to surprise listeners once again toward the end of a movement and to leave a special impression in their hearts (153). The cadence's beauty thus aims to convey a new and touching (*rührend*) astonishment and to bring the desired evocation of feelings to an extreme (157).

45. Other examples of mid-eighteenth-century instructional literature on how to play the flute, the piano, and the violin, respectively, include Johann Georg Tromlitz's *Ausführlicher und gründlicher Unterricht die Flöte zu spielen* (Leipzig: Adam Friedrich Böhme, 1791); Friedrich Wilhelm Marpurg's *Anleitung zum Clavierspielen* (Berlin: Haude und Spener, 1761); and Leopold Mozart's *Versuch einer gründlichen Violinschule* (Augsburg: J. J. Lotter, 1756).

46. The original title is "Vom guten Vortrage im Singen und Spielen überhaupt" (Quantz, *Versuch einer Anweisung*, 100).

47. Ibid.; see also Neubauer, *Emancipation of Music from Language*, 22.

48. Quantz, *Versuch einer Anweisung*, 106: "dass Licht und Schatten dabey ständig unterhalten werden."

49. Ibid., 107; emphases in original.

50. The pronoun for the musician in Quantz's text is consistently in the masculine form, and there is no explicit mention of a gender differentiation among musicians. However, playing the piano and other performing arts were theorized as having different

meaning and directives for men and women. See, for example, Hoffmann, *Instrument und Körper*; and Frieling, *Ausdruck macht Eindruck*, 77–109. Quantz himself does not distinguish between male and female players in his treatise.

51. Quantz, *Versuch einer Anweisung*, 109: emphasis in original.

52. Carl Philipp Emanuel Bach, *Versuch über die wahre Art, das Clavier zu spielen* (Berlin: [beim Autor], 1753), 1:1.

53. Ibid., 82.

54. Ibid., 85.

55. Quantz, *Versuch einer Anweisung*, 2.

56. During a recital, women were not supposed to make faces, for example, or to move their arms in unnecessarily agitated ways. Instruments deemed inappropriate for women included brass instruments and the cello; see Hoffmann, *Instrument und Körper*, 39–49 and 72–79, respectively.

57. Both Freia Hoffmann and Kirsten Frieling discuss these gendered "spaces of ambiguity" as they were conceived in the context of music-making and in the context of related types of social and cultural interaction. See Hoffmann, *Instrument und Körper*, 79–90; and Frieling, *Ausdruck macht Eindruck*, 109–10.

15 An Archaeology of Artificial Life, Underwater

STEFAN HELMREICH

During ethnographic fieldwork I conducted in the 1990s among Artificial Life researchers at the Santa Fe Institute for the Sciences of Complexity, I heard revisited time and again the history of the field, a history that usually started somewhere in the 1980s with Christopher Langton rediscovering John von Neumann's posthumously published 1949 results on self-reproducing automata, the work of McCulloch and Pitts on nerve-net models, and the investigations of Minsky and Papert on perceptrons. Langton, in his edited volume *Artificial Life*, dubbed the latter two of these research programs "common ancestors" of both Artificial Intelligence and Artificial Life.[1] I heard many stories that took this history back further, too. Langton's introductory essay in the proceedings of the first Artificial Life conference, held in 1987, placed ancient Egyptian water clocks, Hero of Alexandria's first-century *Pneumatics*, and Jacques de Vaucanson's eighteenth-century mechanical duck at the "roots" of Artificial Life.[2] Hans Moravec's contribution to the same volume framed the human creation of artificial life on a somewhat grander scale, as a culmination of the "evolution of terrestrial intelligence,"[3] and Doyne Farmer and Alletta d'A. Belin's "Artificial Life: The Coming Evolution," in the proceedings of the second Artificial Life conference, characterized the advent of Artificial Life as a hallmark of a phase shift from the Darwinian evolution of genetic information to the speedier Lamarckian evolution of cultural information.[4] Such narratives allowed Artificial Life scientists to ground their claims for the field not only in a venerable intellectual lineage, but also in a *genealogy*—sometimes even an explicitly evolutionary one—leading up to their own endeavor.[5]

GENEALOGIES, BIOGRAPHIES, AND
ARCHAEOLOGIES OF SCIENTIFIC OBJECTS

This essay offers a commentary on the practice of finding genealogies for Artificial Life. Such an enterprise carries historiographic and epistemological risks. If we simply claim that the notion of synthesizing "life" can be traced back to stories of Rabbi Löw's Golem or Paracelsian homunculi, for example, we overlook Foucault's famous caution about the recent origins of the biological concept of "life,"[6] that theoretical construct upon which Artificial Life arguably depends. If we offer an account of early automata without attention to the language that described them during their own period, we might easily mistake family resemblance to, say, present-day robots for direct family relation or even descent. In other words, as Lorraine Daston writes, "A history of the word without the thing risks degenerating into etymology; a history of the thing without the word risks anachronism."[7] Objects—like the scientific object of "life" with which Artificial Life has been concerned—come into being as at once material and semiotic entities. *Biography*, the term Daston has proposed,[8] might serve as a better concept to capture this process of coming-into-being; making *genealogies* risks essentializing scientific objects and finding ancestors everywhere one looks.

But of course *biography*—the writing of *a life*, after all—might get us into other hot waters when in comes to documenting the coming-into-being of an object like "artificial life." As Foucault argues in *The Archaeology of Knowledge*, obsessions with genesis can fool us into imagining clear historico-genetic lines culminating in consistent objects. Furthermore, a fixation on genesis results in "the analysis of silent births, of distant correspondences, of permanences that persist beneath apparent changes, of slow formations that profit from innumerable blind complicities, of those total figures that gradually come together and suddenly condense into the fine point of the work."[9] One might object to Foucault that any history is inevitably teleological and that keeping the rhizomatic character of possible histories constantly in focus is an impossible task. But I want to suggest that this task of keeping origins productively out of focus is especially necessary when dealing with such a slippery concept as "artificial life."

In *On the Origin of Objects*, Brian Cantwell Smith offers a theory of ontology-in-the-wild that is organized—happily, for our purposes—around the coming-into-being of computational objects. Smith argues that the very existence of computer programs is completely enmeshed in our social, material, and linguistic world, and that how disk drives, Windows programs, file systems, and RAM caches work is the result of many scientific, cultural, and economic decisions.[10] To treat computers as entities that can harbor "life" ignores the work and decisions that have produced them, labor that may not have a single originary moment. "Life" *in silico* is an object that emerges at the intersection of a heterogeneity of practices—in taxonomy, molecular biology, computer science, for starters, and probably also in cinema, the politics of fetal imaging, and psychedelics, to continue this list in a somewhat aleatory fashion. Silicon life is *not*, as Moravec and others would have it, simply the spirit of evolutionary history made digital.

If "life" is a fact on which biology is made, it is one of the sort that Mary Poovey writes of in her argument about the rise of the very idea of the fact: "*A History of the Modern Fact* insists that texts and events generate effects in multiple domains, even those distant from the domain a writer intended to affect."[11] Artificial Life can have *many* histories, not all of which are about vitality, or even machines. I think Foucault's *archaeological* method is congenial to the telling of these histories and allows us not to mistake one story for the true tale of artificial life. Foucault's archaeology does not lapse into genealogy and does not try to restore "original" meaning, for it recognizes that every history is a rewriting, a "regulated transformation of what has already been written."[12] As an ethnographer, I find this approach has the added benefit of preserving the historical specificity and multiplicity of Artificial Life as a late twentieth-century formation.

Archaeologies are always partial and interested, of course.[13] And what I propose here is a shallow dig whose aims are perhaps more anthropological and philosophical than historical. I want to focus on the time depth at which we can begin to see simulations *in silico*. More narrowly, I want to center my attention on simulations that have visual interfaces through which we are invited to "see" into virtual universes and, more narrowly still, those that offer us artificial life forms visible through the computer screen understood as a kind of virtual aquarium. I argue that "life" surfaces in such simulations as an ontologically and epistemologically liminal—and therefore, to people interested in the boundary object that artificial life forms necessarily represent, as a particularly *persuasive*—object. If the reader will follow me deeper into the metaphor of archaeology, what I am interested in here is a sort of *underwater archaeology* of recent Artificial Life—an archaeology that operates though the remote sensing of its object, an object partially obscured by the medium in which it sits, which may in some sense be inaccessible to direct theoretical touch.[14]

ARTIFICIAL WATERWORLDS

Claus Emmeche means in the title of his book *The Garden in the Machine* to call attention to the ways "nature" has been imported into the computer simulations of Artificial Life.[15] Such creations as *SimLife* have offered a kind of green Edenic space in the machine. Conceiving computers as spaces within which "nature" can reside has many historical moorings. Peter Galison has offered the invention of Monte Carlo simulations as a moment when computer simulations came to occupy a floating place between theory and experiment: "data" generated by such simulations could be given the same epistemic status as data from "real" experiments.[16] On a deeper level, these simulations—using pseudorandom numbers as starting points for the emulation of physical processes—could be seen to share a "fundamental affinity" with "the statistical underpinnings of the world itself."[17] Galison writes, "The computer began as a 'tool'—an object for the manipulation of machines, objects, and equations. But bit by bit (byte by byte), computer designers deconstructed the notion of a tool itself as the computer came to stand not for a tool, but for nature itself."[18]

I want to extend this historical analysis with an observation from symbolic anthropology. When computers came to have monitors, and particularly when these began to display computer graphics representing the unfolding of programmed procedures, the idea that computers contained worlds in which life might emerge was provided with a visually persuasive rhetorical armature.[19] Further, as I have argued in *Silicon Second Nature*, simulations that use the computer screen as a "window" into another world have influenced how people have come to "see" artificial life forms in computers. In this section I look more closely at simulations that fashion the computer into a kind of fish tank that users can peer into and in which they can see artificial life forms swimming about.

Let me begin with what I take to be a canonical artifact in this genre: Demetri Terzopoulos, Xiaoyuan Tu, and Radek Grzeszczuk's simulation of fish locomotion, written up in "Artificial Fishes with Autonomous Locomotion, Perception, Behavior, and Learning in a Simulated World" (see figure 15.1).[20] I first saw a presentation of this model at the fourth conference on Artificial Life in 1994. Terzopoulos enraptured the audience when he showed a video of simulated fish acting out their genetic-algorithm-programmed

FIGURE 15.1 | Artificial fishes in a simulated world, from Demetri Terzopoulos, Xiaoyuan Tu, and Radek Grzeszczuk, "Artificial Fishes with Autonomous Locomotion, Perception, Behavior, and Learning in a Simulated Physical World," in *Artificial Life IV*, ed. R. Brooks and P. Maes (Cambridge, MA: MIT Press, 1994). Courtesy of Demetri Terzopoulos.

capacities to "swim" and "hunt." His presentation ended with some of the simulations strung together in an extended (and soundtracked) parody of a Jacques Cousteau nature documentary (thus summoning up, alongside the medium of water, the medium of film, understood to "capture" reality). As the audience laughed at the video and at the movements of the artificial fish on the screen, it became clear that the "lifelike" quality of these simulations produced an unease and a sense of wonder that were themselves precisely the cultural resources that made these creatures seem lively. The laughter bespoke a set of unarticulated intuitions and untheorized thoughts about autonomy and agency. We were meant to see the computer screen as a kind of aquarium, itself a technology that has fashioned for us a way of manufacturing, thinking about, and seeing self-contained ecologies. The aquarium is an exhibitionary technology that creates what Paul Edwards has called a "closed world."[21] The simulation of Terzopoulos and his colleagues summons forth a sense that worlds are just bigger aquaria or terraria.

There are a host of technological features of such computer simulations that can be used to motivate belief in the fidelity of artificial life creations to real life. Most invisible, perhaps, is the fact that computer processing power is used to make images unfold in what looks like real time, in the same temporality in which observers observe.[22] But also important for fishy Artificial Life simulations—and Terzopoulos and his colleagues are not alone: Aqua Zone™ is an Artificial Life toy that invites us to treat our computer as a fish tank (we even feed the artificial fish through the disk drive!)—is the very notion that we are watching "life" *suspended* in another *medium*. Artificial Life scientist Walter Fontana, who in the 1990s paid careful attention to defining computers as worlds, told me in an e-mail interview, "Computers are candidates for *being* worlds, not just describing them. The computer is not just a tool for fast calculations, sequence analysis, data base management, etc. The computer is a medium." The contention that computers are *media* enables the evocation of the *medium* of water as a proxy for the alternative world computers are meant to represent/materialize. Water is a persuasive rhetorical referent in the visualization of these simulations. In addition to the symbolic association of water with life (which has water as the medium within which life began), there is the imagery of floating, about which more below.

Wateriness shows up again in Karl Sims's simulated genetic-algorithm organisms, represented on his computer screen as inhabiting a three-dimensional world and having boxes for arms, legs, torsos, and heads. During a video Sims showed at the same conference at which Terzopolous spoke, his skillful graphics presented a set of clumsy creatures engaging in competitions for the possession of a small cube.[23] As they went about their Darwinian wrestling matches, competing for the right to have their constitutive programs reproduced, these boxy critters again elicited laughter from the audience. I joined the scientists in their pleasure at these images and experienced the activity of the simulated creatures as cute, especially when they could be interpreted as valiantly failing at their tasks. What made the images funny was a sense that Sims was not fully in control; he had programmed a three-dimensional artificial world (and a visual representation of it) that simulated Newtonian physics, gravity, fluid dynamics, and surface friction, and he had introduced into this world a set of virtual creatures visualized as made of three-dimensional, rigid parts that could interact with this world. Different creatures had different characteristics and could do more or less well at the task of capturing the box from competitors. Because

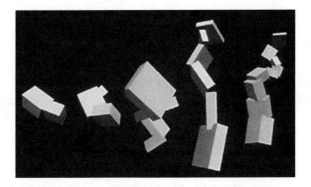

FIGURE 15.2 | An evolutionary sequence of Sims's virtual creatures, selected for "swimming." Courtesy of Karl Sims.

the simulated physics and creatures were programmed together, most behaviors looked realistic and purposeful. But because Sims occasionally made errors in modeling physics, sometimes behaviors came off completely wrong, as when creatures bounced out of the world because of his mistakes in modeling gravity. In a brilliant bit of showmanship, Sims showed videotapes of malfunctioning creatures, explaining that creatures were "exploiting" bugs in the program and were "making fun of [his] physics." Sims's ventriloquism delighted the audience and added a sense that his creatures were not only mimicking familiar behaviors, but also were mimicking behaviors associated with the playfulness of some life forms, a playfulness perhaps most readily compared with that of mammalian babies. This became more effective still when he sought to teach the creatures to "swim," an activity associated with early activity in many mammalian lives (figure 15.2). In this simulation, images of artificial life forms swimming in the medium of an alternative world allow the category and performance of "life" to "float" in front of our eyes. Life becomes both floating signifier *and* signified, something, to borrow from one of the OED's definitions of the intransitive senses of "float," that moves "unsteadily to and fro like an object on the surface of a liquid." Philosophers of artificial life would do well, I think, to attend to such rhetorical productions of "floating" in Artificial Life. "Emergence" is not quite the same thing as this neutrally buoyant submergence. Emergence is meant to direct our attention to how low-level dynamics produce processes that summon realities at higher levels, for which interpretation—the eye of the beholder—is a crucial component. Floating rather shows us how the *effect* of empiricity can be sustained through the medium of interpretation itself.

BRAINS IN VATS AND MINDS IN WATERS

I now want to take an orthogonal detour into a different realm of modeling, that of cognition in Artificial Intelligence. Artificial Life was named on analogy to Artificial Intelligence, and many practitioners see Artificial Life as a field growing from the discipline of Artificial Intelligence, even as Artificial Life means to leave behind the overly representational commitments that have hampered the attempts of Artificial Intelligence researchers to re-create intelligence in *silico*. However, rather than thinking about Artificial Intelligence as "ancestral" to Artificial Life (a "lineage" that has been explored elsewhere—sometimes in ways that assimilate the philosophical problems of Artificial Life to Artificial Intelligence, sometimes in ways attempting to distinguish them),[24] I want to turn to the object of Artificial Intelligence—intelligence—and think about it "underwater," using this meditation to think further about what it means to simulate life in floating worlds. My first example comes from Daniel Dennett's "Brain in a Vat" thought experiment—canonical in philosophy of mind—and my second comes from John Lilly's 1960s work on dolphin intelligence.

In his 1978 essay "Where Am I?" Dennett offered a thought experiment meant to help think through the possibility that a physically instantiated mind could exist separate from the body and, indeed, could be replicated in a computer.[25] In the fable Dennett offers, his brain has been removed from his body and placed in a vat. He recounts seeing his brain, from the vantage point of his newly brainless body, suspended in water: "I peered through the glass. There, floating in what looked like ginger ale, was undeniably a human brain, though it was almost covered with printed circuit chips, plastic tubules, electrodes, and other paraphernalia."[26] Dennett then details successive disorientations as his brain is replaced by a digital double into which his consciousness can switch without signal. Throughout this piece, Dennett's consciousness floats, the briny substance in which his gray matter is floating becoming a kind of pointer to the rhetorical at-sea-ness of his mind. As he asks, "Where am I?" again and again, the answer slides around, suspended between the various material substances—the vat, a computer, and so on—into which his being has been distributed. His "mind," his "soul," his "life" all float.

At around the same time Artificial Intelligence was being consolidated as a named field of inquiry, John Lilly was investigating nonhuman cognition in another world, that of the dolphin. Lilly's work on the consciousness of cetaceans unfolded in the 1960s as he sought to communicate with dolphins. Lilly was engaged with his dolphin charges in what we might call an underwater version of the Turing test. In his 1962 essay "A Feeling of Weirdness," published in Joan McIntyre's 1974 *Mind in the Waters: A Book to Celebrate the Consciousness of Whales and Dolphins*, Lilly writes of "a very peculiar effect which we have noticed in the laboratory while working with the bottlenose dolphin (*Tursiops truncalus*)."[27] He reports that as he and his compatriots navigated the "rough sea of the unknown,"

> we began to have feelings which I believe are best described by the word "weirdness." The feeling was that we were up against the edge of a vast uncharted region in which we were about to embark with a good deal of mistrust concerning the appropriateness of our own equipment. The feeling of weirdness came on us as the sounds of this small whale seemed more and more to be forming words in our own language. We felt we were in the presence of Something, or Someone, who was on the other side of a transparent barrier which up to this point we hadn't even seen.[28]

According to Lilly, his dolphins began to repeat English phrases. Lilly gained access to the "mind in the waters" though tools that rhyme with Dennett's "printed circuit chips, plastic tubules, electrodes, and other paraphernalia": "We first obtained the mimicry effect in 1957 by the use of electrodes implanted deep within the rewarding sites in the brain structures in these animals." Such "direct" access was later complemented by the use of recording devices to replay dolphin sounds. Lilly and his colleagues began to replay tapes, speeding them up, slowing them down, and taking into account what happens to sounds underwater. Direct apprehension was aided by a mediation meant to reveal what dolphins were really up to. Lilly was aware, of course, of the possibility that the communication that he heard emerged from his interpretive relationship with the dolphin. He did not allow that he might be imagining voices, but did permit his interpretative impulse to contribute to his observations. Later, Lilly would turn away from dolphins and move to self-experiment in isolation tanks, where he would float, attempting to achieve the altered states that might give insight into what the mind was made of.

Both Dennett and Lilly are accessing a process, and research object, that floats elusively in water. "Mind" is suspended in relations of interpretation sited in a medium, liquid, that comes to stand for the ambiguities of the object and for the fluid relationships between the empirical and interpretive. We might usefully classify both experimental projects as trysts with what Richard Doyle has called "wetwares," "encounter[s] with flesh as a refrain, a repetition of algorithms or recipes of sufficient complexity that only through instantiation can they be experienced."[29] "Mind" cannot exist apart from its contingent, slippery experience.

Artificial Life, of course, has been no stranger to epistemological debates about the relation between the world and the agent that knows it.[30] The vacillation between strong and weak claims in Artificial Life, between the idea that Artificial Life practice synthesizes or simulates vitality, organizes much of the field. Some of the strong claims for life *in silico* are grounded in the claim that properties like "life," "fire," and "wetness" only exist with respect to patterned relationships and that these can be suitably—and genuinely—instantiated in virtual worlds. As one Artificial Life scientist put it to me in an interview, "'Life' can only be defined *with respect to a particular physics.*" In this view, processes can be alive, on fire, wet, or infectious (like computer viruses) with respect to the computational realm. The "wetness" evoked by simulations such as that of Terzopolous supports the possibility that his artificial fish have artificial life. They and the "life" they represent float—to lift again from the *OED*—"suspended in a liquid with freedom to move." As such, the "wetness" of these entities is not the same as the wetness of the wetwares of which Doyle writes—nor, it would seem, as those Riskin has documented in her "Eighteenth-Century Wetware."[31] Riskin tells us that in their construction of automata, the Jaquet-Droz family used "lifelike materials such as leather, cork, and papier-mâché to give their machines the softness, lightness, and pliancy of living things."[32] For *in silico* artificial life, being "lifelike" requires performative, not textural, simulation; after all, Langton canonically defines "life" as "a property of the organization of matter, rather than a property of matter itself."[33] Intriguingly, however, the ghost of "matter itself" is constantly summoned forth, and perhaps in its most insidious—because invisible—and persuasive form when it appears as virtual water surrounding the fishy software of artificial fish and their kin.

THE UNDERWATER ARCHAEOLOGY OF KNOWLEDGE

Doing a sort of underwater archaeology of knowledge, then, we could transport Peter Galison's arguments about the philosophical status of artificial realities into the virtually aqueous domain of such simulations as Terzopoulos's fish worlds and see that wateriness is an important rhetorical float for the suspension of belief in artificial life itself.

Why underwater archaeology? Engineer and historian of technology David Mindell argues that a special characteristic of deep-water archaeology is that it must be enabled by techniques of remote sensing. For example, high-frequency, narrow-beam, sub-bottom sonar and remotely operated vehicles (ROVs) outfitted with cameras can retrieve data from the seafloor with which archaeologists can build up models of what might be buried down below, at great depth, pressure, and in darkness. Trying to sound out the outlines of "life" using an underwater archaeology of knowledge requires us to admit that we cannot finally lay our hands on the buried treasure of "life itself." Life is a material-semiotic relationship between tools and interpretations, not a thing-in-itself.

To conclude this meditation, let me go off the deep end and suggest, following Mario Biagioli's analysis of the Museum of Jurassic Technology in Los Angeles, that Artificial Life "confabulates" life.[34] "Confabulations," according to Valentine Worth—a writer whose publications seem only to be available from the Museum of Jurassic Technology and who may or may not exist—are "artificial constructions of our own design built around sterile particles of retained experience which we attempt to make live again by infusions of imaginations."[35] Like the Museum of Jurassic Technology, an institution that reproduces, repurposes, parodies, and confuses the very notion of a museum by archiving and displaying knowledge and artifacts that may or may not be part of actual human history, Artificial Life simulations may be "a setting of and for confabulation where hermeneutics is suspended."[36] "Life" floats suspended in the medium in which it swims.

NOTES

1. Christopher G. Langton, "Artificial Life," in *Artificial Life*, ed. Christopher G. Langton (Redwood City, CA: Addison-Wesley, 1989), 1–47, at 19.

2. Ibid., 6–9. And see Jessica Riskin, "The Defecating Duck, or, the Ambiguous Origins of Artificial Life," *Critical Inquiry* 29, no. 4 (2003): 599–633.

3. Hans Moravec, "Human Culture: A Genetic Takeover Underway," in Langton, *Artificial Life*, 167–99, at 178.

4. J. Doyne Farmer and Alletta d'A. Belin, "Artificial Life: The Coming Evolution," in *Artificial Life II*, ed. Christopher G. Langton, Charles Taylor, J. Doyne Farmer, and Steen Rasmussen (Redwood City, CA: Addison-Wesley, 1992), 815–40.

5. See Stefan Helmreich, *Silicon Second Nature: Culturing Artificial Life in a Digital World*, updated with a new preface (Berkeley: University of California Press, 2000).

6. Here is Foucault on the matter:

Historians want to write histories of biology in the eighteenth century; but they do not realize that biology did not exist then, and that the pattern of knowledge that has been familiar to us for a hundred and fifty years is not valid for a previous period. And that, if biology was unknown, there was a very simple reason for it: that life itself did not exist. All that existed was living beings, which were viewed through a grid of knowledge constituted by natural history.

See Michel Foucault, *The Order of Things: An Archaeology of the Human Sciences* (New York: Random House, 1970), 127–28; first published in French as *Les mots et les choses* in 1966.

7. Lorraine Daston, "Objectivity and the Escape from Perspective," in *The Science Studies Reader*, ed. Mario Biagioli (New York: Routledge, 1999), 110–23, at 113.

8. Lorraine Daston, ed., *Biographies of Scientific Objects* (Chicago: University of Chicago Press, 2000).

9. Michel Foucault, *The Archaeology of Knowledge and the Discourse on Language*, trans. A. M. Sheridan Smith (New York: Pantheon, 1972), 138; first published in French in 1969.

10. Brian Cantwell Smith, *On the Origin of Objects* (Cambridge, MA: MIT Press, 1996).

11. Mary Poovey, *A History of the Modern Fact: Problems of Knowledge in the Sciences of Wealth and Society* (Chicago: University of Chicago Press, 1998), 17.

12. Foucault, *Archaeology*, 140.

13. Nadia Abu El-Haj, *Facts on the Ground: Archaeological Practice and Territorial Self-Fashioning in Israeli Society* (Chicago: University of Chicago Press, 2001).

14. David A. Mindell and Katherine Croff, "Deep Water, Archaeology and Technology Development," *MTS Journal* 36, no. 3 (2002): 13–20.

15. Claus Emmeche, *The Garden in the Machine: The Emerging Science of Artificial Life*, trans. Steven Sampson (Princeton, NJ: Princeton University Press, 1994). He also of course plays on Leo Marx's *The Machine in the Garden* (New York: Oxford University Press, 1964), a discussion of how technology has both disturbed and remodeled pastoral ideals in U.S. literature and culture.

16. Peter Galison, "Computer Simulations and the Trading Zone," in *The Disunity of Science: Boundaries, Contexts, and Power*, ed. P. Galison and D. J. Stump (Stanford, CA: Stanford University Press, 1996), 118–57, at 142–43.

17. Ibid., 144.

18. Ibid., 156–57. See also my argument in "The Word for World Is Computer: Simulating Second Natures in Artificial Life," in *Growing Explanations: Historical Perspectives on the Sciences of Complexity*, ed. Norton Wise (Durham, NC: Duke University Press, 2004), 275–300.

19. See Richard Doyle, *On Beyond Living: Rhetorical Transformations in the Life Sciences* (Stanford, CA: Stanford University Press, 1997), for an argument about the "rhetorical software" such imagery has constituted.

20. Demetri Terzopoulos, Xiaoyuan Tu, and Radek Grzeszczuk, "Artificial Fishes with Autonomous Locomotion, Perception, Behavior, and Learning in a Simulated Physical World," in *Artificial Life IV*, ed. R. Brooks and P. Maes (Cambridge, MA: MIT Press, 1994), 17–27.

21. Paul Edwards, *The Closed World: Computers and the Politics of Discourse in Cold War America* (Cambridge, MA: MIT Press, 1996).

22. On time in cinema and simulation, see Christopher Kelty and Hannah Landecker, "A Theory of Animation: Cells, L-Systems, and Film," *Grey Room* 17 (2004): 30–63.

23. Karl Sims, "Evolving 3D Morphology and Behavior by Competition," *Artificial Life* 1, no. 4 (1994): 353–72.

24. Elliott Sober, "Learning from Functionalism: Prospects for Strong Artificial Life," in Langton, Taylor, Farmer, and Rasmussen, *Artificial Life II*, 749–65; and Brian Keeley, "Against the Global Replacement: On the Application of the Philosophy of Artificial Intelligence to Artificial Life," in *Artificial Life III*, ed. Christopher G. Langton (Redwood City, CA: Addison-Wesley, 1994), 569–87.

25. Daniel Dennett, "Where Am I?" in *Brainstorms: Philosophical Essays on Mind and Psychology* (Montgomery, VT: Bradford Books, 1978), 310–23.

26. Ibid., 312.

27. John Lilly, "A Feeling of Weirdness," in *Mind in the Waters: A Book to Celebrate the Consciousness of Whales and Dolphins*, ed. Joan McIntyre (New York: Scribner's Sons and Sierra Club Books, 1974), 71–77, at 71.

28. Ibid., 71.

29. Richard Doyle, *Wetwares: Experiments in Postvital Living* (Minneapolis: University of Minnesota Press, 2003), 186. According to Doyle, The Online Hacker Jargon File defines "wetware" as "Human beings (programmers, operators, administrators) attached to a computer system, as opposed to the system's hardware or software." Doyle offers the following: "'Taint software, 'taint hardware" (181), and "Wetwares are inconceivable not because they sublimely exceed any reduction or representation but because they quite simply cannot be modeled in advance" (186).

30. See, for example, N. Katherine Hayles, "Simulated Nature and Natural Simulations: Rethinking the Relation between the Beholder and the World," in *Uncommon Ground: Toward the Reinvention of Nature*, ed. W. Cronin (New York: W. W. Norton, 1995), 409–25.

31. Jessica Riskin, "Eighteenth-Century Wetware," *Representations* 83 (2003): 97–125.

32. Riskin, "Defecating Duck," 606.

33. Christopher G. Langton, "Toward Artificial Life," *Whole Earth Review* 58 (1988): 74–79, at 74.

34. Mario Biagioli, "Confabulating Jurassic Science," in *Late Editions 2: Technoscientific Imaginaries: Conversations, Profiles, and Memoirs*, ed. George E. Marcus (Chicago: University of Chicago Press), 399–431.

35. Valentine Worth, quoted in Biagioli, "Confabulating," 409.

36. Biagioli, "Confabulating," 426.

16   Booting Up Baby

EVELYN FOX KELLER

In the last few years, a new field has entered the academic fray, and it opens yet another chapter in the history of Artificial Life. Sometimes called Robotic Psychology, at other times Epigenetic, Humanoid, or Developmental Robotics, this field has been developed by researchers who identify themselves simultaneously as psychologists and as roboticists. In a move that is becoming increasingly familiar, they have begun to establish formal affiliations with departments of psychology, neuroscience, and cognitive science. The most visible pioneer in these efforts is of course Rodney Brooks, director of the "The Cog Shop" at MIT—a venture Brooks describes as "a long-term exploration of what developmental psychology can teach robotics, and vice versa." Brooks and his colleagues have been hard at work building robots capable not only of interacting with their environment but also of learning from their interactions with human trainers, much as infants learn from their interactions with their parents. In a shift that reflects recent preoccupations of developmental psychologists with the importance of interactive (intersubjective) dynamics, the robots coming out of the Cog Shop are "sociable."

They are also embodied. Predicated on the assumption that "humanoid intelligence requires humanoid interactions with the world," they are equipped with real-time sensors and actuators designed to approximate the sensory and motor dynamics of a human body. In part, such approximation is inspired by another recent turn of (at least some) psychologists, namely, to the importance of the ways in which cognition is embodied. But another, and perhaps even more critical motivation for the approximation to embodied human beings is to enable the kinds of interactions with humans that social learning requires. As the authors of the Cog Shop Web site explain, "If the robot has humanoid form then it will be both easy and natural for humans to interact with it in a human-like way. In fact it has been our observation that with just a very few human-like cues from a humanoid robot, people naturally fall into the pattern of interacting with it as if it were a human."[1] The hope is that a robot built with the capabilities of six-month-old infant can, by way of such interactions, develop abilities corresponding to that of a two-year-old child. For this project, the model of social dynamics is thus parent-infant interactions, with the robot occupying the role of infant, and the trainer that of parent (Breazeal and Scassellati, 2000).[2]

Cog was the first such "humanoid robot" to engage public attention, and with its deliberately ambiguous gender, he/she/it remains the master model, in which are implemented the designs for more elaborate behavioral systems as they are perfected on subsidiary "platforms." One of the first and most crucial interactive capacities built into Cog was a visual system that can distinguish and zero in on faces, locate the eyes, and direct its gaze onto the directed gaze of the trainer. Furthermore, it does so in ways that its human trainers recognize as eye contact. Using a relatively low-powered, hierarchical series of algorithms, Cog looks you in the eye in real time, creating in you, the observer, the uncanny experience of looking back into an actual living sentience. Next, a system enabling Cog both to track the direction of your visual gaze and to direct your attention by pointing to a third point (object or person) was designed by Brian Scassellati in 1996 and has been under development ever since.[3] Clearly, such a capacity for "shared attention" reinforces one's sense of interacting with a sentient being. But it is not just a matter of creating the impression of a sentient being. By so engaging your attention, the argument goes, the robot draws you unwittingly into a form of interaction that plays a central role in the cognitive development of the infant.

This argument arises from observations reported in the early 1990s by a group of psychologists studying autism. With considerable surprise to cognitive psychologists (even if not to many parents), the British cognitive psychologist Baron-Cohen and his colleagues provided substantial evidence for the critical importance of two specific abilities in the development of a child's ability to recognize others as subjects—that is, as subjects with other minds: (1) the ability to determine whether the eyes of another are directed at oneself or elsewhere and hence to distinguish "seeing" from "gazing" (this requires a mechanism the authors refer to as an "Eye-Direction Detector" [EDD]), and (2) the ability to attend to the object of another's attention (requiring a "Shared-Attention Mechanism" [SAM] permitting triadic representations). In this view, a defect in one or more of these biologically wired (or embodied) mechanisms could be responsible for the failure of autistic children to develop a "theory of mind" or, more generally, for what Baron-Cohen called "mind-blindness."[4] Cog, equipped with mechanisms to approximate these functions, might thus be able to develop a theory of mind; but, perhaps more important, even if not, others would interact with the robot *as* if witnessing the recognition of another subject. Cog's ability to build on such responses was, however, limited. Furthermore, Cog quickly proves wearing to interlocutors, apparently lacking the kinds of responses that work to keep the infant's caretaker properly engaged.

Kismet was designed to do better, and it clearly does. Two of Brooks's students, Cynthia Breazeal and Brian Scassellati, teamed up to chart the next leap forward. The two crucial dimensions that infants have but Cog lacked, or so they reasoned, were, on the one hand, needs and, on the other, a system designed to motivate a caretaker to satisfy those needs. Their understanding of needs (drives or, as they put it, "a motivational system")[5] was strongly influenced by Lorenz's ethology; similarly, their picture of infant interactions drew from the work of infant developmental psychologists (e.g., Trevarthan, Bullowa, Tronick, et al.). From this work they learned that infants interact with their caretakers with the help of an elaborate system of expressions that work to induce strong emotional responses in the parent/caretaker and that can also provide the feedback necessary to let the adult know when he or she is in fact doing the right thing. In other words, the infant must be able not only to respond, but also to initiate interactions and to regulate them. They write, "Soon after birth, an infant is able to display a wide variety of facial expressions. As such, he responds to events in the world with expressive cues that his mother can read, interpret, and act upon. She interprets them as indicators of his internal state (how he feels and why), and modifies her actions to promote his well being."[6] The question is then, what would it take to build a robot that could overcome the next hurdle, that could be transformed from a primarily reactive cognitive agent like Cog into an active participant in an affective and intersubjective dialogue with its caretaker?[7]

Silvan Tomkins taught a generation of psychologists to think of the face as the primary organ of emotion; in line with this view, designing a robot capable of displaying facial expressions seemed like an obvious first step. Kismet was constructed with a face that, while not quite that of a human infant, has the characteristic cuteness of infancy and is capable of exhibiting a range of expressions that we readily interpret as anger, fatigue, fear, disgust, excitement, happiness, interest, sadness, or surprise (see figure 16.1). It is also equipped with a primitive "motivational system"—interpreted here as a set of needs (in particular, the need to interact socially, to be stimulated, and a need for rest) linked to affective "values" that, in turn, trigger the facial expressions corresponding to these values ("sadness," "happiness," "fear," etc.). The motivational system is designed to bias Kismet's behavior toward eliciting the input required to keep those needs met at an appropriate level of "well-being."

The effect is dramatic. Kismet's trainers find themselves immediately hooked—not only do they attend to the robot, but they do so in ways strikingly reminiscent of human parents. Apparently eager to keep Kismet content, they find themselves speaking "motherese," shifting the tenor of their voices upward, exaggerating their prosody, and accentuating their facial expressions. Furthermore, like parents, they are quick to attribute meaning to Kismet's responses and seem delighted to instruct their electronic charge.[8]

FIGURE 16.1 | Kismet engaging a human caregiver and toy using gaze direction and facial expression. Courtesy of Donna Coveney/MIT.

They know they are talking to a robot with only the most primitive hard-wired caricature of a motivational system, but they behave as if interacting with an infant with a fully developed set of emotional and cognitive capacities. Kismet's appearance and simple behavioral repertoire are crucial for invoking this response, and the reaction is, in turn, assumed to be crucial to Kismet's development. As Graham-Rowe puts it, "The robots, like children, will not develop unless their carers read more into their behaviour than is actually there."[9]

So far so good. Kismet has engaged the interest and "motivational system" of a human other, one who is intent on attributing to it human qualities and is eager to instruct. But what new capacities can Kismet actually develop, and exactly how does it learn new kinds of behavior? Engaging the interest of the caregiver is certainly an achievement, but Kismet still needs to make use of this in its development. The question of how Kismet can learn inevitably leads to the question of how human infants learn novel behavior. One answer to this question—that they learn through imitation—has gotten a significant boost in recent years. The support comes from new studies of infant interactions and from recent reports of "mirror neurons" in monkeys (neurons that fire when an action is performed or seen by the monkey). Of particular interest to Breazeal and her colleagues are the observations of the imitation of facial expressions in neonates first reported by Meltzoff and Moore and the subsequent attempts of these researchers to explain their findings.[10] The particular model that Meltzoff and Moore subsequently proposed for the imitative capacity of infants has three key components. First, they hypothesize a kind of "body babbling" (akin to vocal babbling) by which the neonate, perhaps starting even in utero, constructs a mapping between movements and goal states through a combination of trial-and-error movements and proprioceptive feedback. Second, they suggest an innate mapping between the observation and execution of movements (like that suggested by mirror neurons, on the supposition that they are innate). Finally, they describe a representational structure for this mapping that allows for fine-tuning with experience.

Leonardo is a robot that builds on Kismet, incorporating structures capable of performing at least approximations of the functions suggested by Meltzoff and Moore.[11] The hope is that, so equipped, Leonardo will be able to increase its behavioral repertoire through imitation. The training program proceeds as follows: First the trainer, by presenting his face to the robot's visual field, arouses its attention and activates its body babbling system. The trainer then proceeds to imitate Leonardo's facial expression. If Leonardo can recognize the trainer's expressions as imitation, it can then use this feedback to consolidate and refine its representation of the correspondence between the trainer's face and its own. But how is it to recognize imitation before it has become trained? Here (as in many other places), the designers need to fudge: they employ temporal contingency between the robot's expressions and the trainer's responses to signal imitation, and it seems to work. That is, with practice, Leonardo's representation of the correspondence becomes good enough to represent even novel expressions, and Leonardo is now in a position to imitate the trainer and hence, in principle, to learn the specific skills of a trainer's agenda. So far, Leonardo has been able to imitate such simple facial expressions as frowning, alertness (eyes wide open), raised eyebrow, and so forth. It is worth noting here that Leonardo is built to look more like a pet than a baby, and, correlatively, the trainer is no longer figured as a mother. Indeed, the aims inspiring Leonardo's design have shifted from building a baby that can grow into a toddler to building a robot that can serve as a helpmate to a human in need of care. Furthermore, in this context, the goal of simulating developmental processes observed in human infants has given way to that of simulating the skills required of socially competent partners. For these purposes, perhaps more clearly than was true for Kismet, the question of the nature of robotic emotions (are they "real" emotions?) becomes moot if not altogether irrelevant. As Breazeal explains, their interest in exploring "emotion-like" processes is purely pragmatic. Her aim is to endow robots with a "system that will serve the same useful functions [as emotion], no matter what we call it."[12]

Brian Scassellati, Breazeal's former colleague, is now at Yale, and he has taken the development of social robots in slightly different directions. He and his colleagues are building a robot, Nico, designed to engage our attention and further the range of interactions through the initiation of, recognition of, and response to both gestures and vocalizations. Like Kismet and Leonardo, Nico is designed for social competence: it both seeks engagement and learns from it. But while Breazeal's focus has veered toward the development of robots capable of serving as human partners, Scassellati's focus has turned to cognitive science and developmental psychology, with a particular focus on autism. Supported by the National Institute of Mental Health (NIMH) and the Doris Duke Foundation for autism research, Scassellati heads a joint project with the Child Study Center at Yale that is working to develop Nico, first, into an "early detection system for vulnerabilities for autism" and, second, into a therapeutic aid for autism. Infants vulnerable to autism, the argument goes, tend (as early as the first weeks of life) to look at mouths rather than eyes, and therefore are thought to miss essential social information. Nico's eye-tracking capabilities, combined with its interactive competence, enable it to measure the infant's patterns of eye-tracking in a naturalistic setting. Hence, it is believed to provide a diagnostic tool that is, because better standardized and hence more objective, more reliable than human observers and, because more lifelike, more reliable than computerized recordings of infants watching videotapes. More ambitiously, its creators anticipate that Nico will be able to track the development of subsequent social behavior as the infant matures. The therapeutic goals for Nico are of course the most ambitious of all. Motivated by the observation that repeated social interactions of a simple nature are useful in treating autistic children, Scassellati and his colleagues ask, why not use a robot that has been specially designed to engage in such rudimentary social exchanges?

Finally, in one last turn of the screw, many researchers in humanoid robotics, initially basing their designs on current models of development in the psychological literature, have now begun to argue for the use of these same robots to test such models. As Nico's designers explain, "Just as simulations of neural networks have been useful in evaluating the applicability of models of neural function, these robots can serve as a test-bed for evaluating models of human development."[13] Here, Nico is part of a much larger endeavor that is now laying claim to the status of a new field that—whether called "robotic psychology," "epigenetic robotics," or "developmental robotics"—is clearly aimed at informing the discipline from which it drew its initial inspirations. As Luc Berthouze and Tom Ziemke write in a special issue of *Connection Science* dedicated to epigenetic robots, "What is apparent from such studies is that the main challenge for the field will be for robotic models to progress from being merely instructed or inspired by infancy research, to generating testable predictions which, in turn, can stimulate further research in developmental psychology."[14] That at least some psychologists take this hope seriously is attested to by the increasing willingness of neuroscience, psychology, and cognitive science departments to house workers in this new discipline. In fact, some relatively conventional psychology departments (e.g., the Department of Child Development at Tufts University) are establishing new programs requiring training in robotics.

Obviously these developments raise many issues. All of them are challenging, and more than a few provoke serious anxieties. For example, how far can the development of humanoid robots go? How close an approximation of human behavior can be achieved? How close an approximation should we aim for? How close to the edge of "as if" are we prepared to tolerate?[15]

How much and what kinds of traditionally human care are we willing to entrust to robots? How might the reliance on humanoid robots (e.g., in child care) shape the actual development of children? Or how might it shape the ways in which we think of ourselves as human? Or how will the use of such models of development affect our conceptions of human psychology as scientists? And so on. These, like the many other kinds of questions raised by the emergence of a robotic psychology (e.g., economic, technological, epistemological), go well beyond the scope of this essay, but I do want to suggest that a subset—the particular issues raised by the apparently circular trajectory of this endeavor—warrants a few comments by virtue of its resonance with other issues raised in this volume around the topic of simulation. I refer in particular to the trajectory that begins with the implementation (or simulation) of contemporary models of infant development and ends with the use of that implementation or simulation to test, refine, and extend those very models. The circularity of this trajectory bears a striking resemblance to procedures that have become familiar in recent years in the uses of computer simulation in the physical and biological sciences.

Take, for example, the history of computer simulation in physics after World War II and the emerging uses of computer models as experimental tools.[16] The actual term *computer experiments* arose in molecular dynamics to refer to the use of computers to follow the dynamics of simulated systems of idealized particles (e.g., hard spheres) in order to identify the salient features required for physically realistic approximations, and hence to produce equations (or models) that would be both physically realistic and computationally tractable—that is, in order to develop better models.[17] This use of computer simulation seems to me to be quite close to the uses anticipated for Nico (and other humanoid robots). Similarly, the *de novo* construction of computer models (theoretical and/or "practical") of phenomena for which no general theory exists and for which only rudimentary indications of the underlying dynamics of interaction are available (a procedure I have elsewhere described as the third stage in the evolution of simulation as experimental tools)[18] provides another close parallel to the expected uses of humanoid robots in psychology. Or consider some of the recent uses of computational models in contemporary biology in which genetic (and biochemical) data and the schematic models that molecular biologists have until now employed to make sense of that data are used to construct computer simulations that are then, in turn, analyzed to explore the adequacy of the original data and accompanying models. In a number of cases, this procedure has revealed inadequacies in the original models and has accordingly led to the development of better models.[19] My point in raising these examples from other fields is this: If there appears to be a disturbing circularity in the expectations for robotic simulations of human development, and if I am right in suggesting that the same problem arises in the uses of computer simulations, then the issue becomes a more general one. Furthermore, an examination of constructive examples in other fields ought to help us to see how what first appears as circular might be more appropriately characterized as spiral, with a forward momentum quite compatible with the back-and-forth traffic between model and modeled that is more the norm than the exception for so much of techno-scientific practice.

POSTSCRIPT

This is of course not the first time people have tried to build machines that mimic the processes of human development and learning, but earlier efforts were not nearly so successful, nor were they quite so seductive. The best-known example, and an important precursor to much of the current work, was Frank Rosenblatt's effort in the 1950s to build a computer capable of learning to "perceive, recognize and identify its surroundings without human training or control." Rosenblatt, too, had been inspired by developments in psychology, especially by Donald Hebb's claim (1949) that an ensemble of neurons could learn if the connection between neurons were strengthened by excitation, and he attempted to build a machine that could learn in just that way. He called his machine the perceptron, and made large claims for it:

> Our success in developing the perceptron means that for the first time a non-biological object will achieve an organization of its external environment in a meaningful way.... The distinctive characteristic of the perceptron is that it interacts with its environment, forming concepts that have not been made ready for it by a human agent. Biologists claim that only biological systems see, feel, and think, but the perceptron behaves *as* if it saw, felt, and thought.[20]

When asked what practical use such machines might have, Rosenblatt replied, "None, whatever.... In these matters, you know, use follows invention." But unfortunately, others disputed his claims of success, and Rosenblatt's "invention" failed to survive long enough for anything to follow, or at least not immediately.[21]

Hindsight suggests that, relative to the recent efforts of Brooks and his colleagues, Rosenblatt was hampered by at least three deficits: first and most obvious, the computer power available in his day was not up to his needs; second, Hebb's insights, important as they were, lacked an appreciation of the social and intersubjective dimensions of learning; and third, despite his own disavowal of the driving force of need, the demand for such devices were simply not in evidence.

For the most recent generations of humanoid robots, however, use runs hand in hand with, if it does not actually precede, invention. For example, anticipating the needs of a rapidly aging population, the Japanese Ministry of International Trade and Industry and the Agency of Industrial Science and Technology have launched a large humanoid robot research initiative. The hope is that such robots will be able to serve the elderly—as companions, as helpers, as nursemaids.[22] Similarly, the high priority that U.S. funding agencies are now giving to research on autism reflects a growing concern over recent reports of a meteoric rise in the incidence of autism in the general population,[23] coupled with the emergence of therapeutic programs requiring enormous investments of professional one-on-one care. The incentive to find alternatives to the intensive human care that is now required—surrogates capable of meeting at least some part of these growing needs—could hardly be more glaring.

NOTES

1. See the overview at the Cog Shop Web site: http://www.ai.mit.edu/projects/humanoid-robotics-group/cog/overview.html.

2. Cynthia Breazeal (Ferrell) and Brian Scassellati, "Infant-like Social Interactions between a Robot and a Human Caretaker," *Adaptive Behavior* 8, no. 1 (2000): 49–74.

3. Students of infant development have concluded that the capacity for shared attention does not begin to appear in the human infant before the age of nine months and takes another nine months or so to fully develop; see, for example, Philippe-Rochat, *The Infant's World* (Cambridge, MA: Harvard University Press, 2001). Thus the effective age of Cog, once implemented with this capacity, has been advanced significantly beyond six months.

4. See, for example, Simon Baron-Cohen, Helen Tager-Flusberg, and Donald J. Cohen, *Understanding Other Minds: Perspectives from Autism* (Oxford: Oxford University Press, 1993); Simon Baron-Cohen, *Mindblindness: An Essay on Autism and Theory of Mind* (Cambridge, MA: MIT Press, 1995).

5. The term *motivational system* has become widely used in recent years in a wide range of fields (including neuroscience, cognitive science, psychoanalysis, humanoid robotics, political science, and management science) to refer to any (usually unspecified) system responsible for generating motivated behavior in an individual.

6. Breazeal and Scassellati, "Infant-like Social Interactions," 54.

7. It is noteworthy that in Breazeal and Scassellati's thinking about the engagement of an affective and intersubjective exchange, the adult "caretaker" has taken on the distinct identity of the "mother."

8. Cynthia Breazeal and Lijin Aryananda, "Recognizing Affective Intent in Robot Directed Speech," *Autonomous Robots* 12, no. 1 (2002): 83–104.

9. Duncan Graham-Rowe, "Booting Up Baby," *New Scientist*, 22 May 1999, 42–43. I take the title of this essay from his article.

10. See A. N. Meltzoff and M. K. Moore, "Imitation of Facial and Manual Gestures by Human Neonates," *Science* 198 (1977): 75–78; and "Explaining Facial Imitation: A Theoretical Model," *Early Development and Parenting* 6 (1997): 179–92. These claims of neonatal imitation have been enthusiastically received in many quarters, but they have also been challenged on statistical grounds; see, for example, M. Anisfeld, G. Turkewitz, S. A. Rose, F. R. Rosenberg, F. J. Sheiber, D. A. Couturier-Fagan, J. S. Ger, and I. Sommer, "No Compelling Evidence That Neonates Imitate Oral Gestures," *Infancy* 2 (2000): 111–22. Indeed, some critics suggest that the very enthusiasm with which they have been received is suspect: they confirm what adults are predisposed to believe about infants.

11. See C. Breazeal, D. Buchsbaum, J. Gray, D. Gatenby, and B. Blumberg, "Learning from and about Others: Towards Using Imitation to Bootstrap the Social Understanding of Others by Robots," *Artificial Life* 11, nos. 1–2 (2005): 31–62.

12. Cynthia Breazeal, "Function Meets Style: Insights from Emotion Theory Applied to HRI," in *IEEE SMC Transactions on Systems, Man, and Cybernetics*, part C, 34, no. 2 (2004): 187–94, at 187.

13. See the Web site of the Social Robotics Lab@Yale University: http://gundam.cs.yale.edu/Projects.htm.

14. Luc Berthouze and Tom Ziemke, "Epigenetic Robotics: Modeling Cognitive Development in Robotic Systems," *Connection Science* 15, no. 4 (2003): 151–90.

15. In "Learning from and about Others," Breazeal cautions against fashioning robots that bear too close a physical resemblance to humans because of the danger of raising false expectations and hence inviting frustration. But Breazeal (like many others) is also influenced by the concerns raised by the Japanese roboticist Masahiro Mori in the late 1970s. On the basis of empirical studies of human responses to lifelike androids, Mori observed that while the emotional engagement of people increases with the similitude of the android, it does so only to a point. Too close a resemblance, and the positive response gives way first to disinterest and soon to revulsion. Exact resemblance to humans restores the positive response. Mori warned roboticists to steer clear of this chasm that he dubbed "the valley of the uncanny," and not only robotocists but also the entire media industry have taken careful note ever since. See Masahiro Mori, *The Buddha in the Robot* (Charles E. Tuttle, 1982); see also the "Written Nonfiction" page at Dave Bryant's Web site: http://www.arclight.net/~pdb/nonfiction/uncanny-valley.html.

16. See, for example, Fritz Rohrlich, "Computer Simulation in the Physical Sciences," *Philosophy of Science Association* 2 (1990): 507–18; Peter Galison, *Image and Logic* (Chicago: University of Chicago Press, 1997); Eric Winsberg, "Sanctioning Models: The Epistemology of

Simulation," *Science in Context* 12, no. 2 (1999): 275–92; Evelyn Fox Keller, "Models, Simulation, and 'Computer Experiments,'" in *The Philosophy of Scientific Experimentation*, ed. Hans Radder (Pittsburgh, PA: University of Pittsburgh Press, 2002).

17. For further discussion, see Keller, "Models."

18. Ibid.

19. See, for example, George von Dassow, Eli Meir, Edwin H. Munro, and Garrett M. Odell, "The Segment Polarity Network Is a Robust Developmental Module," *Nature* 406 (2000): 188–92; Dennis Bray and Thomas Duke, "Conformational Spread: The Propagation of Allosteric States in Large Multiprotein Complexes," *Annual Review of Biophysics and Biomolecular Structure* 33 (2004): 53–73.

20. Frank Rosenblatt, quoted in the *New Yorker*, 6 December 1958, 44–45. See also Donald O. Hebb, *The Organization of Behavior* (New York: John Wiley, 1949).

21. For a brief discussion of the rise and fall of the perceptron and of what is often regarded as the stillbirth of "connectionism," see Evelyn Fox Keller, "Marrying the Pre-Modern to the Post-Modern: Computers and Organisms after WWII," in *Growing Explanations*, ed. Norton Wise (Durham, NC: Duke University Press, 2004).

22. C. A. Breazeal, A. Brooks, J. Gray, G. Hoffman, C. Kidd, H. Lee, J. Lieberman, A. Lockerd, and D. Mulanda, "Tutelage and Collaboration for Humanoid Robots," *International Journal of Humanoid Robotics* 1, no. 2 (2004): 315–48.

23. See, for example, Peter Chowka, "Autism: As the Incidence Rises, Possible Answers and Solutions Emerge," on the Web site of La Leva di Archimede: http://www.laleva.cc/choice/autism.html.

17 Body Language: Lessons from the Near-Human JUSTINE CASSELL

The story of the automaton had struck deep root into their souls and, in fact, a pernicious mistrust of human figures in general had begun to creep in. Many lovers, to be quite convinced that they were not enamoured of wooden dolls, would request their mistresses to sing and dance a little out of time, to embroider and knit, and play with their lapdogs, while listening to reading, etc., and, above all, not merely to listen, but also sometimes to talk, in such a manner as presupposed actual thought and feeling. | E.T.A. Hoffmann, "The Sandman," 1844

INTRODUCTION

It's the summer of 2005, and I'm teaching a group of linguists in a small Edinburgh classroom. The lesson consists of watching intently the conversational skills of a life-size virtual human projected on the screen at the front of the room. Most of the participants come from formal linguistics; they are used to describing human language in terms of logical formulae and usually see language as an expression of a person's intentions to communicate that issues directly from that person's mouth. I, on the other hand, come from a tradition that sees language as a genre of social practice, or interpersonal action, situated in the space between two or several people, emergent and multiply determined by social, personal, historical, and moment-to-moment linguistic contexts, and I am as likely to see language expressed by a person's hands and eyes as by mouth and pen. As a graduate student pursuing a dual Ph.D. in linguistics and psychology in the 1980s, I had felt profoundly inadequate in the presence of these scholars: their formalized theories belong to a particular kind of technical discourse that is constructed in opposition to everyday language and that had seemed more scientific than my messy relational and embodied understanding of how language looks.[1] Those feelings of inadequacy—along with the experience of having articles rejected from mainstream journals and conferences—led me to try to formalize or "scientify" my work. I undertook a collaboration with computer scientists in 1993 to build a computational simulation of my hypotheses that took the form of virtual humans who act on the basis of a "grammar" of rules about human communication. In turn, that simulation has, in the manner of all iconic representations, turned out to both reveal and obscure my original goals, depending on what the technical features of the model can and cannot handle. And the simulation has, like many scientific instruments, taken on a life of its own—almost literally in this instance—as the virtual human has come to be a playmate for children, a teaching device for soldiers, and a companion on cell phones—a mode of interacting with computers as well as a simulation that runs on computers.

But back to the classroom in Edinburgh. In the intervening fifteen years since graduate school, I have armed myself with a "sexy" demo to show other scientists, and times have changed so that the notion that language is embodied is somewhat more accepted in linguistics today. And so these formal linguists have chosen to attend the summer school class on "face-to-face pragmatics" that I am co-teaching. In the conversation today I'm trying to convince them of two points: that linguists should study videotapes and not just audiotapes, and that we can learn something important about human language by studying *embodied conversational agents*—fake humans who are capable of carrying on a (very limited) conversation with real humans—such as the one we call NUMACK, shown in figure 17.1.

I show the students a new video of NUMACK (the Northwestern University Multimodal Autonomous Conversational Kiosk) interacting with a real human, a simulation of our latest work on the relationship between gesture and language during direction-giving. On the basis of an examination of ten people giving directions to a particular place across campus, my students and I have tried to extract generalities at a fine enough level of detail to be able to understand what the humans are doing, and to use that understanding to program our virtual humans to give directions in the same way as humans do. The work exemplified in this particular video has concentrated on the shape of the people's hands as they give directions and on what kind of information they choose to give in speech and what kind in gesture. I'm excited to share this work, which has taken over a year to complete—moment-by-moment investigations into the minutiae of human gesture and language extracted from endless examinations of videotapes that show four views of a conversation (see figure 17.2), followed by complicated and novel implementations of a computer system that can behave in the same way. In fact, this is my own first view of the newly updated system; I've been traveling, and my graduate students finished up the programming and filmed the demo.

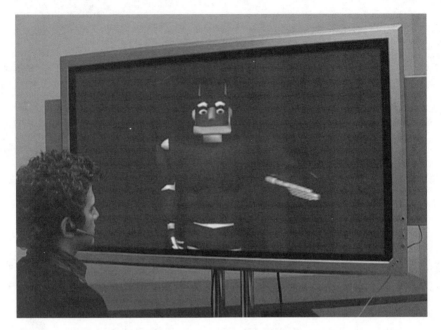

FIGURE 17.1 | NUMACK, the Northwestern University Multimodal Autonomous Conversational Kiosk, giving directions to a real human.

The Edinburgh linguists and I watch the video of NUMACK giving directions to a person, and it looks terrible! The small group of students tries to look down so as not to reveal that they don't think this is a fitting culmination of one year of work. I break the silence and say, "It looks ridiculous! Something is really off here. What is wrong with it? Can anybody help me figure out why it looks so nonhuman?" The students look surprised—after all, NUMACK looks nonhuman along hundreds of dimensions (starting with the fact that it is purple). Accustomed as they are to seeing impeccably animated characters in movies and on Web pages, they have expected to hurt my feelings by criticizing the virtual human's poor rendering of reality. But, as we watch the video over and over again, what stands out is how NUMACK's interaction violates our intuitions about how direction-giving should look. After three or four viewings, one student notes that NUMACK's two hands operate independently in giving directions. The virtual human says, "Take a right," and gestures with his right hand. He then says, "Take a left," and gestures with his left hand. I've never thought about this before, but in looking at the robot I am struck by the fact that we humans don't do that—we must

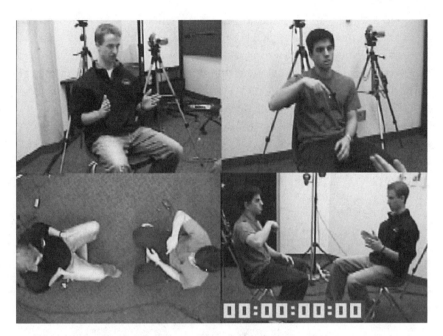

FIGURE 17.2 | Analysis of videotapes allows us to draw generalizations about human-to-human direction-giving.

have some kind of cohesion in our gestures that makes us use the same hand for the same set of directions. Another student points out that the virtual human describes the entire route (in roughly fourteen "turn left," "turn right," "go straight ahead" kinds of segments) at once, with only an "uh-huh" on the part of the real human. No real human would do that—the directions are too long and couldn't possibly be remembered in their entirety.

I am thrilled and once again amazed at how much I learn about human behavior when I try to re-create it, especially when, and because, my imitations are partial and imperfect. Only when I try to reproduce the processes in the individual that go into making embodied language, do I get such a clear picture of what I do not yet know. For example, here I have realized that we will need to go back to our ten human direction-givers and look at their choice of hands—can I draw any generalizations about the contexts in which they use their right hand or their left? When is the same hand used repeatedly, and when do they switch to a different hand? Likewise, we will need to look further at the emergent properties of their directions. What behaviors signal to the direction-giver when to pause and when to continue, when to elaborate and when to repeat? What embodied and verbal actions serve to alert the two participants to that the message has been taken up and understood, and that the next part of the message can be conveyed? I am also struck once again at the extent to which people are willing to engage with the virtual human, both as participants in a conversation about how to get to the campus chapel, and as participants in a conversation about the holes in our theory of the relationship between verbal and nonverbal elements in conversation.

I have learned something about the particularities of human communication here despite the fact that what I am viewing is a freak of artificial nature—a virtual human that is both generic and very particular, general and very detailed. In fact, for the experiment to work, we depend in part on the not-so-laudable schemas and expectations of our viewers and ourselves—that there can be such a thing as a generic human, which probably entails, for a direction-giving robot, that it is male and humanoid (albeit purple) and that its voice is Caucasian and American. As Clifford Nass and Scott Brave point out, violating cultural assumptions about expertise and gender or race produces distrust in users.[2] In the art world, Lynn Hershman Leeson, among others, has violated exactly these assumptions by synthesizing an infinitely smart female robot (or "bot") whose body is present only in certain contexts, and who reproduces herself. But, in the current case, these largely unconscious assumptions on the part of the linguists examining the simulation are what allow them to identify as failings not a lack of personality or cultural identity in the virtual human, but simply that the hands are not synchronized. Thus I have learned something about human communication despite all of the ways in which this virtual human is not very human at all. I question these assumptions below, but for the moment let us return to the fundamental questions that guide this work.

Artificial Intelligence (AI) investigators and their acolytes, like the creators of automata before them, ask, "Can we make a mechanical human (or, in the weaker version, a human-like machine)?" I would rather ask "What can we learn about humans when we make a machine that *evokes* humanness in us—a machine that acts human enough that we respond to it as we respond to another human (where I mean both respond to us in our status of interlocutor or of scientist)?" Some researchers are interested in stretching the limits and capabilities of the machine or in stretching the limits of what we consider human by building increasingly human machines. Such is the case for the work described by Evelyn Keller in this volume. In my own work, at the end of the day I am less interested in the properties of machines than in the properties of humans. For me there are two kinds of "aha!" moments: those in which I learn from my successes by watching a person turn to one of my virtual humans and unconsciously nod and carry on a conversation replete with gestures and intent eye gaze; and those in which I learn from my failures by watching the ways in which the real human is uncomfortable in the interaction, or the interaction looks wrong, as I illustrated in the Edinburgh classroom. These simulations serve as sufficiency proofs for partial theories of human behavior—what Keller has described as the second historical stage in the use of simulation and computer modeling[3]—and thus my goal is to build a virtual human to which people cannot help reacting *as if it were actually human*, to which people cannot prevent themselves from applying native-speaker intuitions. And key to the enterprise is the fact that those theories of human behavior and those native-speaker intuitions refer to the whole body, as it enacts conversations with other bodies in the physical world.

In the remainder of this chapter, I discuss my work on one particular kind of virtual human called an embodied conversational agent (ECA) in terms of its dual function as a simulation and as an interface. That is, I describe how these virtual humans have allowed me to test hypotheses about human conversation and what they have taught me by their flaws. I also describe the life that ECAs have acquired when they leave the lab—the uses to which companies and research labs have put them. In this way, I hope to illuminate the kinds of conversations that these virtual humans engage in when scientists use them as tools to study conversational phenomena and when ordinary people use them as tools to access information, dial phone numbers, learn languages, and so forth.

EMBODIED CONVERSATIONAL AGENTS AS CONVERSATIONAL SIMULATIONS

Just to be clear about our terms, *embodied conversational agents* are cartoonlike, often life-size, depictions of virtual humans that are projected on a screen. They have bodies that look more or less human, they are capable of initiating and responding in (very limited) conversations (in preset domains) with real humans, and they have agency in the sense that they behave autonomously, in the moment of their deployment, without anybody pulling the strings. Of course, this agency relies on a prior preset network of interactions among their inventors, their users, and the sociotechnical context of their deployment. As a point of contrast, consider chat bots or chatterbots.

Chat bots (such as the popular Alice, at http://www.alicebot.org/, which readers can try out for themselves) rely on a mixture of matching input sentences to templates, stock responses, and conversational tricks (such as "What makes you say X [where X is what the user typed in]?" or "I would need a more complicated algorithm to answer that question" when they don't understand). Chat bots are increasingly employed by artists such as Lynn Hershman Leeson, STELARC, or Kirsten Geisler because they are relatively easy to program and thus allow the artist to concentrate on the aesthetic experience she or he wishes to provoke in the viewer. Chat bots often communicate with viewers only through text, but when embodied, they usually have only a head, which displays only the most rudimentary of behaviors (blinking, looking left and right). Embodied conversational agents, on the other hand, are by definition models of human behavior, which means that at least along some dimension they must function in the same way humans do. Thus, the pedagogical agent of Wang and his colleagues and the virtual actor created by Walker, Cahn, and Whittaker both rely on Brown and Levinson's theory of politeness and language use.[4] Poggi and Pelachaud base the facial expressions of their ECA on Austin's theory of performatives.[5] Likewise, ECAs are fully functioning artificial intelligence systems in the sense that they understand language by composing meanings for sentences out of the meanings of words; they deliberate over an appropriate response, deliver the response, and then remember what they said so as to make the subsequent conversation coherent. They mostly have both heads and bodies, and their behavior is based on observation of human behavior.

Figure 17.3 shows an ECA named REA (for Real Estate Agent) who was programmed on the basis of a detailed examination into the behavior of realtors and clients. Over a period of roughly five years, various graduate students, postdocs, and colleagues in my research group studied different aspects of house-buying talk and incorporated their findings into the ECA. Hao Yan looked at what features of a house description were likely to be expressed in hand gestures and what features in speech. Yukiko Nakano discovered that posture shifts were correlated with shifts in conversational topic and shifts in whose turn it was to talk. Tim Bickmore examined the ways in which small talk was employed to establish trust and rapport between realtor and client. Earlier work by Scott Prevost on intonation and by Obed Torres on patterns of eye gaze also went into the implementation. As our research into human conversation progressed, we also came to better understand some of the overall properties of human conversation, and those insights were also incorporated.[6]

FIGURE 17.3 | REA, the virtual real estate agent.

The result was a virtual woman who tried to sell a home to whomever approached her. A small camera on top of the screen allowed her to detect the presence of real humans and initiate a conversation with them. Her knowledge of appropriate realtor conversation led her to ask questions about a person's housing needs and then nod, seem to reflect, and pull up data on appropriate properties, describing them using a combination of descriptive hand gestures, head movements, and spoken language.

A SAMPLE INTERACTION WITH AN ECA

Figure 17.4 illustrates an actual dialogue between a human user and REA.

Conversational Properties

As well as demonstrating the use of several conversational modalities, such as speech, hand gestures, and head movements, in this example REA is engaging in some very subtle, humanlike behavior that demonstrates four of the key properties of human face-to-face conversation. Those four properties are (1) the distinction between interactional and propositional functions of language and conversation; (2) the distinction between conversational behaviors (such as eyebrow raises) and conversational functions (such as

Mike approaches the projection screen. Rea is currently
turned side on and is gazing idly about. As Mike moves
within range of the two cameras mounted above the screen,
Rea turns to face him and says:
REA> Hello. How can I help you?
MIKE> I'm looking to buy a place near MIT.
Rea nods, indicating that she is following.
REA> I have a house to show you.
*A picture of a house appears on-screen behind Rea who
blinks and looks at the house and then at Mike.*
REA> It's in Somerville.
MIKE> Tell me about it.

Rea briefly looks up and away while she plans what to say.
REA> It's big.

Rea makes an expansive gesture with her hands.
*Mike brings his hands up as if he is about to speak and so
Rea does not continue; instead waiting for him to speak.*
MIKE> Tell me more about it
REA> Sure thing. It has a nice garden.

**Rea sketches a curved gesture with her hands indicating
that the garden extends along two sides of the house.**
MIKE> How far is it?
REA> It is five minutes to the Porter Square T station.
*Rea makes it clear that it is five minutes **on foot** from the
T station by making a walking gesture with her fingers.*
MIKE> How big is the house?
REA> It has four bedrooms, three bathrooms. . .
And the house tour continues.

FIGURE 17.4 | Transcript of an interaction between REA and a person.

turn-taking); (3) the importance of timing among conversational behaviors; (4) the deployment of each modality to do what it does best. Our insights into each of these properties has come through the cycle of watching real humans, attempting to model what we see in virtual humans, and observing the result or observing people interacting with the result.

DIVISION BETWEEN PROPOSITIONAL AND INTERACTIONAL FUNCTIONS | Some of the things that people say to one another move the conversation forward, while others regulate the conversational process. Propositional information corresponds to the content (sometimes referred to as transmission of information) and includes meaningful speech as well as hand gestures that represent something, such as punching a fist forward while saying "she gave him one" (indicating that the speaker's meaning is that she punched him, not that she gave him a present). *Interactional* information regulates the conversational process and includes a range of nonverbal behaviors (e.g., quick head nods to indicate that one is following, bringing one's hands to one's lap and turning to the listener to indicate that one is giving up one's turn), as well as sociocentric speech ("Huh?" or "Do go on"). It should be clear from these examples that both functions may be filled by either verbal or nonverbal means. Thus, in the dialogue excerpted above, REA's nonverbal behaviors sometimes contribute propositions to the discourse, such as the gesture that indicates that the house in question is five minutes *on foot* from the T stop, and sometimes they regulate the interaction, such as the head nod that indicates that REA has understood Mike's utterance.

DISTINCTION BETWEEN FUNCTION AND BEHAVIOR | When humans converse, few of their behaviors are *hard-coded*. That is, there is no mechanism or database "look-up table" that gives the appropriate response for every possible conversational move on the part of one's partner. Every day we hear thousands of phrases that we have never heard before, assembled through the infinite creativity of language use, and we reply to each of these phrases in just a couple of milliseconds, with an equally creative response. Gestures and head movements are no more likely to be routinized—head nods will look different if we are looking up at a taller interlocutor or down at somebody short, if we are wearing a hat or bareheaded. And other than the small number of culturally meaningful gestures (such as "V for victory" or "Up yours"), gestures display a great variety across people and even for one person across time. In observing human-human conversation, our group discovered that speakers do not always nod when they understand. Instead they sometimes signal that they are following along by making agreement noises such as "uh-huh." In our simulation of this behavior, then, instead of hard-coding, the emphasis is on identifying the high-level structural elements that make up a conversation. We describe these elements in terms of their role or function in the exchange. Typical discourse functions include *conversation invitation, turn-taking, providing feedback, contrast and emphasis,* and *breaking away.* Each function can be filled through a number of different behaviors, in one or several modalities. The form given to a particular discourse function depends on, among other things, current availability of modalities such as the face and the hands, type of conversation, cultural patterns, and personal style.

REA generates speech, gestures, and facial expressions based on the current conversational state, the conversational function she is trying to convey, and the availability of her hands, head, and face to engage in the desired

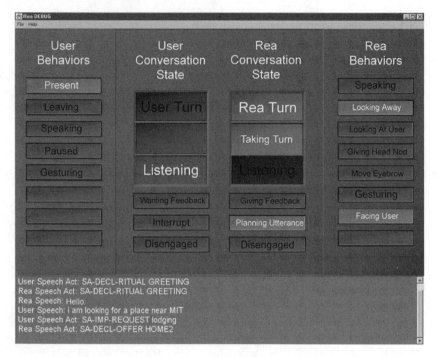

FIGURE 17.5 | Visualization of ECA and human conversational state.

behavior. For example, when the user first approaches REA ("user present" state), she signals her openness to engage in conversation by looking at the user, smiling, and/or tossing her head. Figure 17.5 shows a visualization of REA's internal state with respect to conversational behaviors and conversational states.

IMPORTANCE OF TIMING | The relative timing of conversational behaviors plays a large role in determining their meaning. That is, for example, the meaning of a nod is determined by where it occurs in an utterance, all the way down to the 200-millisecond scale; consider the difference between "you did a [great job]" (square brackets indicate the temporal extent of the nod) and "you did a [. . .] great job"). Thus, in the dialogue above, REA *says* "it is five minutes from the Porter Square T station" at exactly the same time as she *performs* a walking gesture. If the gesture occurred in another context, it could mean something quite different; if it occurred during silence, it could simply indicate REA's desire to take the turn.

Although it has long been known that the most effortful part of a gesture occurs with the part of an utterance that receives prosodic stress,[7] it wasn't until researchers needed to generate gestures along with speech in an ECA—and therefore needed to know the details of the context in which meaningful gestures were most likely to occur—that they discovered that a gesture is most likely to occur with the *rhematic*, or new contribution, part of an utterance.[8] This means that if a speaker is pointing to her new vehicle and saying, "This car is amazingly comfortable. In fact, this car actually has reclining seats," the phrase "amazingly comfortable" would be the rheme in the first sentence, because "car" is redundant (since the speaker is pointing to it) and "reclining seats" would be the rheme in the second sentence, because "car" has already been mentioned. Therefore, the speaker would be most likely to produce hand gestures with "amazingly comfortable" and "reclining seats."

USING THE MODALITIES TO DO WHAT THEY DO BEST | E-mail obliges us to compress all of our communication goals into textual form (plus the occasional emoticon). In face-to-face conversation, on the other hand, humans have many more modalities of expression at their disposal, and they depend on each of them, as well as various combinations of them, to communicate what they want to say. They use gestures to indicate things that may be hard to represent in speech, such as spatial relationships among objects,[9] and they depend on the simultaneous use of speech and gesture to communicate quickly. In this sense, face-to-face conversation may allow us to be maximally efficient or, in other instances, to use conversation to do other kinds of work than information transmission (for example, we may use the body to indicate rapport with others, while language is getting task work done). In the dialogue reproduced above, REA uses the hands' ability to represent spatial relations among objects and places by indicating the shape of the garden (sketching a curved gesture around an imaginary house) while her speech gives a positive assessment of it ("it has a nice garden"). However, in order to produce this description, the ECA needs to know something about the relative representational properties of speech and gesture, something about how to merge simultaneous descriptions in two modalities, and something about what her listener does and does not already know about the house in question.

The need to understand how speech, gestures, and movements of the head and face can be produced together by ECAs has forced me to design experimental and naturalistic methodologies to examine the nature of the interaction between modalities, and has resulted in significant advances in my theorizing about the relationship between speech and gesture in humans. Thus, for example, in my current work with the purple virtual robot NUMACK as a simulation, Paul Tepper, Stefan Kopp, and I have become interested in the seeming paradox of how gesture communicates, given that there are no standards of form in spontaneous gesture—no consistent form-meaning mappings. Some gestures clearly depict visually what the speaker is saying verbally, and these gestures are known as *iconics*. What is depicted on the fingers, however, and its relationship to what is said can be more or less obvious. And two speakers' depiction of the same thing can be quite different.

For example, we compared two people describing the same landmark on Northwestern University's campus: an arch at the entrance to the campus that lies at the intersection of Sheridan Road and Chicago Avenue. In order to collect these data, we hid prizes in various spots on campus and asked one student, who knew where the prize was hidden, to give directions to the prize to a second student. If the second student succeeded in finding it, the two shared the prize (and both were entered into a drawing for an iPod, probably the most motivating feature of the experiment!). We used four cameras to videotape the direction-giving, training them on different parts of the bodies of the two speakers, as described above (and shown in figure 17.2), and then we transcribed each gesture and the speech that accompanied it for further study. One speaker in the experiment, describing directions to a church near the arch, said "go to the arch" and, with his fingertips touching one another and pointing upward, made a kind of tepee shape. In this instance, the gesture seemed to indicate a generic arch. Another participant, referring to that same arch, said, "You know the arch?" but this person, although his fingertips were touching one another, pointed his fingers toward the listener with his thumbs up, making the shape of a right angle. In this instance, the gesture seems to indicate . . . what? An arch lying on its side?! It makes, in fact, no sense to us as observers—unless we know that the arch is located at the *right angle* formed by Sheridan Road and Chicago Avenue. And the speaker's next utterance, "It's located at the corner of Sheridan," supports this interpretation of the gesture. So, in the absence of the relatively stable form-meaning pairing that language enjoys (the same image may not be evoked for both of us, but when I say "right angle" I can be relatively sure that we will both imagine something similar), how do gestures communicate? The answer to this question (which is outside the scope of this chapter but involves the interpretive flexibility of gestures, which have meanings only in situated contexts) resulted both in a new computational architecture, through which gesture and speech are computationally generated together, and a new way of understanding of how gestures communicate among humans.

Translating Conversational Properties into Computational Architectures

The four conversational properties discussed in the previous section gave rise in 2000 to the computational architecture represented in figure 17.6. As this diagram makes clear, and like many systems in Artificial Intelligence, ECAs are largely linear and devoid of contingent functionality—a person asks a question that is collected by the input modules of the system (cameras to view the speaker's gestures and posture, microphones to hear the speech) and then interpreted into a unified understanding of what the speaker meant. In turn, that understanding is translated into some kind of obligation to respond. That response is planned out first in "thought" or communicative intention, and then in speech and movements of the animated body, face, and hands through the use of a speech synthesizer, computer graphics engine, and various other output modes. Meanwhile, so as not to wait for all of that processing to be completed before a response is generated, a certain number of hardwired responses are sent to be realized: head nods, phatic noises (mmm, uh-huh), and shifts of the body.

The linear nature of this architecture is one of the constraints imposed by the scientific instrument—like trying to cut out circles with straight blades.

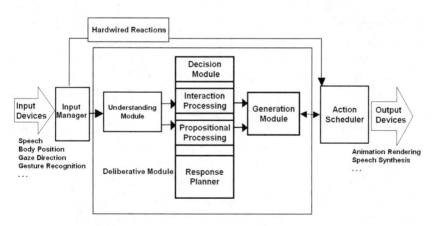

FIGURE 17.6 | Computational architecture of an ECA.

When I first began to collaborate with computer scientists in 1993–94 to build a virtual human, I asked them to build one that was responsive to itself and to its interlocutor in a number of ways. I told them that I wanted the virtual human to be able to "see" its own hands gesturing, and from what it saw decide what it wanted to say in the moment—the way humans often do, for example, when they cannot recall a word until they make the gesture for it. And I told them I wanted some kind of entrainment or accommodation between the different participants in the conversation, such that their language and gestures grew increasingly alike as they came to mirror one another. The response was incredulity and a request for me to be better informed before I went asking for features. The goal, I was told, was autonomy and not codependence. Of course, as Suchman has pointed out about other work in Artificial Intelligence, this means that we have not produced a truly conversational agent, since "interaction is a name for the ongoing, contingent co-production of a shared social/material world."[10] But the kinds of interdependence that we wish to simulate are hard to achieve with our current models.

In general terms, however, building ECAs has forced researchers in human behavior to attend to the integration of modalities and behaviors in a way that merges approaches from fields that usually do not speak to one another: ethnomethodological interpretive and holistic studies of human communication merge with psycholinguistic, experimental, isolative studies of particular communicative phenomena. To build a human entails understanding the context in which one finds each behavior—and that context is the other behaviors.

During that first collaboration with computer scientists, when we were building the very first of these animated embodied conversational agents, a different researcher was implementing each part of the body. Catherine Pelachaud was writing the algorithms to drive the character's facial movements (head nods, eye gaze, etc.) based on conversational parameters such as who had the turn. Scott Prevost was writing rules to generate appropriate intonation—the prosody of human language—on the basis of the relationship between the current utterance and previous utterances. I was working on where to insert gestures into the dialogue. After several months of work, we finally had a working system. In those days, ECAs needed to be "rendered"—they were not real-time—and so with bated breath we ran the simulation, copied it to videodisc, and then watched the result. The result was an embodied conversational agent who looked like he was speaking to very small children or to foreigners. That is, the resultant virtual human used so many nonverbal behaviors that signaled the same thing, that he seemed to be trying to explain something to a listener who didn't speak his own language or was just very stupid. This system, called Animated Conversation, was first shown at SIGGRAPH, the largest computer graphics conference, for an audience of four thousand researchers and professional animators (the folks who build cartoons and interactive characters), and they found it hilarious. To my mind, on the other hand, we had made a huge advance. We had realized that the phenomena of hand gestures, intonation, and facial expressions were not separate systems, nor was one a "translation" of the others; instead, they had to be derived from one common set of communicative goals. That was the only explanation for the perception of overemphasizing each concept through a multiplicity of communicative means. The result fundamentally changed the way we build embodied conversational agents, but it was an advance in understanding human communication as well. It also led to a design methodology that I have relied on ever since (see figure 17.7). Iteratively, my students and I collect data on human-human conversation, interpret those data in such a way as to build a formal model, implement a virtual human on the basis of the model, confront the virtual human with a real human, evaluate the results, and collect more data on human-human communication

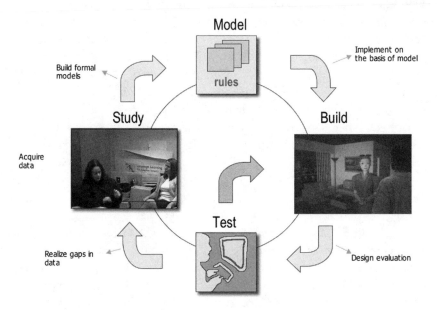

FIGURE 17.7 | Methodology for modeling human conversation and building ECAs.

if needed. (A side effect of this methodology is the need to confront the response of lay viewers to the necessary flaws and lacunae in the implementation, but I try to think of that as character building.)

I should repeat that building a computational system has traditionally demanded a formal or predictive model. That is, in addition to being able to *interpret* why a particular experience occurs in a particular context, one must also be able to *predict* in the future what set of conditions will give rise to a particular experience so that one can generate those behaviors in the ECA in response to the appropriate conditions.[11] Unfortunately, predictive models come with their own baggage, as they tend to underscore the way in which fixed sets of conditions give rise to fixed outputs, as opposed to highlighting the very contingent, coproduced nature of human conversation where, on the fly, hearers and speakers influence one another's language and indeed their very thinking patterns, as Suchman has forcefully argued.[12] In this sense, I sometimes worry that building computational simulations of this sort may set back the computational study of language; that phenomena that cannot yet be modeled in virtual people will be ignored. On the other hand, for the most part, before the advent of ECAs, computational linguistics and work on dialogue systems (which arose from the cognitive sciences—psychology, linguistics, philosophy, computer science) concentrated on the *propositional* functions of language, which many linguists saw as the primary if not the only functions of language. Before ECAs, computational models of language were, for the most part, capable only of simulating task talk bereft of social context and bereft of nonverbal behavior. And given the power of these computational models, perhaps the arrival of ECAs, with their attention to the non-informational and socially contextualized functions of language, has played some positive role in the cognitive sciences.

Now that there has been a decade of research on ECAs, several researchers, including myself, are beginning to explore other kinds of computational architectures and techniques that do not require deterministic, formal, input-output style models of conversation. Probabilistic techniques—such as spreading activation, Bayesian reinforcement learning, and partially observable Markov decision processes—are being applied to the newest phenomena to be modeled with ECAs. These phenomena, which tend to have more to do with social context than local linguistic context, include the effect of emotion on verbal and nonverbal behavior in conversation, the role of personality and cultural differences, social influence, etiquette, and relationship-building.[13]

In all of these implementation experiments, ECAs are *tools to think with*, much like other computer software and hardware that allow us to evaluate our own performance in the world.[14] They allow us to evaluate our hypotheses about the relationship between verbal and nonverbal behavior, and to see what gaps exist in our knowledge about human communication, by seeing ourselves and our conversational partners in the machine. How do we go about evaluating our hypotheses? As described above, we watch the ECAs and observe our own reactions. But we also put others in front of the ECAs and examine the differences between their behavior with ECAs and their behavior with other humans. This second kind of experiment relies on the supposition that correctly implemented virtual humans evoke natural human responses. Mechanisms that seem human make us *attribute* humanness and aliveness to them and make us react in natural human ways. Successful virtual humans evoke distinctly human characteristics in our interaction with them. The psychological approach to artificial life leads to functional bodies that are easy to interact with, "natural" in a particular sense: they evoke a natural response.

In an early experiment, for example, Kris Thorisson and I compared reactions to three versions of an ECA called Gandalf (this was 1996, and the ECA consisted of a head with one disembodied hand, as shown in figure 17.8). Our goal was to demonstrate, in those early days, that interactional behaviors—ones that did not move the conversation forward—could be simulated computationally, and that those behaviors in virtual humans would elicit similar behaviors on the part of human interlocutors. An additional goal was to demonstrate that if we had to choose only certain nonverbal functions to implement computationally, they should be interactional (what we called "envelope" functions) rather than emotional functions. We felt

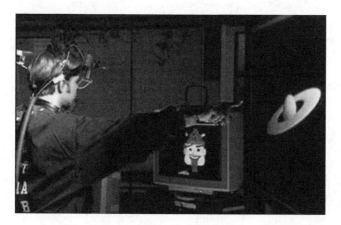

FIGURE 17.8 | Person interacting with Gandalf.

that emotional reactions should be studied only after these very ubiquitous interactional behaviors had been simulated.

In the first version, called "content-only," the ECA spoke but used no nonverbal expressions of any kind. An example of an interaction with such an ECA follows:

> Gandalf: "What can I do for you?" (*Face looks at user. Eyes do not move.*)
>
> User: "Will you show me what Mars looks like?" (*User looks at Gandalf.*)
>
> Gandalf: "Why not—here is Mars." (*Face maintains orientation. No change of expression. Mars appears on monitor.*)
>
> User: "What do you know about Mars?" (*User looks at map of solar system.*)
>
> Gandalf: "Mars has two moons." (*Face maintains orientation. No change of expression.*)

In the second version, called "content + envelope," the ECA spoke and also used eye gaze, eyebrow and head movements, and an occasional hand wave to moderate turn-taking, feedback, and other conversational-envelope processes. An example of an interaction with such an ECA follows:

> User: "Is that planet Mars?"
>
> Gandalf: "Yes, that's Mars." (*Gandalf raises eyebrows and performs beat gesture while saying "yes," turns to planet, and points at it while saying "that's Mars," and then turns back to face user.*)
>
> User: "I want to go back to Earth now. Take me to Earth." (*User looks at map of solar system, so Gandalf looks at solar system.*)
>
> Gandalf: "OK. Earth is third from the sun." (*Gandalf turns to planet as he brings it up on the screen, then turns to user and speaks.*)
>
> User: "Tell me more." (*Gandalf takes about 2 seconds to parse the speech, but he knows within 250 ms when the user gives the turn, so he looks to the side to show that he's taking the turn, and his eyebrows go up and down as he hesitates while parsing the user's utterance.*)
>
> Gandalf: "The Earth is 12,000 kilometers in diameter." (*Gandalf looks back at the user as he speaks.*)

In the final version, called "content + emotional," the ECA spoke and also smiled, frowned, and looked puzzled as the occasion warranted. An example of an interaction with such an agent follows:

> Gandalf: "What can I do for you?" (*Gandalf smiles when user's gaze falls on his face, then stops smiling and speaks.*)
>
> User: "Take me to Jupiter." (*User looks at screen and then back at Gandalf, and so Gandalf smiles.*)
>
> Gandalf: "Sure thing. That's Jupiter." (*Gandalf smiles as he brings Jupiter into focus on the screen.*)

User: (*Looks back at Gandalf. Short pause while deciding what to say to Gandalf.*)

Gandalf: (*Looks puzzled because the user pauses longer than expected. Waits for user to speak.*)

User: "Can you tell me about Jupiter?"

The study consisted in asking people to interact with Gandalf and then examining the real human's conversational-envelope and emotional behaviors during the interaction, as well as asking subjects to fill out a questionnaire assessing "lifelikeness." What we discovered was that participants tended to mimic the virtual human: if he stood rigid, so did they; if he was animated, so were they. In fact, the people standing in front of the content-only version of Gandalf were most animated in their expressions of frustration—sighs and the occasional request for signs of life ("Gandalf, are you there?"). People interacting with the content + envelope version, on the other hand, started off wary as Gandalf's head and single hand began to describe the solar system, and then, after an utterance or two, became more animated, gesturing and nodding to Gandalf in much the same way as they had to the experimenter before the experiment started.[15] Finally, we discovered no difference in the people's interaction with, nor in their assessment of, the ECA between the content-only version and the content + emotion version.

More recently, Yukiko Nakano and I carried out a study of the role of nonverbal behaviors in grounding and how these behaviors could be implemented in a virtual human.[16] *Common ground* is the sum of mutual knowledge, mutual beliefs, and mutual suppositions necessary for a particular stage of a conversation.[17] *Grounding* refers to the ways in which speakers and listeners ensure that the common ground is updated, such that the participants understand one another. Grounding may occur by nodding to indicate that one is following, by asking for clarification when one does not understand, or by asking for feedback, as in "You know what I mean?" Here, too, an extensive study of human-human behavior in the domain of direction-giving paved the way for the implementation of an ECA that could ground while giving directions using a map and hand gestures. And here, too, we evaluated our work by comparing people's reactions to two versions of the ECA: one that demonstrated grounding behaviors and one that had the grounding "turned off." When the ECA's grounding behaviors were turned off, the person simply acted as if she were in front of a kiosk and not another human—not gazing at the ECA or looking back and forth between him and the map. When the ECA did engage in grounding behaviors, the human acted strikingly . . . human, looking back and forth between the map and the ECA, as shown in figure 17.9.

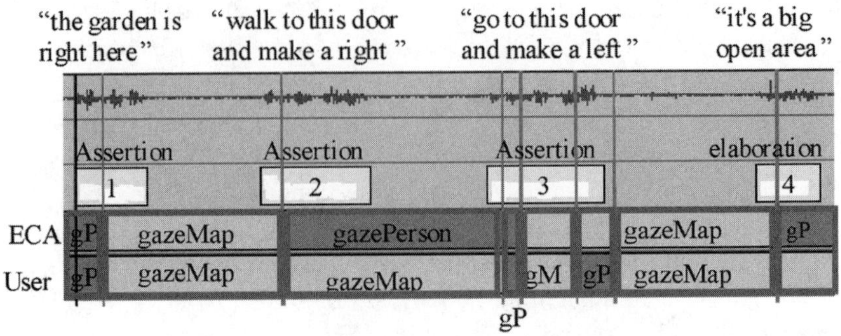

"the garden is right here"	"walk to this door and make a right"	"go to this door and make a left"	"it's a big open area"
Assertion	Assertion	Assertion	elaboration
1	2	3	4

ECA	gP	gazeMap	gazePerson			gazeMap	gP
User	gP	gazeMap	gazeMap	gM	gP	gazeMap	

gP

FIGURE 17.9 | Analysis of grounding behaviors in human-ECA conversation.

A final example, and perhaps the one that best illustrates how ECAs—properly constructed, on the basis of theories elaborated from human observation—can elicit natural behavior from humans who interact with them (and how this behavior can be illuminating along both positive and negative dimensions), is an experiment in which we endowed REA with social chitchat skills.[18] As mentioned above, Tim Bickmore carried out an extensive study of small talk among realtors and traveling salesmen. The results indicated that small talk was not randomly inserted into conversation, but served specific purposes (e.g., to minimize the potential face threat of personal questions such as "How much do you earn?"). These functions of small talk could be simulated in a way that would allow us to implement a small-talking realtor who used chitchat to smooth the progress of a house-selling transaction with a human. In order to test our model of human conversation, we asked people to interact with one of two versions of the ECA. One used task-talk only, while the other added social chitchat at key places in the interaction. Figures 17.10 and 17.11 show the actual conversational contributions by REA under the two conditions.

The people who interacted with each ECA were asked to evaluate their experience: how natural they felt the interaction to be, how much they liked the ECA, how warm they felt she was, and how trustworthy. We also tested the subjects on their own social skills, dividing them into extroverts and introverts by using a common psychological scale. The results showed that extroverts preferred the small-talk version of the ECA, while introverts preferred the ECA to keep to the task (we also discovered that it was difficult to find extroverts among MIT students, but that's another story).

An introvert working with the small-talk version remarked:

```
1   How about this weather?
2   I think winters in Boston are awful.
3   How do you like Boston?
4   I have lived in Boston all my life. Come to think of
    it, I have lived inside this room all of my life. It is
    so depressing.
5   Boston is certainly more expensive than it used to be.
6   So, where would you like to live?
7   How many bedrooms do you need?
8   Do you need access to the subway?
9   Is one bath enough?
10  You know, I keep showing the researchers here the same
    houses, over and over again. Maybe one day I will get
    lucky.
11  Have you been in the Media Lab before?
12  Do you know that the Media Lab is going to expand into
    another building. Things are really going well for the
    researchers here.
13  It is pretty cool do you think?
14  They are doing some crazy things in here.
15  I have shown houses to lots of students and faculty
    from MIT. But I always enjoy talking to them.
16  Anyway, What can you afford?
17  What kind of down payment can you make?
18  Let me see what I have available.
```

FIGURE 17.10 | Small talk in ECAs.

REA exemplifies some things that some people, for example, my wife, would have sat down and chatted with her a lot more than I would have. Her conversational style seemed to me to be more applicable to women, frankly, than to me. I come in and I shop and I get the hell out. She seemed to want to start a basis for understanding each other, and I would glean that in terms of our business interaction as compared to chitchat. I will form a sense of her character as we go over our business as compared to our personal life. Whereas my wife would want to know about her life and her dog, whereas I really couldn't give a damn.

An extrovert working with the same version had a very different response: "I thought she was pretty good. You know, I can small-talk with somebody for a long time. It's how I get comfortable with someone and how I get to trust them and understand how trustworthy they are, so I use that as a tool for myself."

Clearly, the people in this experiment are evaluating the ECA's behaviors in much the same way as they would evaluate a flesh-and-blood realtor. And clearly, our unexamined implementation of the realtor as a woman instead of a man has played into those evaluations, as much as have any

```
1   So, where would you like to live?
2   What can you afford?
3   What kind of down payment can you make?
4   How many bedrooms do you need?
5   Do you need access to the subway?
6   Is one bath enough?
7   Let me see what I have available.
```

FIGURE 17.11 | Task-only talk in ECAs.

of our carefully examined decisions about small talk, hand gestures, and body posture. Although our goal was to obtain input for a theory of the role of small talk in task talk, this response from one of REA's interlocutors effectively demolishes the claim that human identity can be denuded of its material aspects. Much of the previous work on responses to ECAs as interfaces has concentrated on exactly this sort of effect, with some researchers advising industry executives to use a female ECA to sell phone service, but a male ECA to sell cars.[19] In response to this unintended research finding in our small-talk study, my students and I have begun to use the virtual human paradigm to investigate explicitly which linguistic, nonverbal, and visual cues signal aspects of identity. Some have suggested that the race of ECAs be paired to the putative race of the user; my students and I have begun to look at the complex topic of racial identity and how a person's construction of his/her own race, and recognition of the racial identity of others, may be conveyed not just by skin color, but (also) by aspects of linguistic practice, patterns of nonverbal behavior, and narrative style.[20]

Embodied Conversational Agents as Interfaces

I've alluded to other ways in which ECAs are used, where they serve not as scientific instruments or tools to think with, but as interfaces to computers. In this function, ECAs might take the place of a keyboard, screen, and mouse—the user speaks to them instead of typing. Or they might represent the user in an online chat room. ECAs can also serve as teachers or tutors in combination with educational software—so-called "pedagogical agents." Research in this applied science examines whether ECAs are preferable to other modalities of human-computer interaction such as text or speech; what kinds of behaviors make the ECAs most believable and most effective (as tutors, information retrievers, avatars); and what personas the ECA should adopt in order to be accepted by their users. My students and I have also conducted some of this research, looking' at whether virtual children are effective learning companions for literacy skills, whether people are willing to be represented by ECAs in online conversations, and whether tiny ECAs—small enough to fit on a cell phone—still evoke natural verbal and nonverbal responses in the people speaking with them.[21] Even here, however, our research on virtual peers has led us back to an exploration of human-human communication, as we attempt to identify the features that signal to children that somebody else is a peer, is good friendship material, is worth listening to and telling stories with. Our related exploration of the pragmatics of the body has led us to some key features of social interaction—how rapport and friendship are negotiated—which, in turn, have led us to a better understanding of peer learning. Figure 17.12 shows one of our virtual peers.

Most recently, Andrea Tartaro and I have begun to look at how children with autism can play the role of scientist—learning about the gaps in their knowledge of communication and social interaction by authoring virtual people and watching them interact with others.[22] Mostly, however, our work is focused on the minutiae of human interaction and is therefore sometimes less useful to interface designers. In fact, computer scientists sometimes respond to my talks about NUMACK, the direction-giving robot, by asking, "But wouldn't it just be more effective to display a map on the computer screen and skip the virtual human?" When I respond that such an interface

FIGURE 17.12 | A child playing with Sam, the virtual peer.

wouldn't teach us anything about human communication, those same questioners often nod sagely, as if they knew all along that my interest was only in humans. Others have taken the ECA much further as an interface—the most extreme work (and most studied by historians of science and technology) is being done at the Institute for Creative Technologies (ICT) at the University of Southern California. Funded in equal parts by the U.S. Army and Hollywood, the ICT has created a vast, immersive, video-game-like room intended to teach soldiers before they enter the field—what Tim Lenoir has called a "military entertainment complex."[23]

The development of the ECA from a scientific instrument that simulates human behavior to an attractive interface bears interesting parallels to the history of mechanical automata. Automata makers of the sixteenth century, such as the one who built the perpetually praying monk described so elegantly by Elizabeth King in this volume, depended on the gaze of the perceiver to confer lifelikeness on the machine. Automata makers of the eighteenth century intended to find out in what way the human activities of drawing and writing and playing an instrument differed, if at all, when machines performed them.[24] Droz's writing boy, whose pen moves across the page just as real writers' pens move, is an example. The ECAs that I build today are likewise a way to compare conversation among humans with conversation between a human and a humanlike machine in order to discover what we know and do not know about human communication, and that simulation only works because of the "life" conferred on the virtual human by the interlocutor. Mechanical automata of the later nineteenth century, however, were meant to entertain, and not illuminate. An example of such a pretty virtual body as entertainment is the Pierrot automaton doll that writes—but simply by moving an inkless pen smoothly across a page—while sighing deeply and progressively falling asleep by the lamplight. These latter examples of mechanical humans did sustain relationships with real humans in that people wanted to own the pretty mechanical toys and were fascinated by them. But in these instances, the gaze of the viewer was one of concupiscence, not that of an interlocutor. Likewise, the tiny virtual human on a cell phone is meant to evoke the greedy desire of the collector more than the unconscious gaze of a partner in conversation.

CONCLUSIONS

These exercises in building virtual people have led to advances in what we know about the interaction between verbal and nonverbal behavior in humans, about the role of small talk in task talk, about the kinds of functions performed in conversation by the different modalities of the body, and about how learning is linked to rapport in children. Learning what must be implemented in order to make embodied conversational agents evoke a natural response and learning what the technology can and cannot do at the present time have also given me a sense of the meaning of humanness through human behavior. It is the ensemble of behaviors in all of their minuteness and unconscious performance that create the sense of humanness. Flaws and lacunae in that ensemble of behaviors give the scientist interlocutor a sense of what we do not know about human communication. Strengths and continuities in the theory that underlies the implementation lead people to respond to the virtual human as if it were another human being. The sufficiency criterion in cognitive science consists of explaining human cognitive activity by showing how a computer program may bring about the same result when the computer is provided with the same input.[25] In virtual human simulations, cognitive activity is not sufficient. I know that my model successfully explains human behavior when it evokes human behavior, because human communicative behavior is intrinsically relational and cannot be understood without two humans.

To come back to the anecdote with which this essay began, it is important to note the essential role of the physical body in the study of both language and social experience (insofar as those might be distinguishable). Language has traditionally been relegated to taking place purely in the head. But I hope the examples of communicative functions given above have made it clear that language is spread throughout the whole body—the hands, the torso, the eyes—and across two bodies in interaction. My original goal in building virtual humans was to focus attention on the whole-body aspects of language and on its intrinsically relational nature. As Descartes points out, the difference between real men and artifacts that only have the shape of men exists both in word and movement: imitation and gesture are as constitutive of humanness and social interaction as spoken language:

> And suppose there existed machines built in the image of our bodies, and capable of imitating our actions, as far as morally possible, there would still remain two certain tests by which to know that they were not really men. The first is that these automata could never use words or other signs in conversation, as we are able to do in order to convey our thoughts to others; for even if we can easily conceive of a machine that can emit the sounds of speech, or that can respond to external action such that, for example, if touched in one particular place it

may ask what we wish to say to it; if touched in another it may cry out that it is hurt, and so forth; we nevertheless cannot imagine a machine that can answer to what is said in its presence, as even fools can do. The second test is that even though such machines may carry out actions as well or even more perfectly than we humans can, they still will fail in executing other actions, by which we can discover that they did not act from knowledge, but from a particular arrangement of their organs.[26]

NOTES

The research reported in this chapter would not have been possible without the hard work, persistence, and insight of graduate students who were so generous with their time and knowledge that they quickly became my colleagues and teachers. Sincere thanks go to the members of Animated Conversation, Gesture and Narrative Language, and the ArticuLab. Thanks to Jessica Riskin for including me in this volume, and abiding gratitude to Ken Alder, Pablo Boczkowski, Sid Horton, Jessica Riskin, Dan Schwartz, Matthew Stone, and two reviewers for careful and perceptive comments that greatly improved the quality of the manuscript.

1. Philip Agre, "Formalization as a Social Project," Quarterly Newsletter of the Laboratory of Comparative Human Cognition 14, no. 1 (1992): 25–27.

2. Clifford Ivar Nass and Scott Brave, Wired for Speech: How Voice Activates and Advances the Human-Computer Relationship (Cambridge, MA: MIT Press, 2005).

3. Evelyn Fox Keller, "Models, Simulation, and 'Computer Experiments,'" in The Philosophy of Scientific Experimentation, ed. H. Radder (Pittsburgh: University of Pittsburgh Press, 2003).

4. See N. Wang, W. L. Johnson, P. Rizzo, E. Shaw, and R. E. Mayer, "Experimental Evaluation of Polite Interaction Tactics for Pedagogical Agents," in Proceedings of the International Conference on Intelligent User Interfaces (New York: ACM Press, 2005); Marilyn A. Walker, Janet E. Cahn, and Stephen J. Whittaker, "Improvising Linguistic Style: Social and Affective Bases for Agent Personality," in Proceedings of the First International Conference on Autonomous Agents (New York: ACM Press, 1997); and Penelope Brown and Stephen C. Levinson, Politeness: Universals in Language Usage, vol. 4 of Studies in Interactional Sociolinguistics (New York: Cambridge University Press, 1987).

5. Isabella Poggi and Catherine Pelachaud, "Performative Facial Expressions in Animated Faces," in Embodied Conversational Agents, ed. J. Cassell, J. Sullivan, S. Prevost, and E. Churchill (Cambridge, MA: MIT Press, 2000); John Austin, How to Do Things with Words (Oxford: Oxford University Press, 1962).

6. See Hao Yan, Paired Speech and Gesture Generation in Embodied Conversational Agents (M.S. diss., MIT Media Lab, Massachusetts Institute of Technology, 2000); Justine Cassell, Yukiko Nakano, Timothy Bickmore, Candy Sidner, and Charles Rich, "Non-Verbal Cues for Discourse Structure," in Proceedings of the Thirty-ninth Annual Meeting of the Association of Computational Linguistics (Philadelphia: Linguistics Data Consortium, 2002); Timothy Bickmore and Justine Cassell,

"Small Talk and Conversational Storytelling in Embodied Conversational Characters," in *Narrative Intelligence: Papers from the Fall 1999 AAAI Symposium* (Menlo Park, CA: AAAI Press, 1999); Scott Allan Prevost, "Modeling Contrast in the Generation and Synthesis of Spoken Language," in *Proceedings of the Fourth International Conference on Spoken Language Processing [ICSLP '96]* (New York: IEEE, 1996); and Obed E. Torres, Justine Cassell, and Scott Prevost, "Modeling Gaze Behavior as a Function of Discourse Structure," in *Proceedings of the First International Workshop on Human-Computer Conversation* (n.p., 1997).

7. Adam Kendon, "Some Relationships between Body Motion and Speech," in *Studies in Dyadic Communication*, ed. A. W. Siegman and B. Pope (Elmsford, NY: Pergamon Press, 1972).

8. M. A. K. Halliday, *Intonation and Grammar in British English* (The Hague: Mouton, 1967).

9. Justine Cassell, Matthew Stone, and Hao Yan, "Coordination and Context-Dependence in the Generation of Embodied Conversation," in *Proceedings of the INLG, 2000* (Mitzpe Ramon, Israel: Association of Computational Linguistics, 2000).

10. Lucy Suchman, "Writing and Reading: A Response to Comments on Plans and Situated Actions," *Journal of the Learning Sciences* 12, no. 2 (2003): 299–306.

11. Daniel L. Schwartz and Taylor Martin, "Representations That Depend on the Environment: Interpretative, Predictive, and Praxis Perspectives on Learning," *Journal of the Learning Sciences* 12, no. 2 (2003): 285–97.

12. Lucy Suchman, "Do Categories Have Politics? The Language/Action Perspective Reconsidered," in *Human Values and the Design of Computer Technology*, ed. B. Friedman (Cambridge: Cambridge University Press, 1997).

13. See, respectively, Cristina Conati and Xiaoming Zhou, "A Probabilistic Framework for Recognizing and Affecting Emotions," in *Proceedings of the AAAI Spring Symposium on Architectures for Modeling Emotions* (Stanford, CA: Stanford University Press, 2004); Fiorella de Rosis, Catherine Pelachaud, Isabella Poggi, Valeria Carofiglio, and Berardina Nadja De Carolis, "From Greta's Mind to Her Face: Modelling the Dynamics of Affective States in a Conversational Embodied Agent," in "Applications of Affective Computing in HCI," special issue, *International Journal of Human-Computer Studies* 59 (2003): 81–118; Gene Ball and Jack Breese, "Emotion and Personality in a Conversational Agent," in Cassell et al., *Embodied Conversational Agents*; Stacy Marsella, David V. Pynadath, and J. Stephen Read, "PsychSim: Agent-Based Modeling of Social Interactions and Influence," in *Proceedings of the Sixth International Conference on Cognitive Modeling* (Mahwah, NJ: Lawrence Erlbaum Associates, 2004); Timothy Bickmore, "Unspoken Rules of Spoken Interaction," *Communications of the ACM* 47, no. 4 (2004): 38–44; Justine Cassell and Timothy Bickmore, "Negotiated Collusion: Modeling Social Language and Its Relationship Effects in Intelligent Agents," *User Modeling and Adaptive Interfaces* 12 (2002): 1–44; Bas Stronks, Anton Nijholt, Paul van der Vet, and Dirk Heylen, "Designing for Friendship: Becoming Friends with Your ECA," in *Proceedings of Embodied Conversational Agents: Let's Specify and Evaluate Them!* (New York: ACM Press, 2002).

14. Sherry Turkle, *Life on the Screen: Identity in the Age of the Internet* (New York: Simon & Schuster, 1995).

15. Justine Cassell and Kristinn R. Thorisson, "The Power of a Nod and a Glance: Envelope vs. Emotional Feedback in Animated Conversational Agents," *Applied Artificial Intelligence* 13 (1999): 519–38.

16. Yukiko I. Nakano, Gabe Reinstein, Tom Stocky, and Justine Cassell, "Towards a Model of Face-to-Face Grounding," in *Proceedings of the Forty-first Annual Meeting of the Association for Computational Linguistics* (Philadelphia: Linguistics Data Consortium, 2004).

17. Herbert H. Clark, *Arenas of Language Use* (Chicago: University of Chicago Press, 1992).

18. Cassell and Bickmore, "Negotiated Collusion."

19. Nass and Brave, *Wired for Speech.*

20. Justine Cassell, Andrea Tartaro, Vani Oza, Yolanda Rankin, and Candice Tse, "Virtual Peers for Literacy Learning," in "Pedagogical Agents," special issue, *Educational Technology,* forthcoming.

21. See, respectively, Kimiko Ryokai, Catherine Vaucelle, and Justine Cassell, "Virtual Peers as Partners in Storytelling and Literacy Learning," *Journal of Computer Assisted Learning* 19, no. 2 (2003): 195–208; Justine Cassell and Hannes Vilhjálmsson, "Fully Embodied Conversational Avatars: Making Communicative Behaviors Autonomous," *Autonomous Agents and Multi-Agent Systems* 2 (1999): 45–64; and Timothy Bickmore, "Towards the Design of Multimodal Interfaces for Handheld Conversational Characters," in *Proceedings of the Conference on Human Factors in Computing Systems* (New York: ACM Press, 2002).

22. Andrea Tartaro and Justine Cassell, "Using Virtual Peer Technology as an Intervention for Children with Autism," in *Towards Universal Usability: Designing Computer Interfaces for Diverse User Populations,* ed. J. Lazar (Chichester, UK: John Wiley and Sons, in press).

23. Timothy Lenoir, "All but War Is Simulation: The Military-Entertainment Complex," *Configurations* 8, no. 3 (2000): 289–335.

24. Jessica Riskin, "Moving Anatomies" (paper presented at the annual meeting of the History of Science Society, 3–7 November 1999, at Pittsburgh, PA).

25. Allen Newell and Herbert A. Simon, *Human Problem Solving* (Oxford, UK: Prentice-Hall, 1972).

26. René Descartes, *Discours de la Méthode,* in *Oeuvres et Lettres* (1637; Paris: Librairie Gallimard, Bibliothèque de la Pléiade, 1953), 164–65; my translation.

INDEX

Page numbers followed by "f" refer to figures.